Sustainable Work and the Environmental Crisis

Compared to 20 years ago, the jobs many people do today are increasingly characterised by low pay and insecurity, while countless others cope with workplace stress and ill-health. At the same time the consequences of our current model of economic activity are creating dangerous and critical changes in the planet's climate. Until recently debates around these two issues have had little contact with each other.

This book demonstrates that there are definite and complex connections between degraded jobs and a degraded environment, that neither the dominant economic model nor the rate at which we exploit the planet's resources are sustainable and that the limits for both may be reached sooner rather than later. By bringing together insights from critical thinkers in a range of disciplines, the book discusses the requirements and characteristics for work to be at the same time economically, socially and environmentally sustainable and examines the potential for alternative routes to sustainable work in policies and actions that support both the natural environment and worker well-being.

The book will be of interest to researchers, academics and students in the fields of HRM, labour studies, employment relations, sociology, environmental studies and sustainability. It is particularly relevant for those focusing on the link between labour and climate change. It is also highly relevant to policymakers, trade unions and NGOs looking at decent work and sustainability.

Chris Baldry is Professor Emeritus in the Stirling Management School, University of Stirling and a Fellow of the Royal Society of Arts.

Jeff Hyman is Professor Emeritus of Employment Relations at the University of Aberdeen and Honorary Professor of Management at the University of St Andrews.

Sustainable Work and the Environmental Crisis

The Link between Labour and Climate Change

Chris Baldry and Jeff Hyman

Routledge
Taylor & Francis Group

LONDON AND NEW YORK

First published 2022
by Routledge
2 Park Square, Milton Park, Abingdon, Oxon OX14 4RN

and by Routledge
605 Third Avenue, New York, NY 10158

Routledge is an imprint of the Taylor & Francis Group, an informa business

British Library Cataloguing-in-Publication Data
A catalogue record for this book is available from the British Library

Library of Congress Cataloging-in-Publication Data
Names: Baldry, Christopher, author. | Hyman, J. D. (Jeffrey D.), author.
Title: Sustainable work and the environmental crisis : the link between labour and climate change / Chris Baldry and Jeff Hyman.
Description: Abingdon, Oxon ; New York, NY : Routledge, 2021. | Includes bibliographical references and index.
Identifiers: LCCN 2021001196 (print) | LCCN 2021001197 (ebook) |
ISBN 9780367322090 (hardback) | ISBN 9780429317286 (ebook)
Subjects: LCSH: Sustainable development. | Labor–Environmental aspects. |
Labor market–Environmental aspects. | Climatic change–Economic aspects.
Classification: LCC HC79.E5 B3478 2021 (print) | LCC HC79.E5 (ebook) |
DDC 331–dc23
LC record available at https://lccn.loc.gov/2021001196
LC ebook record available at https://lccn.loc.gov/2021001197

ISBN: 978-0-367-32209-0 (hbk)
ISBN: 978-1-032-03340-2 (pbk)
ISBN: 978-0-429-31728-6 (ebk)

Typeset in Sabon
by Taylor & Francis Books

For our grandchildren in the hope that they will inherit a better world.

Contents

Acknowledgements

We would like to thank Felix FitzRoy for his informed and valuable comments, Ian Douglas for his helpful discussion of things technological and Arne Kalleberg for his supportive response to our queries. Any errors are entirely our own.

Abbreviations

AI	Artificial intelligence
CAT	Centre for Alternative Technology
CBI	Confederation of British Industry
CEO	Chief Executive Officer
CME	Coordinated Market Economy
CCC	Committee on Climate Change
CSR	Corporate Social Responsibility
ETUC	European Trade Union Confederation
EU	European Union
EV	Electric vehicles
EWC	European Works Council
GDP	Gross Domestic Product
GHG	Greenhouse gases
GND	Green New Deal
HR(M)	Human Resource (Management)
HSE	Health and Safety Executive
ILO	International Labour Organization
IPCC	Intergovernmental Panel on Climate Change
IPBES	Intergovernmental Science-Policy Platform on Biodiversity and Ecosystem Services
JT	Just Transition
LME	Liberal Market Economy
NGO	Non-governmental Organisation
NMW	National minimum wage
OECD	Organisation for Economic Co-operation and Development
ONS	Office for National Statistics
RSA	Royal Society for Arts
SDG	Sustainable Development Goals
SHP	Shareholder primacy
SIB	State Investment Bank
SOE	State-Owned Enterprise
SUV	Sports utility vehicle
TUC	Trades Union Congress

UBI	Universal Basic Income
UN	United Nations
USEPA	United States Environment Protection Agency
WHO	World Health Organization
WMO	World Meteorological Organization
WTR	Working time reduction

1   Introduction
Degraded work in a degraded planet

What's going wrong?

We are close to burning up our planet. Planet Earth will of course survive but continued human existence will be increasingly difficult. This may at first glance read like scaremongering hyperbole but there is now virtually no scientific disagreement on the mounting evidence of the dangerous impact of our economic activities on the planet's ecosystem. Not simply the exhaustion of finite resources (everything from fish to forests), but the steady increase in global temperatures and resultant climate change, to the extent that some authorities (including the World Bank) now predict that we are close to a tipping point that will make it almost impossible to reverse these processes. In 2012 the World Bank warned that, although there is repeated international agreement that global temperatures should not increase more than 2°C, we were on track for a 4°C warmer world by the century's end and 'there is no certainty that adaptation to a 4° world is possible' (quoted in Klein, 2014: 13). On the same day in 2019 that it was reported that the Greenland ice sheet was melting at a much faster rate than had been predicted (an increase of 400 percent since 2003), the veteran environmentalist David Attenborough was warning the annual World Economic Forum at Davos, 'the Garden of Eden is no more' and that 'we need to get beyond guilt or blame and get on with the practical tasks in hand'.

The chief villain, of course, has been fossil fuels. There has been no shortage of warnings about the impact of fossil fuel dependency on the welfare of the planet and its inhabitants. In a 2018 speech at the Global Climate Summit (held in San Francisco where the effects of climate change are already being experienced), Lord Stern repeated the warnings contained in his authoritative 2006 report (Stern, 2006), which presented detailed evidence for anthropogenic climate change in which carbon-dependency was heavily implicated and was unequivocal in its conclusions: if there is no action, the overall costs and risks of climate change will be equivalent to a loss of five percent of *global* GDP annually, 'now and forever'. If a wider range of risks is considered, estimates of damage rose to a disastrous annual loss of 20 percent of global GDP or more. The 2018 Report of the Intergovernmental Panel on Climate Change was even more stark in its warnings, concluding that following a business-as-usual pathway would result in global mean temperatures increasing by three to five degrees over pre-industrial levels by the end of the century (IPCC, 2018).

Many parts of the planet are having to confront the more immediate consequences of climate change through increases in severe weather incidents which have, for example, seen houses devastated by flooding, as in the UK in 2019–2020; wildfires in New South Wales and California in 2020; record high temperatures in Australia and the USA in 2020; above-average size of hurricanes and tropical storms; and even wildfires within the Arctic Circle. Outside these areas of immediate climatic impact, the average citizen may well ask, as they watch the news or heed the summer warnings on UV radiation and rub on their sunblock, 'What on Earth is happening?'

However, as Tim Jackson has reminded us, for most people 'rain forests are a long way away' and the Arctic ice sheet is similarly remote (Jackson, 2011). Unless you are a Pacific Islander watching the annually rising sea levels threatening to obliterate not just your home but your whole society, warnings about global heating are not your most immediate problem. Indeed, for many people, pandemic aside, the daily source of stress and worry comes from the only means they have of sustaining life in the here and now – selling their labour power.

For some time it has become clear that work in both developed and newly industrialising countries has become increasingly degraded, by which we mean not just a low and static level of pay in return for your labour, but the loss of most of those aspects of work which might compensate for low pay – security, sociability, pride in doing a good job, the chance of advancement and in having a say in how work is undertaken. In the developed economies there has been almost weekly evidence of a steady depreciation of working conditions such that increasing numbers of people are performing low-paid, insecure jobs. There is widespread acknowledgement that the UK is steadily becoming a low-wage economy, with many employment sectors (21 percent of all employed jobs) characterised by remuneration below the living wage so that 55 percent of people in poverty are actually in a working family (Shafique, 2016). In addition, observers across the world have chronicled the spread of precarious and insecure jobs through working in the so-called 'gig' economy (Taylor, 2017), sub-contract and agency working and bogus self-employment (Kalleberg, 2018; Weil, 2014). Even in sectors that retain conventional employment standards, organisational responses to heightened competition have led to the intensification of effort demanded of employees, with ever-increasing performance targets and draconian sickness absence policies. There has also been an increase in reports of organisational cultures characterised by bullying and harassment.

Whilst in the West, workplace psychological and spiritual deprivation are becoming increasingly prevalent, in emergent economies this is surmounted by more extreme physical and physiological demands enforced through working long hours, under threatening supervisory practices and often in unhealthy conditions. For assembly-line workers in China (often sub-contracted by prestige manufacturers like Apple) working under pressurised, toxic and sometimes inhumane battery-hen conditions (Litzinger, 2013) or women workers in Bangladesh's weakly regulated garment industry experiencing brutality from both employers and public officials keen to protect the country's vulnerable export markets (Murphy, 2019) or global agricultural workers, more likely to die at work than in any other sector with three to four million pesticide poisonings and some 40,000 fatalities reported annually (Rossman, 2013: 61), those debates in the West on 'the meaning of work' seem laughably irrelevant.

Whereas the global South has long been the chosen location for the most blatant exploitation, what seems puzzling to workers in mature industrial economies is what exactly has been happening to their own jobs. For someone working today in one of Amazon's 'fulfilment centres', the pressures seem unending: they will be trying to meet a punishing performance standard in the 'pick rate' per hour, on a zero-hours contract and worrying in case they have accrued a critical number of disciplinary points from under-performing or sickness absence. They worry too that Amazon will no longer request their services from the agency that is their nominal employer (Bloodworth, 2018). He or she may, at the end of their gruelling shift, compare their situation from what they remember of their parents' jobs. Those jobs, back in the 1970s say, might have involved a long period of service with the same company, with paid holidays, sick pay and maybe even a company pension. Wages and hours were predictable, as a result of collective bargaining between company and trade union, and the union might also have had oversight of health and safety through a joint company safety committee. Our Amazon workers may indeed wonder to themselves, 'What went wrong?'

An unequal planet

In the following chapters we examine the above twin global crises: degradation of the planet and degradation of work. For most people, and indeed most politicians and policymakers, connections between the two may seem hard to see. Despite increased acknowledgement by some trade unions that work and climate are closely interconnected (Gingrich, 2013), until the IPCC report in October 2018, debates around the issues of work and environmental degradation tended to have little contact with each other. However, the IPCC dramatically warned that to avert global warming from reaching catastrophic levels, there must be 'rapid and significant change' in four major global systems: 'energy, land use, cities and industry' or as one co-chair of IPCC adds, 'this is not about remote science: it is about where we live and work'. The first part of this book will demonstrate the complex ways in which the seemingly inexorable downward pressure on working conditions in both developed and developing economies interconnects with the relentlessly increasing threat to the planet's health.

One trend linking both crises is growing economic inequality both within and between countries, sustained by the ideological triumph in the major economies of a model of 'free market' economics that eschews regulation, and promotes cost minimisation and profit maximisation. This in turn engenders a notable short-term perspective in economic decision making, which contrasts starkly with the long-term vision necessary to understand what is happening to the planet. Notably, the Anglo-American model of capital structure based on share issue rather than bank finance gives primacy to the interests of shareholders. Whereas the interests of employees, who are likely to be, or would wish to be, with the company for the long term are routinely ignored, shareholders, who have little interest in the company apart from its ability to improve share price, are given legal preferential consideration in corporate decision making. These short-term priorities also encourage competition strategy based on cost minimisation (for example through paying low or static wages) rather than the longer-term horizon required for capital investment.

The twin priorities of cost minimisation and maintaining shareholder value have fuelled development of the low-wage precarious economies of the UK and US while at the same time increasing the rewards given to the economic elite: the share of national income going to the top one percent has doubled over the past 30 years while workers have seen their real wages flatline or even fall. This concentration of income is a major contributor to both of the present crises. The upward transfer of income from wage earners to the wealthy means that the two most effective routes out of crisis are closed: consumer demand is choked off, or reliant on increasing levels of domestic debt, and wealth is not being reinvested. The ability of the rich to avoid taxation on most of their wealth reduces the resources of the state which in turn further emiserates the most vulnerable. Authors such as Wilkinson and Pickett (2010) and Dorling (2010) have demonstrated a close correlation between greater social and economic inequality and a wide range of other social issues, including increased ecological degradation as inequality repeatedly fuels demand for conspicuous consumption in what Stiglitz has called 'trickle-down behaviourism' (Stiglitz, 2013). A study of 50 countries found that the more unequal a country is the more likely is its biodiversity to be under threat (Raworth, 2017).

As we shall examine in Chapter 4, in the conventional theory of the economics of the firm the environmental consequences of the firm's activities such as carbon emissions or polluted waterways are treated as 'externalities', essentially meaning that the cost of putting them right will be borne by someone else or not borne at all, leading to the accumulating poisoning of the environment. Relying on someone else to cope with your externalities is, as we shall see, one of the major causes of the disparity between economic activity in wealthier nations and the environmental impact felt by developing countries, competing desperately to secure inward investment and to sustain economic growth.

It should also be borne in mind that it is not just workers who suffer from the devastating effects of pressurised work and polluted workplaces, it is their communities also; studies by Greenpeace point to a quarter of a million premature deaths attributable to reliance on coal power in China (Monbiot, 2017: 169). This again is not a new phenomenon as testified by earlier environmental and safety disasters such the 1984 Bhopal incident when lethal methyl isocyanate gas spilled out from Union Carbide's Indian subsidiary in Bhopal, poisoning thousands of local inhabitants straight away and many more subsequently through unchecked environmental contamination. As with similar and more recent disasters such as the 2009 Deepwater Horizon blowout, employer responsibility proved hard to pin down.

Which path out of the jungle?

In the following chapters we shall first examine the twin crises of environment and work in more detail and the economic mindset that continues to weave inextricable links between them. We shall then look at which of the major actors and institutions we can expect to offer solutions, and the limitations of their abilities to do so. For example, for business to take on board the costs of its current 'externalities' would lead to a drop in shareholder confidence, not to mention the opposition of powerful economic interests, linked to the fossil fuel giants, who continue to fund campaigns of climate change denial. For governments to take unilateral action would threaten their short-term economic competitiveness. Many trade unions are in the double bind position of being in the business of protecting their members' jobs, even if those jobs are in environmentally damaging industries. Non-governmental agencies such as the United Nations possess the breadth of vison for international cooperation in tackling the climate crisis but lack economic and political power, while relying on technological solutions to the crisis of work or the climate crisis will inevitably turn out to be a mixed blessing.

There are obvious limits to the neo-liberal mode of running the world economies, as increased competition from emerging economies and the organisational priority of cost minimisation result in a global 'race to the bottom'. For example, in the UK, a high percentage of benefit claimants are in working families, the state in effect subsidising employers' low wages through payments such as tax credits and housing benefit. As declining money and real wages result in a lower tax take, which lends support to government calls for austerity and further diminishing social support for the vulnerable, it is clear that the current model for working life is socially and economically unsustainable. Yet, the response on the part of political and economic orthodoxy is to restate the goal of more economic growth as the panacea that will restore society to the sunlit uplands of prosperity.

Clearly, however, we are at the limit of what is environmentally possible; already, for every world citizen to enjoy the current standard of living of Swedes or Canadians would require three more planets. Global population is currently 7.3bn and expected to reach 10bn by 2050, while GDP, in the aftermath of the pandemic shocks, is soon expected to resume growth, doubling the global economy by 2037 (Raworth, 2017). More growth would inevitably mean significantly increasing carbon emissions, yet simply to claim that the answer is zero growth would not be to the benefit of those in the global South desperate to raise the living standards of their people.

Thus, it would seem that the conventional answer to one crisis area – more growth – compounds problems in the other. The solutions most usually offered to the twin crises are pulling in opposite directions. This makes a coherent policy strategy to meet these twin crises extremely difficult to gain political or global business acceptance. Such a policy should not save the planet by condemning workers to low-paid drudgery, nor add value to work by wrecking the planet. What is needed, in short, is sustainable work.

What would sustainable work look like?

In 2009 the Sustainable Development Commission identified three operational principles for economic activity to satisfy: a positive contribution to well-being, the provision of decent livelihoods and low material and energy throughput (SDC, 2009). Similar principles formed the background to the UN's sustainable development goals, formally agreed by 193 Heads of Government in 2015 (United Nations, 2015).

Nearly 200 years of debating the 'meaning of work' has shown that, in addition to its primary function as a source of economic livelihood, a range of other non-monetary factors are valued by employees – security, a recognition of worth, pride in work and a degree of control or agency (either individual or collective), the presence of an occupational community, autonomous teamwork and membership of social groups. Many of these qualities can be found in areas of non-employed work ignored by conventional measures of GDP as they are not allocated a market value, such as domestic, voluntary and community work.

Current priorities of short-term profit maximisation create low-paid and insecure jobs yet are not disposed to include long-term issues such as the environment. This suggests that a different kind of economic organisation is needed to deliver sustainable jobs, more akin to redistributive organisations such as co-ops, not for profits and community interest companies. Whilst large and global interests are certain to be immune from and unreceptive to policy shifts to more inclusive organisations, there are many steps, as this book will demonstrate, that established corporations could adopt to offer more sustainable modes of employment.

It has been suggested that as employment in the rich nations moves from manufacturing to services this will have a positive effect through reducing environmental pollution and emission of greenhouse gases. However, all this does is export environmental 'externalities' overseas, letting someone else, somewhere else, do the dirty work. Clearly meeting the criteria for sustainability will require sectoral transformation – the contraction or elimination of extractive industries of coal, oil and gas, reform of industrial agriculture and curtailing speculative finance; to be balanced by the expansion of long-term investment in renewable energy, public transport and circular manufacturing (Jackson, 2011; Klein, 2014). The focus in all sectors would have to be on delivering a product or service that does not decrease the well-being of others or the health of the planet: a transformation that has come to be known as a 'just transition'.

In her critique of economic orthodoxy Kate Raworth has proposed a 'Doughnut Model' of the necessary future economy: below the inner ring of the doughnut, representing human well-being, are shortfalls in access to health, food, education and housing. Above the outer ring, representing the planet's ecological ceiling, are overshoots on pressures on the Earth's systems, such as climate change, soil degradation and ocean acidification. Only between these limits is a space that is ecologically safe and socially just (Raworth, 2017). In order to identify what work in this space would look like we may have to redefine work to include not only that undertaken in a labour market for pay but also unpaid forms of work such as domestic labour, and also examine the monetary and non-monetary rewards conducive to socio-economic sustainability and how sustainable jobs could also contribute to environmental stability.

Degraded work and polluted communities are of course not new phenomena. However, while nineteenth-century mill and factory owners could escape environmental havoc caused by their operations by living above the pollution in higher parts of town or even further away (Freeman, 2018: 27), today environmental disaster threatens us all and there will be no escape for the hyper-rich from the escalating environmental crises caused by global warming. There will be no point fleeing to a tropical island that is in danger of disappearing beneath raised sea levels, or moving to the most exclusive forested areas of California which are regularly being devastated by uncontrollable wild fires. While for 200 years it has been possible to be ignorant of or deny the systemic environmental damage created by the dominant model of industrial activity, that excuse is no longer viable.

In this book: goals and limitations

In the chapters that follow we examine these critical issues in detail and attempt to navigate our way through the complexities of change. What this book will attempt to do is initially to show connections between work and environment; this will include critically revisiting the characteristics of exactly what goods and services we produce and how we produce them including materials, technology and organisational forms. We scrutinise evidence that indicates the consequences these characteristics have for environmental damage. Secondly, we look at the social and economic actors and institutions who we might look to for solutions. From our diagnoses of the human and environmental damage attributable to 'growth at any cost' policies allied to short-term profit maximisation, we examine the potential for alternative courses of sustainable employment located in policies and actions which support both the natural environment and worker well-being. Lastly, we ask what it would take to make work sustainable, both in terms of its human, social and ethical values and in terms of minimising its environmental impact.

Daly and Cobb point out that Adam Smith, the 'father' of economics, saw the economy as part of the entirety of human activity, of both society and history. They argue that, in contrast, how we go about understanding human activity today is largely dictated by the disciplinary organisation of knowledge that maintains that each discipline has a subject matter that distinguishes it from others, with boundaries set by discipline-specific methodologies and conceptual apparatus (Daly and Cobb, 1994). The background of the writers of the current book is in the area of work and employment. However, as the coming environmental crisis will affect everyone, regardless of the particular work they may do or whether they are in work at all, the very nature of the book's focus means that we have attempted to put aside the boundary conventions of academic disciplines and tried to show the connectedness between work, management, economics and the environment. In doing so we have tried to minimise the use of discipline-specific concepts which may not be readily accessible to others and tried to explain as clearly as possible terms and ideas in a way that can be read by anyone. In doing this, by straying into others' areas of expertise we have probably made some mistakes, for which we apologise, and we will probably be accused of over-simplifying things or missing out discussion of key debates in this or that subject area. However, we want our ideas on work to be understood by environmentalists and our thoughts on the environmental crisis to be understood by social scientists; above all we want it to be read by a concerned public. While our analysis and recommendations are unlikely to find favour in all quarters, we attempt to found these on recognition of the complexities of the social, political and environmental problems facing us. We have used academic research and official sources of data wherever possible but, because of the speed of developments in both environmental issues and the world of work we have also used reports from campaigning groups where these support existing data and investigative journalism where published data is lacking, such as the experienced realities of contemporary working conditions.

The environmental crisis, of course, affects the whole planet and, as we try to show, so increasingly do our patterns of work, production and consumption. Nevertheless, we have mainly confined our examination of structures and policies to the mature industrial nations, particularly the UK, Europe and North America. We make no apologies for this as it is their economic activities over two centuries that have been the predominant cause of global heating and its attendant inequalities and thus have the prime responsibility for making the necessary economic and ecological changes. We have, however, indicated where the costs of our economic way of life are currently being borne by countries in the global South, in the form of wildfires, flooding, waste pollution or hazards to health.

One key area that we have deliberately omitted as falling outside our scope is agriculture and land management, both in developed and developing countries; the nature of agricultural employment shares many of the characteristics of other modes of precarious labour and the consequences of industrialised agriculture undoubtedly have to be addressed in any coordinated response to the climate crisis but to deal with this adequately would probably require another book and more expertise in this area than we possess.

Finally, in the middle of writing this book the world was hit by the Covid-19 pandemic, the responses to which, by national governments and international agencies, have served to raise even more questions over future patterns of work and their environmental consequences. Where it has seemed appropriate, we have indicated some of these in the relevant chapters but, like virtually everyone else, we cannot foresee the long-term effects of this catastrophic plague.

Bibliography

Bloodworth, J. (2018), *Hired: Six Months Undercover in Low-Wage Britain*, London, Atlantic.

Daly, H. and Cobb, J. (1994), *For the Common Good*, Boston Mass., Beacon Press.

Dorling, D. (2010), *Injustice: Why Social Inequality Persists*, Bristol, Policy Press.

Freeman, J. (2018), *Behemoth: A History of the Factory and the Making of the Modern World*, New York, Norton.

Gingrich, M. (2013), 'A comparative study of blue-collar unions' reactions to the climate change threat in the United States and Sweden' in N. Räthzel and D. Uzzell (eds), *Trade Unions in the Green Economy*, London, Routledge; 214–226.

IPCC (2018), *Summary for Policymakers of IPCC Special Report on Global Warming of 1.5°C Approved by Governments*. Intergovernmental Panel on Climate Change. www.ipcc.ch/sr15/chapter/spm.

Jackson, T. (2011), *Prosperity Without Growth: Economics for a Finite Planet*, London, Earthscan.

Kalleberg, A. (2018), *Precarious Lives: Job Insecurity and Well-Being in Rich Democracies*, London, Polity.

Klein, N. (2014), *This Changes Everything: Capitalism Versus the Climate*, London, Penguin.

Litzinger, R. (2013), 'The labor question in China: Apple and beyond', *South Atlantic Quarterly*, 112 (1); 172–178.

Monbiot, G. (2017), *How Did We Get into this Mess?*, London, Verso.

Murphy, S. (2019), 'Factory that supplied Tesco compensated abused worker', 22 January, *The Guardian*, https://www.theguardian.com/world/2019/jan/22/bangladeshi-factory-that-supplied-tesco-and-marks-and-spencer-compensates-abused-worker

Raworth, K. (2017), *Doughnut Economics*, |London, Penguin.

Rossman, P. (2013), 'Food workers' rights as a path to a low carbon agriculture' in N. Räthzel and D. Uzzell (eds), *Trade Unions in the Green Economy*, London, Routledge; 58–63.

SDC (2009), *Prosperity without Growth*, Sustainable Development Commission, London, Earthscan.

Shafique, A. (2016), *Addressing Economic Insecurity*, London, Royal Society of Arts/ Nottingham Civic Exchange.

Stern, N. (2006), *Stern Review on the Economics of Climate Change*, London, Cabinet Office/HM Treasury.

Stiglitz, J. (2013), *The Price of Inequality*, London, Penguin.

Taylor, M. (2017), *Good Work: The Taylor Review of Modern Working Practices*, London, Department for Business, Energy and Industrial Strategy.

United Nations (2015), *Transforming our World: The 2013 Agenda for Sustainable Development*, A/RES/70/1, www.sustainabledevelopment.un.org.

Weil, D. (2014), *The Fissured Workplace: Why Work Became so Bad for so Many and What Can be Done to Improve It*, Cambridge Mass., Harvard University Press.

Wilkinson, R. and Pickett, K. (2010), *The Spirit Level*, London, Penguin.

2 Bad day at work

Henry Mayhew, examining the state of work and poverty in London in 1851, gave a vivid description of the desperation of dock workers attempting to gain the attention of the foreman at the start of each day, waving hands in the air, jumping up, shouting and calling out their names, and remarked how sad it was to see thousands of men struggling for only a day's hire. He was describing the system of casual labour that characterised the docks industry until the middle of the twentieth century; dock workers would have to assemble at a 'pen' at the start of the day and hope to be picked by work-gang leaders for that day's work. For workers in a casual labour system life was governed by insecurity and perpetual uncertainty about whether you were going to work that day and, if so, what you might earn. Casualisation continued throughout the twentieth century in particular sectors such as construction and agriculture but a combination of unionisation and employers not wishing to lose employees with company-specific skills in an increasingly complex industrial economy meant that, by the second half of the century, most people's idea of a job was with an identified employer for a given number of hours for a known rate of pay.

Back to the future?

However, since the millennium there have been noticeable and fundamental changes in how a significant sector of the labour market works, which often seem to indicate a return to conditions of the nineteenth century. The term used today is 'precarious' labour, a term that includes workers on zero-hours contracts, temporary agency workers and those in the so-called 'gig' economy whose work is directed through a digital platform. What all these have in common is lack of certainty about when, or if, work will be available, for how long the worker's labour will be required and, consequently, what the reward for selling his or her labour power (i.e. income) will be that week. Increasingly, reports from trade unions, academic and independent research bodies indicate the existence of harsh and disturbing conditions in sectors most prone to using precarious labour. As a recent review of such work concluded: 'Rather than providing the basis for full and flourishing lives, for too many people our economy is inducing anxiety and uncertainty' (Shafique, 2018: 10).

Though we focus on the UK as a leader in precarious working, similar labour market dynamics are evidenced across Europe (Prosser, 2016) and indeed worldwide (Choi, 2018; Manky, 2018; Lee and Kofman, 2012). With precarious working broadly defined as 'a situation of living and working without stability or safety net' (Kalleberg, 2018: 13; Pettinger, 2019), it has been argued that neoliberalism and globalisation (terms that we explore in more detail in subsequent chapters) have undermined post-war gains made by labour and social democratic parties (Standing, 2011), with young workers, immigrants, women and poorly educated workers most at risk of exposure to insecure and overly demanding working pressures (de Ruyter and Warnecke, 2008). At the same time, it should be recognised that whilst precarious work represents a reversal for many workers in economically advanced countries, for many in less developed countries, precarity has always been present. In his detailed study of six 'rich democracies', Kalleberg demonstrates that while precarious work is most strongly associated with liberal market economies, patterns of response by the countries to global and technological developments are not identical as the extent and impact of precarity vary according to political differences in approach to labour market liberalisation. British and American workers experienced the highest degree of market deregulation and loss of social protection; Germany, Japan and Spain adopted a more protectionist and dualist approach, in which there are divisions between workers enjoying a decent job and those lacking stability, security or good working conditions while greater societal risk sharing was demonstrated by Denmark (*ibid.*: 195; Prosser, 2016).

As a response to growing concern and lack of information about the changing labour market, in 2016 the UK Government set up the Independent Review of Employment Practices in the Modern Economy under the chairmanship of Matthew Taylor with the aim of helping to ensure that 'all work in the UK should be fair and decent with realistic scope for development or fulfilment'. However, the report (Taylor, 2017) has to be viewed as a missed opportunity to investigate fundamental changes in how a significant minority of the workforce engage with paid work. Firstly, while Taylor maintained a general optimism about the UK's flexible labour market (in Taylor's words, 'the British Way'), the composition of the review panel has led critics to question just how independent it was, noting that other members consisted of the CEO of a platform-based company, an employment lawyer and the head of the Gangmasters Licensing Authority, with no one to represent trade union or worker voice (Thompson, 2019). Although the TUC and individual unions did make submissions to the review, Taylor's list of 'stakeholders' to be consulted comprises 'entrepreneurs and business, the legal profession, and enforcement agencies' (Taylor, *op. cit.*: 5), whereas it might be thought that one of the more significant stakeholders would be those actually doing the work. Secondly, Taylor's focus was very much limited to the confusing and differential entitlement to rights at work caused by the different categories of employment status currently recognised in UK employment law.

Increasing fragmentation of the labour market certainly has created a confusing situation, with workers themselves sometimes unsure of their contractual status. To understand why this might be the case we need to distinguish between different UK legal employment categories. The *self-employed* independent contractor or freelancer runs his or her own business and contracts with clients or customers to provide work or services; examples would be plumbers and electricians, or designers and others in the creative industries. The *worker*, who Taylor recommended should be seen as a 'dependent contractor', is registered as self-employed but provides a service as part of someone else's business – their contract is not with a client or customer but with the directing business or platform; examples would be parcel delivery drivers and couriers for firms like Deliveroo. These two categories of both real and notional self-employment are sometimes referred to collectively as 'the gig economy'. The *agency worker* has a contract with a labour supply agency, who pays them and administers other employment rights, and from whom they are hired by another company; examples include significant numbers in the social care sector and workers in delivery centres for online retail companies such as Amazon. The *employee* has an employment contract directly with a company with stipulated agreed pay, hours of work and leave entitlement; this includes both conventional full-time employment and zero-hours employment (Balaram, 2018); however, to confuse matters still further, as Taylor points out, in UK tax law there are only employees and the self-employed and the latter pay lower National Insurance (NI) contributions and their hirers pay no NI. To understand the implications of these legal distinctions, we need to unpack and explain some of these terms.

Zero-hours

A zero-hours contract is where the employer requires the worker to be available full-time but does not guarantee any fixed number of hours of work; the worker does not know each day whether they will be working until they get a text or phone call from their manager and they only get paid for the hours actually worked.

As Figure 2.1 demonstrates, the number of workers reporting that they are on zero-hours contracts has been steadily increasing since 2004, with a significant surge since 2013 when, between October to December 2013 and October to December 2018, the number of zero-hours contracts for workers aged 16 years and over increased by 44.2 percent, from 585,000 to 844,000 (ONS, 2019). By the period April to June 2019 there were 896,000 workers on zero-hours contracts, representing 2.7 percent of total employment (compared to the figure for 2013 of 1.9 percent) (ONS, 2020a). By February 2020 the number of those in employment reporting that they are on zero-hours contracts had reached a record high of 974,00 (ONS, 2020b). Total numbers of zero-hours contracts recorded are often higher than numbers of zero-hours workers as some people may hold multiple contracts.

The practice is fairly widespread but is a key characteristic of particular sectors of the economy, being especially prevalent among low-skill workers and in low-wage sectors; more than half the businesses in the UK hotel and catering sector use zero-hours working and it is also a large feature in retailing and

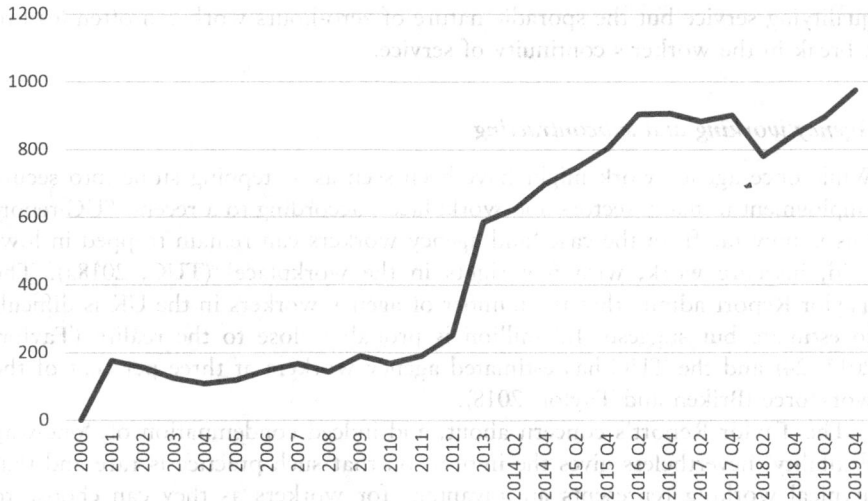

Figure 2.1. Numbers ('000s) reporting zero-hours contracts, UK, 2000–2019.
Source: ONS

education. In 2019 the sectors with the highest number of zero-hours workers were 'accommodation and food services', 'health and social work services' and 'transport, arts and other services'. Over half of those working on such contracts are women, 36.8 percent were aged 16 to 24 years (many in full-time education) while 23.1 percent were 50 to 64 years old. Estimates from the Labour Force Survey show that in April to June 2018 workers on zero-hours contracts usually worked an average of 25 hours per week, compared to 36 hours for all people in employment, while 16 percent of people on a zero-hours contract had worked no hours in the survey reference week (ONS, 2019).

The Taylor Report was broadly positive about zero-hours working, agreeing both with employer organisations' defence of the practice as essential to meet fluctuating demand and with their argument that it has contributed to keeping levels of employment high. The report was however critical of what it describes as 'one-way flexibility', where the choice of what hours are worked, and when, is entirely in the hands of the employer. While in theory workers on zero-hours contracts have the right to turn down work, in reality the risk of the contract being terminated makes zero-hours workers very unlikely to complain. Among other negative aspects to such contracts there is the question of worker status. As outlined in a House of Commons Briefing Paper (Pyper and McGuinness, 2018), zero-hours working has been the subject of several cases where the Employment Appeals Tribunal has had to decide whether or not those working under such contracts are 'employees', 'workers' or self-employed, with direct consequences for employment rights. Even where employment rights seem to be clear, the nature of zero-hours working may jeopardise them: for example, the right not to be unfairly dismissed is only available to employees with two years' qualifying service but the sporadic nature of zero-hours work can often lead to a break in the worker's continuity of service.

Agency working and subcontracting

While once agency work might have been seen as a stepping stone into secure employment in many sectors and workplaces, according to a recent TUC report this is now far from the case 'and agency workers can remain trapped in low-paid, insecure work, with few rights in the workplace' (TUC, 2018a). The Taylor Report admits that the number of agency workers in the UK is difficult to estimate but suggests 1.2 million is probably close to the reality (Taylor, 2017: 24) and the TUC has estimated agency workers at three per cent of the workforce (Briken and Taylor, 2018).

The Taylor Report's concern about, and indeed condemnation of, 'one-way flexibility' nevertheless gives the impression that such practice is rare and that atypical working represents an advantage for workers as they can choose to work in a range of different ways to fit around other responsibilities, such as studies or caring responsibilities. This view, maintained throughout the report, that atypical working is 'supply-led' (Moore *et al.*, 2019), in other words a response to workers' desire for more choice, is not reflected in the findings of most academic research on these forms of working; research that the review panel seem to have largely ignored (the Report's 105 references contain not a single academic source) (Thompson, 2019).

Indeed, Briken and Taylor's (2018) research into agency working in an Amazon fulfilment centre in South Wales found that the most common entry route, in an area of de-industrialisation and high unemployment, was that of unemployed workers facing benefit sanctions if they declined job offers. Although it was widely felt in the local community that working at Amazon was thoroughly unpleasant, many of their respondents reported being forced into working for Amazon's labour supply agency by threat of sanction. In this example the state job centre became a fourth actor, intervening between worker, agency and final hirer, to compel unemployed claimants to take unwanted jobs on insecure contracts through agencies and other labour providers. The authors conclude that narratives about 'choice' must include the conditions under which such choices are made and recognise that 'for millions the prevalence of *constrained choice* is fundamental to "modern work"'.

Recent work on agency workers in the homecare sector finds a good example of another aspect of one-way flexibility in which employers shift the cost of unproductive time onto the workers (Moore and Newsome, 2019). Because pay in homecare is usually on the basis of contact time only (where workers are in clients' homes), travel time between clients, supervision training and attending staff meetings go unpaid, meaning care workers may dip below the National Living Wage when time and pay are averaged out. Homecare workers in this study also reported the episodic nature of work, where there is so-called 'down-time' or 'waiting time' during the day where workers may not be able to go home and must wait between visits, effectively available to, but unpaid by, the employer. Hours can change from week to week (Biggs, 2016: 168) and, perhaps unsurprisingly, turnover rate in the care sector is running at around 25 percent (and higher in the private sector) (Bloodworth, 2018: 85).

The gig economy

The so-called 'gig' economy has been defined as work that is found and accessed online using a laptop, smartphone or other internet-connected device. Typical jobs include taxi driving, deliveries, office work, design, software development, cleaning and household repairs. In many cases companies providing the apps, and thus the allocation of tasks, define themselves as an internet platform rather than an employer and the workers are usually classed as self-employed (Taylor, 2017). Estimates of numbers in the gig economy vary: an RSA survey estimated 1.1 m (Balaram *et al.*, 2017) while the CIPD has put it at 1.3 million, 58 percent of whom are estimated to be permanent employees in their 'main' job, suggesting that a lot of gig work is to top-up unsatisfactory income (CIPD, 2017). The TUC, who prefer the term 'platform work', used a slightly broader definition than the Office for National Statistics, to include paid tasks found via a website or app, finding that the number of working age adults who have worked for an online platform at least once a week had doubled from 2.3 million workers three years ago to 4.7 million in 2019; six in ten platform workers in their survey were between 16 and 34 (TUC, 2019a).

The above surveys reveal some key features of gig workers: around 40 percent are in the 18–29 age group, are predominantly male, more concentrated in London and have a much broader range of ethnic backgrounds than other workers (CIPD, 2017; Taylor, 2017; Balaram *et al.*, 2017). Gig workers tend to have higher educational qualifications (44 percent have university degrees): this may reflect some platforms requiring high levels of skills, or alternatively it may indicate under-employment (i.e. workers being overqualified for the jobs they are doing). One in four of the RSA's respondents were in gig work because they could not find sufficient work by other means; over half cited better conditions in terms of flexible working and pay, and nearly two-thirds have been using gig work to supplement their income (Balaram *et al., op. cit.*).

The confusing picture created by these characteristics suggests that the widely used term 'gig economy' is not particularly analytically useful as it lumps together those platforms for professional services or skilled trades, catering for a high level of genuine self-employment, with those platforms with low entry levels and specialising in low-skilled work, such as driving or delivery and often characterised by bogus self-employment. The only thing these employment sectors have in common is the technology. This makes it difficult to evaluate survey respondents' comments about being attracted to gig work because of the flexibility, freedom and control.

Although the criterion used in employment law to distinguish between employed and self-employed status is usually that of the degree of control over the job exercised by worker and company respectively, it is clear from a recent study of 'self-employed' owner-drivers that, while they may lease or own their vans, their work is in fact totally controlled by the delivery companies (Moore and Newsome, 2018). A driver is paid by the number of deliveries or drops completed and may be contracted to be available for delivery 52 weeks and 365 days of the year. If they take a holiday or are sick it is their responsibility to find a substitute to cover the work or they could be fined by the company to cover the cost of hiring an agency worker. The advantage of the system for the company, in addition to not having to cover sick pay, holiday pay or pension contributions, is the removal of the costs of non-delivery.

In response to the Taylor Review the UK Government scrapped the agency worker loophole that allowed agency workers to be paid less than others doing the same job. They did, however, refuse to ban zero-hours contracts as that would 'negatively impact' more people than it helped. In comparison the New Zealand government abolished zero-hours contracts in 2016: from 2017 all workers had to be offered a fixed minimum number of hours a week and could not be penalised for refusing to accept extra work at short notice.

The challenge to employment status has been the main focus of trade unions who have several times tried to take employers like Uber, Hermes and Deliveroo to court to get staff re-categorised as workers in order to gain protections such as sick pay. In February 2019 the GMB union concluded a compromise deal with the courier company Hermes, which gives those couriers who choose it a guarantee of the minimum wage and holiday pay, but while continuing to be classed as self-employed (M. Taylor, 2019). In September 2019 the State of California passed a bill that will make it harder for platform-based users of gig economy labour to classify workers as independent contractors rather than employees and may enable gig economy workers to get holiday and sick pay.

Performance management and control

The fear of not getting work also means that the main employer in a precarious labour sector can impose ever more stringent performance standards and expectations, as the company's control over the allocation of work, and thus a worker's potential income, serves as a disciplinary mechanism: an aspect of 'modern working practices' largely ignored by the Taylor Report. For the owner-drivers mentioned above, failure to meet set delivery targets can mean the driver might be taken off easier routes with lots of short-distance drops (such as a city route) or have his contract terminated altogether. To complete an allocated workload a driver might be in the van for 12 to 13 hours daily and subject to close monitoring and surveillance through hand-held Personal Digital Assistants that electronically track and trace the movement of goods, but also the movement of drivers and vans (Moore and Newsome, 2018).

Amazon's 'fulfilment centres' operate a well-documented points-based disciplinary system in which penalty points are given in the event of a worker being late in starting a shift or returning from a break, or for sickness or absence; an accumulation of six points can result in the agency terminating your employment (Briken and Taylor, 2018; Bloodworth, 2018). In the South Wales centre there was a system of dual labour control where, in addition to Amazon managers, agency staff with an on-site office acted as workplace supervisors and were responsible for day-to-day discipline and performance management.

Similar disciplinary regimes have been revealed by investigative reporters working under cover for some of the most well-known employers of precarious labour. James Bloodworth (2018) spent several months as an agency worker working in jobs such as a 'picker' for Amazon, an Uber driver and a care worker and describes in each case a punishing work culture in which you could be penalised for not working fast enough or for being ill. Similarly, two *Guardian* reporters described what it's like to work in the warehouse of sports clothing retailer Sports Direct (Goodley and Ashby, 2015) where the agency workers are expected to walk up to 20 miles a day picking products from the shelves, are given a list of 802 sports and clothing brands they are not allowed to wear (and have to undergo visual checks of their socks and underwear to enforce this ban) and again, can accrue penalty points for 'crimes' that include 'excessive toilet breaks', 'time wasting', 'excessive chatting' and 'using a mobile phone'. As Bloodworth found at Amazon, times allocated for picking tasks are almost impossible to hit without running and each employee's performance is published on league tables outside the canteen showing the percentage of targets achieved. In all cases, the undercover journalists' fellow-workers were almost entirely migrants, often from Eastern Europe, with limited command of English. There are two more things to note in these accounts of low-wage precarious working. One is that both companies are personally owned by millionaires (and, in the case of Amazon's Bezos, reputedly one of the richest men in the world), second is that while online web-based technology is the basis for Amazon's whole *raison d'être* and a large part of Sports Direct's business, the actual starting point of the delivery chain is reliant almost entirely on low-tech manual work.

Thus, the day-to-day working reality in these low-wage, precarious sectors of the labour market seems far from the ideal where flexibility offers the worker freedom of choice and the ability to fit work around other social demands:

> From this perspective, the 'British way' becomes a de-humanised, punitive sanctions-based workfare regime that criminalises the poor and dragoons the unemployed, lone parents and disabled into fruitless hours seeking often non-existent decent work with, for many, no choice but to take the degrading jobs on offer.
>
> (Briken and Taylor, 2018)

The precarious position of these workers, daily living 'on the edge' of survival, was starkly illustrated during the coronavirus epidemic. The UK Government's policy responses of 'if you can, work at home' and of extending the period of sick pay, ignored the working reality of the nominal 'self-employed' who, firstly, did not qualify for sick pay and, secondly, could not stop working as they had no other means of support. To opt for the other government suggestion, that they should apply for benefits, would mean an average five-week wait for the payments to come through (even if their claims were successful). Meanwhile, those on zero-hours contracts have seen their hours dwindle or dry up altogether as firms shed staff in a desperate attempt to keep solvent through the economic crisis induced by the epidemic and government reaction.

Work intensity, bullying and the organisational culture

The above concentration on the negative dimensions of 'precarious' work should not be interpreted as implying that the quality of work for those in regular employment remains satisfactory. Phil Taylor (2019) has criticised the Taylor Report for limiting examination of 'modern working practices' to contractual precarity in the labour market, ignoring the deterioration in conditions for full-time permanent employees who form the bulk of the labour force and who experience work intensification through higher workloads or tighter deadlines, and work-related physical and mental health issues.

There is growing evidence that in many workplaces the organisational culture frequently features bullying and harassment. At the end of the 1990s a national workplace survey covering over 5,000 employees and managers in 70 different organisations spanning most of the major employment sectors found that one in ten respondents reported being bullied within the previous six months and that when the reporting period was extended to the previous five years, the incidence rose to a quarter of all correspondents (Hoel and Cooper, 2000). Bullying is a long-term process (a single argument does not constitute bullying) and usually involves a power imbalance, which makes it difficult for the bullied person to defend themselves. Typically, two-thirds of reported bullying cases involve a manager or supervisor as the perpetrator, while a third involves bullying by colleagues or peers (often, however, connected to the presence of a bullying manager). Hoel and Beale (2006) make the further distinction between 'victimisation' in which the target of abusive behaviour is a given individual, and 'oppressive work regimes' where everyone may be subject to the experience. These studies are now 20 years old and the evidence is that the prevalence of such work cultures has not diminished: for example, Graeber has recently collected employees' accounts of what he terms 'ritual humiliation', which has no purpose other than to keep the underling in her place and remind her of the supervisor's superior power position (Graeber, 2019: 119). In May 2019, an independent legal review of allegations of widespread bullying in a Scottish NHS Board found that 'hundreds' of staff had endured years of fear and intimidation by senior managers so severe that some were driven to depression and alcoholism, many had resigned moved to other jobs or retired as a result of their experiences (Scottish Government, 2019).

Health and well-being

The consequences of these intensive and uncertain work regimes can be drastic for workers' physical and mental health. In his 2020 review Sir Michael

Marmot revisited the themes of his comprehensive report of ten years earlier on 'health equity' in the population and concluded that, despite recommendations he made in 2010, the situation regarding inequalities of health and well-being had in fact worsened. In his review of the data on employment his team concluded that there had been an increase in poor quality work, including part-time, insecure employment, real pay was still below 2010 levels, there had been an increase in the proportion of people in poverty living in a working household and the incidence of stress caused by work had increased since 2010. The main causes of work-related stress were found to be workload pressures, including tight deadlines, long hours, too much responsibility, a lack of managerial support and fear of losing the job. The effect of these trends on the overall health of workers and their families had been significant. As stated in their summary of the findings:

> A poor quality or stressful job can be more damaging to health than being unemployed. Unemployment and poor-quality work are major drivers of inequalities in physical and mental health. Being in poverty and working in poor quality employment have marked effects on physical and mental health, including on children in the families concerned.
>
> (Marmot *et al.*, 2020)

As examples of the direct effect of the factors identified by Marmot, Amazon's warehouses have seldom been out of the news. For workplace injuries to be reported to the Health and Safety Executive, they must be severe enough to prevent someone from performing their usual work duties for at least seven days or be from a specified list of injuries that includes fractures, amputation, crushing, scalping and burning. Between 2016–2017 and 2018–2019 a total of 622 legally required Reporting of Injuries, Diseases and Dangerous Occurrences Regulations injury (RIDDOR) reports were made from Amazon warehouses to the HSE – with the number rising each year (152 in 2016–2017, 230 in 2016–2017 and 240 in 2018–2019) (TUC, 2020); these have been only one company's contribution to the figure of 0.5 million work-related musculoskeletal disorder cases in 2018–2019 (HSE, 2019).

In the same period, 2018–2019, there were 0.6 million reported work-related stress, depression or anxiety cases (new or longstanding) (HSE, 2019). Mental health in the workplace has, over the past decade, risen to near the top of current occupational health concerns and could be affecting half of all employees, according to a survey of 44,000 workers carried out by the mental health charity Mind. Due to fear, shame and job insecurity (300,000 people lose their job each year due to a mental health problem) only half of those surveyed had talked to their employer about it (TUC, 2018b). While some employers have responded with mental health first aid initiatives, training 'mental health champions' to support those with stress related problems, unions have pointed to a lack of evidence that such policies have led to any improvements and have emphasised the need for preventive action. In many workplaces even becoming physically ill can have severe consequences for employees due to draconian sickness absence policies, which all too often result in workers continuing to work while ill for fear of losing their jobs (Taylor *et al.*, 2010). In 2018 it was reported that Newcross Healthcare Solutions, one of Britain's biggest providers of agency care workers, was fining staff £50 if they phoned in sick.

Of even greater concern is where the pressures of the workplace become so unbearable that the employee feels they have no option but to end it all in the most final way. A recent meta-analysis of international research in this area concluded that 'regardless of the pathways through which it occurs, it is clear that job stressors are associated with increased risk of suicide' (Milner *et al.*, 2018). In 2007 the Renault Technocentre in France experienced at least three suicides in the space of four months. What is significant is that the suicides were not those of car assembly workers but those of highly qualified engineers and technicians working in a €800m research and design plant set in rolling countryside, with its own restaurants, banks and other facilities. The primary cause of the tragic deaths, according to French unions, was the imposition of quantitative production targets to the creative process. An even more severe series of 18 suicides and 13 suicide attempts were recorded at French telecoms group France Telecom/Orange between 2008 and 2010. In a case going through the French courts at the time of writing, former executives of the company are accused of 'moral harassment' and prosecutors have outlined how, following the privatisation of the former state-owned monopoly, the company aimed to cut its workforce by 22,000 and redeploy 10,000 workers, forcing them into new jobs and giving them unattainable performance objectives.

In case such incidents should be interpreted as reflecting over-badly on the French it should be noted that in the UK employers do not have to report workplace suicides as they are explicitly excluded from the RIDDOR workplace and injury reporting regulations. However, the Unite union has reported that construction on the Hinkley Point nuclear power station was associated with ten suicide attempts by workers in the first four months of 2019 and a 'mental health crisis' (TUC, 2019b).

Low pay

The concentration of zero-hours working in low-paying sectors implies that workers on such contracts (who are more likely to have fewer than 12 months' employment duration compared to other workers) may be experiencing income-insecurity and in-work poverty, a conclusion reached by the House of Commons Briefing Paper (Pyper and McGuinness, 2018). Earnings in gig work seem to support the suggestion that most gig workers are supplementing their income: roughly a third earn less than £4,500 per year from gig work. Clearly, if workers are topping up their income in such large numbers, it suggests a pervasive degree of low pay in full-time employment (Balaram *et al.*, 2017). A survey of the evidence on the degree of precarious employment in the UK found that 21 percent of all employee jobs (5.5 million jobs), including 46 percent of part-time jobs, pay less than the Real Living Wage (a measure of the amount of money that employees and their families need to live) (Shafique, 2018).

A recent report by the Joseph Rowntree Foundation found that, even though official pre-Covid employment rates had reached their highest ever levels, in-work poverty had risen from ten percent of workers 20 years ago to 13 percent and that 56 percent of people in poverty were now in a working family, a significant change from 20 years ago when 39 percent were. This change has been particularly dramatic for children, with seven in ten children in poverty now in a family where at least one person is working (JRF, 2020). Some of the sectors with the highest levels of in-work poverty are familiar to us from reviews of precarious work: the accommodation and catering sector, retail and residential care.

The Joseph Rowntree Foundation measures poverty based on a family's total earnings in work (after taxes), plus any income from benefits and other sources, after they have paid their housing costs. This means the risk of poverty in working families could rise because low-income families' earnings are growing slowly, because changes to the tax and benefits system leave them worse off, or because of a rise in housing costs. Although the National Living Wage was introduced in 2015, pushing up the hourly wage of the lowest-paid employees, this has not reduced in-work poverty. The first reason is that low-income families do not keep that much of any extra income they get from work, because they see their social security payments reduced sharply as they earn more. Secondly, all minimum wage levels are set at an hourly rate and we have already seen several examples of precarious workers not being able to secure the hours they would like to work. JRF found that the main reasons for this were being unable to find work with more hours, a lack of affordable, flexible childcare and transport barriers. Low-paid workers are more likely to work non-standard hours such as evenings and weekends or have irregular shift patterns than other workers and fitting childcare around these work patterns is very difficult (JRF, *ibid.*: 40–41).

If systemic low pay in sectors of the labour market was not enough, many workers also have to contend with what the Citizens' Advice Bureau has termed 'wage theft' – underpayment for work done or unauthorised deductions from pay. A comprehensive report by Clark and Herman (2017) on 'Unpaid Britain' used a number of official data sources to investigate the extent to which wages currently go unpaid in Britain and identify the mechanisms and reasons for non-payment. They found that unpaid wages may present in several forms simultaneously (combinations of deductions from wages, often as 'penalties' for alleged misconduct, unpaid overtime, withholding holiday pay, failing to pay statutory pay, or paying allegedly 'self-employed' workers below the national minimum wage). All these may be accompanied by other abuses of workers' rights such as failure to provide pay slips or written terms of employment. Other common causes of underpayment included employers disappearing while owing wages (known as 'knocking' in the construction industry) or dissolving a company that owes wages in order to start up afresh with a new company ('phoenixing').

The researchers concluded that such practices have been around for a long time and, in the absence of better data, it is unclear whether they are increasing. One reason for their continued persistence, however, may lie in under-reporting by workers due to their giving various forms of consent to the practice: 'uninformed consent', where a worker might agree to the offered terms of employment because they are not aware of their legal entitlement (such as holiday pay for part-time work), 'coerced consent', where a worker accepts the terms offered to avoid losing the job or to avoid benefits sanctions for refusing an offer of employment, or lastly, 'collusion', where the worker accepts inferior terms to avoid tax, keep receiving in-work benefits or evade immigration-related curbs on the right to work (Clark and Herman, 2017).

The new organisation

The Taylor Report and other commentators (Balaram *et al.*, 2017) assert correctly that the world of work is changing but fail to investigate or suggest *why* this might be. So how did this steady erosion of job security for so many workers come about? Why is there such a marked contrast between the degraded nature of many jobs now and the conditions enjoyed in the most mundane of jobs 30 years ago? One underlying reason for the prevalence of insecure and low-paid work is a fundamental shift in employers' organisational strategies. Whereas previously most large employers undertook all functions with in-house staff, so that for example an engineering works would employ its own cleaners and canteen staff, the current prevailing organisational model is to outsource everything but the company's 'core' activity.

While this strategy of working 'lean' has its origins in post-war Japan and the development by Toyota of its 'just-in-time' production system, the organisational model offered such an advance in cost-saving that the idea quickly spread to other manufacturing sectors and subsequently to the service sector in every major industrial economy. Although hyped by management as 'doing more with less', as it has been developed as the current organisational paradigm, 'lean' basically means the ability of figurehead firms to shift the costs of uncertainty to someone else; this could be a supplier, a subcontractor, a franchise and, more importantly for our focus, the employees working for any of these. The French suicides mentioned above represent the most drastic consequence of the uncritical spread of 'lean' production techniques into areas of white-collar work and professional work that were formerly characterised by a high degree of autonomy. A study of the effects of the introduction of lean management into the UK Revenue and Customs service found that the imposition of both quantitative and qualitative performance targets and the monitoring of performance and the public rebuking of officers who failed to meet set performance levels resulted in 63 percent of staff feeling 'very pressurised' compared to one percent prior to 'lean' (Carter *et al.*, 2011).

In 1984 the economist John Atkinson highlighted what he saw as the key features of the emerging 'flexible firm': a core of permanent staff engaged on the essential work of the company and outer zones of peripheral employment, of increasing levels of impermanence, of temporary and part-time workers, outsourced and agency workers, subcontractors and sub-subcontractors and the self-employed. Despite much debate at the time, Atkinson's conceptual model proved to be extremely prescient, with the result that in many sectors of the advanced economies we now have what Weil (2014) terms 'the fissured workplace' in which the hiving off of operations to tiers of contractors and subcontractors, results in the 'fragmentation of the employment relationship' so that agency and temporary workers may be unclear as to which organisation they actually work for. They may be paid by a labour supply agency but obliged to wear the uniform and adhere to the production or service delivery standards of the flagship company: 'Employment is no longer the clear relationship between a well-defined employer and a worker. The basic terms of employment – hiring, evaluation, pay, supervision, training and coordination – are now the result of multiple organisations' (Weil, 2014: 7).

The splitting up of economic activity into layers of responsibility each handled by a different subcontracting organisation also means that 'certain important decisions that do not directly affect the cost of any of the employers, fall through the cracks' (Weil, *ibid.*: 19): among the most important of these are occupational health and safety and environmental impact.

This is the much-hyped 'UK flexible labour force' in which companies can shed some fixed costs by getting someone else to bear them. As we have seen, this has two immediate consequences: the outsourced supplier of labour such as an agency has to bid for the contract by offering the most competitive price and such low-price bids are achieved by paying low wages and dispensing with holiday pay, sick pay or, in the case of bogus 'self-employment', NI contributions.

The second consequence is that the main company can achieve flexibility by laying off or not hiring agency labour when demand is slack without having to worry about paying redundancy money, thus ensuring that such agency work and its rewards are essentially insecure and unpredictable. In the original version of 'lean' at Toyota, one 'waste' to be eliminated was waste time; this was achieved by achieving a continuous flow of components arriving 'just in time', an emphasis on the build quality of the components and a punishing work rate (Kamata, 1983). As Graeber (2019) has suggested, many jobs contain portions of working time where nothing productive is happening; as the ideology of lean has permeated management thinking employers have found they can eliminate those elements by exporting them outside the company. Here the worker bears the cost of unproductive time without financial reward.

The Taylor Report commented that 'a business that can design a workforce model that relies almost exclusively on self-employed labour has the potential to gain a significant market advantage' (Taylor, 2017: 67), to which one might respond that this exactly why so many companies do this, particularly given the UK's short-term business strategies, which prioritise cheap labour over long-term investment.

So, what has this got to do with the planet?

We admit that this chapter paints a fairly gloomy picture of some recent trends that, while arguably more pronounced in liberal market economies such as the UK and USA, nevertheless are affecting many workplaces in the developed world. Of course, not everyone's workplace will display every one of these characteristics but we anticipate enough readers will recognise some of them to perceive current trends in the degradation of work. Were we to add the conditions faced by those in the developing economies, the picture would be even gloomier as an era of unrestricted global business has allowed corporations, through outsourced and agency operations, to take advantage of countries with low environmental standards and few employment protections. And here is the link: as we shall see, the dominant 'free-market' economic model views any sort of government intervention as a hindrance to the effective working of the market and free-market evangelists in political think-tanks and consultancies constantly press for 'de-regulation' (euphemistically described as 'cutting bureaucratic red tape'). This applies as much to environmental regulation, such as limits on emissions, as it does to the regulation of employment and the workplace. In the next chapters we will look at the extent to which a parallel diminution and spoliation has been taking place in the natural environment and then suggest reasons why the two trends are not unconnected.

Bibliography

Atkinson, J. (1984), 'Manpower strategies for flexible organisations', *Personnel Management*, August, 28–31.

Balaram, B., Warden, J. and Wallace-Stephens, F. (2017), 'Good gigs: a fairer future for the UK's gig economy', London, Royal Society of Arts, www.thersa.org/discover/p ublications-and-articles/reports/good-gigs-a-fairer-future-for-the-uks-gig-economy.

Balaram, B. (2018), 'Three stumbling blocks to reforming the gig economy', London, Royal Society of Arts, www.thersa.org/discover/publications-and-articles/rsa-blogs/ 2018/09/three-stumbling-blocks-to-reforming-the-gig-economy.

Biggs, J. (2016), *All Day Long: A Portrait of Britain at Work*, London, Serpent's Tail.

Bloodworth, J. (2018) *Hired: Six Months Undercover in Low-Wage Britain*, London, Atlantic Books.

Briken, K. and Taylor, P. (2018), 'Fulfilling the "British way": beyond constrained choice – Amazon workers' lived experiences of workfare', *Industrial Relations Journal*, 49 (5–6); 438–458.

Carter, B., Danford, A., Howcroft, D. *et al.* (2011), 'All they lack is a chain', *New Technology, Work and Employment* 26 (2); 83–97.

Choi, S. (2018), 'Masculinity and precarity: male migrant taxi drivers in South China', *Work, Employment & Society*, 32 (3); 493–508.

CIPD (2017), 'To gig or not to gig? Stories from the modern economy: survey report', March, www.cipd.co.uk/knowledge/work/trends/gig-economy-report.

Clark, N. and Herman, E. (2017), *Unpaid Britain: Wage Default in the British Labour Market, Project Report*, London, Middlesex University/Trust for London.

De Ruyter, A. and Warneke, T. (2008), 'Gender, non-standard work and development regimes: a comparison of the USA and Indonesia', *Journal of Industrial Relations*, 50 (5); 718–735.

Goodley, S. and Ashby, J., (2015), 'Cheap labour, low-tech work: undercover in Ashley's "gulag"', *The Guardian*, 10 December.

Graeber, D., (2019), *Bullshit Jobs: The Rise of Pointless Work and What We Can Do About It*, London, Penguin/Random House.

Hoel, H. and Beale, D. (2006), 'Workplace bullying, psychological perspectives and industrial relations', *British Journal of Industrial Relations*, 44 (2); 239–262.

Hoel, H. and Cooper, C. (2000), *Destructive Conflict and Bullying at Work*, Manchester, Manchester School of Management, UMIST.

HSE (2019), Health and Safety Executive, 'Health and safety at work: summary statistics for Great Britain 2019', London, HSE, www.hse.gov.uk/statistics/overall/hssh1819.pdf.

JRF (2020), Joseph Rowntree Foundation, 'UK poverty 2019/20', www.jrf.org.uk/report/uk-poverty-2019-20-work.

Kalleberg, A. (2018), *Precarious Lives: Job Insecurity and Well-Being in Rich Democracies*, Cambridge, Polity.

Kamata, S. (1983), *Japan in the Passing Lane: An Insider's Account of Life in a Japanese Auto Factory*, New York, Random House.

Lee, C. and Kofman, Y. (2012), 'The politics of precarity: views beyond the United States', *Work and Occupations*, 39 (4); 388–408.

Manky, O. (2018), 'Resource mobilisation and precarious workers' organisations: an analysis of the Chilean subcontracted mineworkers' unions', *Work, Employment & Society*, 32 (3); 581–598.

Marmot, M., Allen, J., Boyce, T., Goldblatt, P., and Morrison, J., (2020), *Health Equity in England: The Marmot Review 10 Years On*, London, Institute of Health Equity. www.instituteofhealthequity.org/resources-reports/marmot-review-10-years-on/the-marmot-review-10-years-on-executive-report.pdf.

Milner, A., Witt, K., La Montagne, A., Niedhammer, I., (2018), 'Psychosocial job stressors and suicidality: a meta-analysis and systematic review', *Occupational and Environmental Medicine*, 75 (4); 245–253.

Moore, S. and Newsome, K. (2018) 'Paying for free delivery: dependent self-employment as a measure of precarity in parcel delivery', *Work Employment and Society*, 32 (3); 475–492.

Moore, S. and Newsome, K. (2019), 'Management by exception? The Taylor Review and workforce management', *New Technology Work and Employment* 34 (2); 95–99.

Moore, S., Tailby, S., Antunes, B. and Newsome, K., (2019), 'Fits and fancies: the Taylor Review, the construction of preference and labour market segmentation', *Industrial Relations Journal*, 49 (5–6); 403–419.

ONS (2019), 'Labour market economic commentary: July 2019', https://www.ons.gov.uk/employmentandlabourmarket/peopleinwork/employmentandemployeetypes/articles/labourmarketeconomiccommentary/july2019.

ONS (2020a), 'Labour market economic commentary: January 2020', https://www.ons.gov.uk/employmentandlabourmarket/peopleinwork/employmentandemployeetypes/articles/labourmarketeconomiccommentary/latest#analysis-of-zero-hours-contract-working-in-the-economy.

ONS (2020b), 'Employment in the UK: February 2020', https://www.ons.gov.uk/employmentandlabourmarket/peopleinwork/employmentandemployeetypes/bulletins/employmentintheuk/february2020.

Pettinger, L. (2019), *What's Wrong with Work?* Bristol, Polity Press.

Prosser, T. (2016), 'Dualization or liberalization? Investigating precarious work in eight European countries', *Work, Employment & Society*, 30 (6); 949–965.

Pyper, D. and McGuinness, F. (2018), 'Zero-hours contracts: House of Commons Briefing Paper 06553', 17 August.

Scottish Government (2019), www.gov.scot/publications/scottish-government-response-report-bullying.

Shafique, A. (2018), *Addressing Economic Insecurity*, London, Royal Society of Arts/ Nottingham Trent University.

Standing, G. (2011), *The Precariat: The New Dangerous Class*, London, Bloomsbury.

Taylor, M. (2017), *Good Work: The Taylor Review of Modern Working Practices*, https://assets.publishing.service.gov.uk/government/uploads/system/uploads/attachment_data/file/627671/good-work-taylor-review-modern-working-practices-rg.pdf.

Taylor, M. (2019), *Hermes and the GMB – the taxman cometh?*, London, Royal Society of Arts. www.thersa.org/publications-and-articles/matthew-taylor-blog/2019/01/hermes-and-the-gmb-the-taxman-cometh.

Taylor, P., Cunningham, I., Newsome, K. and Scholarios, D., (2010) '"Too scared to go sick": reformulating the research agenda on sickness absence', *Industrial Relations Journal*, 41 (4); 270–288.

Taylor, P. (2019), 'A band aid on a gaping wound: Taylor and modern working practices', *New Technology, Work and Employment*, 34 (2); 95–99.

Thompson, P. (2019), 'The Taylor Review: a platform for progress?', *New Technology, Work and Employment*, 34 (2); 106–110.

TUC (2018a), Trades Union Congress, 'Ending the Undercutters' Charter: why agency workers deserve better jobs', London, TUC. www.tuc.org.uk/research-analysis/reports/ending-undercutters-charter.

TUC (2018b), Trades Union Congress, 'Poor mental health at work "widespread"', *Risks* 866, 15 September 2018.

TUC (2019a), Trades Union Congress, 'UK's gig economy workforce has doubled since 2016', www.tuc.org.uk/news/uk's-gig-economy-workforce-has-doubled-2016-tuc-and-feps-backed-research-shows.

TUC (2019b), Trades Union Congress, 'Union action call on suicide linked site', *Risks* 911, 24 August 2019.

TUC (2020), Trades Union Congress, 'Risks 936', 29 February 2020.

Weil, B. (2014) *The Fissured Workplace: Why Work became so Bad for so Many and What Can be Done to Improve It*, Cambridge Mass., Harvard University Press.

3   A threatened environment

Introduction

Since ancient times humanity has lived in uneasy and ambivalent relations with its environment (Thiele, 2016). For 12,000 years, during the geological epoch known as the Holocene, the Earth's climate was stable and warm (Raworth, 2017: 48), sustained by modest and balanced human demands on the environment. The present epoch, the Anthropocene, a name derived from Greek for 'recent time' and 'humans' (Lewis and Maslin, 2018: 5), reflects the environmentally disruptive impacts of industrial and associated commercial activities, together comprising 'all the immense and far-reaching impacts of human actions on Earth' (*ibid.*: 6; Lovelock, 2019). At one environmental extreme a sustainable, albeit fragile, harmonic balance can exist between nature and satisfying man's consumptive and productive needs, though Harari points out in his popular history that vigorous depletions of animal species from the earliest days of humanity to the present 'makes *Homo sapiens* look like an ecological serial killer' (Harari, 2011: 74; 83). Nevertheless, as environmental activist George Monbiot reflects: 'the world lives within us; we live within the world. By damaging the living planet we have diminished our existence' (Monbiot, 2017: 91). In other words, what we take from the environment, we need to return to the environment, which is the essence of sustainability (FitzRoy and Papyrakis, 2016: 54). Indeed, a standard definition of sustainable development rests on this fine balance: 'humanity has the ability to make development sustainable to ensure that it meets the needs of the present without compromising the ability of future generations to meet their own needs' (Helm, 2015, in Pilling, 2018: 210), though, of course, interpretations of 'needs' can vary significantly.

We can identify two diametrically opposed ways of seeing our relationship to the environment. Balance is still pursued, though increasingly precariously, among many indigenous communities, who aim to protect traditional fishing, hunting, gathering and agricultural practices in the face of incursions from invasive modern methods of industrial extraction, transport and production (Klein, 2014: 443). The words of Raoni Metuktire (2019), chief of the indigenous Brazilian Kayapó tribe, addressed directly to rainforest developers but also to the wider community of backers and beneficiaries, are chillingly apposite:

To live you must respect the world, the trees, the plants, the animals, the rivers and even the very earth itself. Because all of these things have spirits, all of these things are spirits, and without the spirits the Earth will die, the rain will stop and the food will wither and die too.

We all breathe this one air, we all drink the same water. We live on this one planet. We need to protect the Earth. If we don't, the big winds will come and destroy the forest.

Then you will feel the fear that we feel.

The other, modernist, extreme, is typified by humanity's attempts in pursuit of economic progress and efficiency to dominate nature and environment, supported by political imperatives and ideological dogma and underpinned by advances in technological and commercial development. Tensions between tradition and economic modernity are manifest today with many indigenous communities attempting to maintain customary sustenance activities in unequal competition with international conglomerates whose profit-seeking pursuits extend commercial activities to the most vulnerable and inaccessible areas of the planet, often supported by aggressive local political and military actions deployed against host communities (Klein, 2019: 189). A prime example is the case of rainforests being forcefully depleted to make way for commercial products such as palm oil and soybeans.

Two routes to crisis

The threat to the planet's environment from two centuries of cumulative economic activity arrives via two routes. The one we usually think of, under the rough heading of 'pollution', consists of all the waste materials that are expelled, dumped and pumped into the earth, seas and atmosphere because they have no economic market value: chemicals, heavy metals, plastics, furnace slag, mining spoil and gases. The long-held assumption was that the seas and atmosphere had an infinite capacity to absorb these wastes without incurring any substantial change to how the planet worked; we now know how mistaken that was. The environmental crisis that is precipitating global heating is attributable primarily to the gas emissions caused by burning fossil fuels (CDP, 2017). Concentrations of CO_2 and other greenhouse gases (GHGs) such as methane and nitrous oxide have increasingly accumulated in the atmosphere since the start of the coal-powered industrial revolution some two hundred years ago, a process that is still occurring today according to observations undertaken by the World Meteorological Organization (WMO, 2019). Global emissions of the potent and persistent GHG, nitrous oxide, a by-product of nitrogen-based fertilisers, also appear to be rising much faster and are twice as large as previously estimated (Thompson *et al.*, 2019). The greatest abuse, of course, against nature, and paradoxically, against humanity itself, is in the continued, and indeed, expanding extraction and exploitation of carbon-derived products, such that despite all the warnings of its contribution to global heating and catastrophic climate change, in the twenty-first century, coal represents the fastest-growing energy source (Goldstein and Qvist, 2019: 4).

Another major source of environmental concern that has received recent attention is the extent of plastic pollution, especially in marine and waterways environments, with estimates of over five trillion plastic pieces, weighing over 250,000 tons located in the world's oceans, then fragmenting and entering the extensive marine food chains. Microplastics, with traces found in food and air samples, have become another potential pollutant, though long-term health implications are not yet known with certainty. Nevertheless, a key review article identifies possible hazards to health through particle, chemical and microbial contamination, inhalation and ingestion, stimulating potential allergic and immune responses (Wright and Kelly, 2017).

The second route to damaging the planet comes from the initial processes of extraction of raw materials, demonstrated by the systematic deforestation referred to above. The importance of the role played by our forests in acting as a sink absorbing carbon dioxide from the air and thus mitigating GHG emissions is now understood, despite which they are being cut down at an ever more alarming rate. One major cause is the mass market for beef in the diets of developed nations, which has outstripped the ability of domestic agricultures to produce enough cattle food, this gap being filled by soybean cultivation. According to the World Wildlife Foundation, in the past 70 years there has been a massive expansion of soybean production, especially in North and South America, which has contributed to deforestation, and hence reduced carbon reservoirs, displacement of local farmers and their communities. The other major cause of deforestation is the creation of palm oil plantations. Almost all oil palm, an ingredient found in consumer products such as soap and cosmetics, has been planted in areas populated by rainforests. A recent study shows that nearly half of oil palm plantations in South East Asia and almost a third in South America came from areas that were virgin forests 30 years ago (Vijay *et al.*, 2016). Many plantations in South East Asia are operated and owned by North American food and agri-business concerns.

Deforestation operations help to eliminate biodiversity and deprive local communities of both employment and means of sustenance. Worse, it has been calculated that deforestation causes four-fifths of Indonesia's CO_2 emissions, making the country one of the world's largest emitters of greenhouse gases (Rainforest Action Network, 2019). According to the World Resources Institute's *Global Forest Watch*, some 12 million hectares of tropical forest cover were lost in 2018 (Showstack, 2019) with nearly half the loss in Indonesia and Brazil. Between 2018 and 2019 there was a reported 84 percent increase in the number of fires across Brazil, with more than half in the Amazon region. There were nearly 2,500 outbreaks in Acre state, representing a 176 percent increase from 2018 (Garvey *et al.*, 2019). And worse is expected; the Brazilian president, Jair Bolsonaro, is establishing steps to 'develop' the Amazon rainforest, including a hydroelectric dam, intrusive highways and plans to integrate the previously unspoiled Amazon region into the national economy (Woodward, 2019).

There are additional environmental problems associated with the extraction of fossil fuels themselves. Transportation, increasingly originating from inaccessible extraction sites, requires lengthy pipelines, risking spillage and leakage, often in environmentally fragile areas. Recent examples include the 210,000 gallons of oil leaked from the Keystone Pipeline in South Dakota in 2017. According to Greenpeace (2018), since 2002, in Minnesota alone seven pipeline operators have reported '132 hazardous liquids incidents, 17 of which were larger than 2,100 gallons'. Serial offender pipeline operator Enbridge has been cited for multiple incidents involving leaks and failure to fully inspect its pipeline network. Also, massive oil tankers require dedicated harbour facilities, which risk multiple disruption to local habitats. Negative environmental impacts of shipping destinations derive from the vicinity of the port, through the infrastructure of service roads and railways, to the impact of transfer activities themselves. According to the OECD (2011), impacts include noise pollution from engines and loading/unloading; particles from nitrogen, sulphur and carbon exhausts; and dust from handling coal, sand, grain and other products. Tankers themselves are at risk of spilling huge amounts of deadly pollutants over wide areas, as demonstrated tragically with the *Exxon Valdez* disaster in 1989, when 37,000 metric tonnes of crude oil spilled into remote Prince William Sound in Alaska, devastating local wildlife and bankrupting fishing families. Thirty years later, oil residues still remain. Whilst the incidence of spills has declined in recent years, as recently as 2018 the tanker *Sanchi* sank east of Shanghai following a collision with a cargo ship. Crew members died following the collision, after which some 113,000 tonnes of natural gas condensate was released, causing widespread contamination of neighbouring coasts. Extraction, processing and transport of tar-sands deposits from Alberta demonstrate both the contribution to GHGs and the frightening multiplicity of powerful interests that underpin and control fossil fuel projects. Canada's tar-sands contain vast quantities of trapped carbon, with potential yield even higher than that of Saudi Arabia. The extraction process has been described in graphic detail: these huge pits

> are torn in state-sized chunks out of the earth's mantle by monstrous wrecking machines ripping out the boreal forest lands by the roots, destroying the carbon sinks and water-hold stabilization they provide in the Northern region as well as pumping out ever more climate-changing gases.
> (McMurtry, 2019)

The tar then has to be steam-driven from the open-pit mines using massive quantities of water, which becomes contaminated in the process. Then comes the problem of transport to the USA of the crude product. Transport by massive tankers through the environmentally sensitive coast of British Columbia or by pipeline over the Sierra mountain ranges becomes a major potential ecological and logistical problem and whilst the pipeline is currently protected by legislation, opposition to the statute has been remorseless from Conservative politicians, 'Big Oil' corporations, supported by financial and ideological underpinning from the Koch brothers and other libertarian interests (*ibid.*, 2019).

An old problem

Recognition of the dangers posed by anthropogenic, and especially industrial, activity is not new. In his comprehensive history of the industrial factory system, Freeman quotes a contemporary account of the poisoned environment of mid-nineteenth-century Manchester as 'the entrance to hell realized', the whole city being 'a chimney'. Water contamination from textiles waste and sewage resulted in one main water source being described as 'less a river than a flood of liquid manure, in which all life dies' (Freeman, 2018: 27). Freeman also comments on the environmental damage associated with cotton mills in America and Britain, whose location and operation demanded deforestation, resulting in soil depletion and reduced carbon capture (*ibid.*: 29). Early recognition of the detrimental impact of industrialisation on society led to limited and fairly conservative attempts to localise pollution, so that mill owners and a growing middle class could live above and away from the urban nightmare that their industrial activities had provoked (Akatsu, 2015). While such overt industrial despoliation is now relatively rare in developed countries, the problems have been exported to newly industrialising countries, keen to capture markets and sustain economic growth, irrespective of negative environmental impact, often making the understandable argument that the mature economies of the North have exploited the environment for at least two centuries, during which time their carbon emissions have persisted in the atmosphere and it is unjust to blame and penalise newly industrialising countries for adopting similar features of production and patterns of consumption (Klein, 2014: 416).

Nevertheless, not only vulnerable developing countries face life-threatening consequences of industrial despoliation: poorer areas of rich countries are equally susceptible. Overlooking Richmond in environmentally aware California, the massive Chevron refinery processes a quarter of a million barrels of crude oil daily. The rate of children's asthma is reported to be twice that of bordering communities and those living closest to the refinery report high levels of cardiovascular disease and cancer (Cagle, 2019). In parts of Louisiana, highly toxic chemicals, including carcinogenic benzene, chloroprene and ethylene oxide emitted from chemical companies are reported to contaminate the atmosphere. According to the United States Environmental Protection Agency (USEPA), the Ponchartrain facility, once operated by DuPont and now owned by Japanese company Denka, along with other local petrochemical operations, delivers the highest risk of cancer from airborne pollution in North America, up to 50 times the national average (Hersher, 2018). These claims are strongly denied by local petrochemical thinktank Lake Area Industrial Alliance, with support from the libertarian Cato Institute, which has described the alleged pollution as an 'environmental myth' (LAIA, undated).

Notwithstanding persistence of these 'myths', negative sustainability consequences have become evident, alongside growing realisation and acceptance that survival of the planet is now under existentialist threat (Thiele, 2016; Leichenko and O'Brien, 2019; Romm, 2018). This growing emergency was highlighted in a recent UN report: 'Climate change is now affecting every country on every continent. It is disrupting national economies and affecting lives, costing people and countries dearly ... Without action, the world's average surface temperature is likely to surpass 3 degrees centigrade this century' (United Nations, 2019). If unchecked, scientists warn that this level of warming would be globally calamitous (Stern, 2006; IPCC, 2018). Impending catastrophe provided the focus for the 2015 Paris Agreement, which sought to limit global temperature rise to below 2°C and to aim for 1.5°C. World leaders were united in signing the agreement to cut GHG emissions largely responsible for the rise in global temperatures. Other key targets aim to maintain balance between human-sourced GHG emissions and their natural absorption; to provide five-yearly reviews of each country's progress in cutting emissions; and for wealthy countries to provide financial incentives to poorer ones to switch to renewable energy sources (IPCC, 2018).

The Paris agreement received a considerable setback when Donald Trump declared his intention to withdraw from the accord, a decision that was finalised in November 2020, but certain to be reversed by incoming President Joe Biden. Nevertheless, as Klein reports (2014: 409), the USA, with less than five percent of the world's population, contributes at least 14 percent of total carbon emissions, levels that if unchecked are unlikely to reduce in the absence of emission controls and in a country that has positively promoted fossil fuel extraction (Parker and Davenport, 2016). According to data reported by researchers at *Global Energy Monitor*, from a total of 302 oil and gas pipelines under development in 2019, over half are located in the USA; $232.5bn in capital spending is being directed into these North American pipeline projects, with at least $1tn allocated to total oil and gas infrastructure. In the USA, these developments in gas output would lead to an additional 559 million tons of CO_2 emissions annually by 2040, above 2017 levels (electricityinfo, 2019). Ted Nace, co-author of the report, maintains that 'this is a whole energy system not compatible with global climate survival. These pipelines are locking in huge emissions for 40 to 50 years when the scientists say we have to move in 10 years' (Ambrose, 2019).

The Paris agreement followed an earlier attempt, the 1997 Kyoto Protocol, to establish emission-reducing targets, directed specifically at economically mature countries. China and India, two of the main producers of GHGs but then regarded as developing economies, were not required to participate, other countries failed to comply with the protocol and the USA subsequently withdrew support. Criticisms were aimed both at the limited scope of the Protocol and its inherent 'institutional design failures' all of which contributed to its comparative failure (Rosen, 2015). The 2018 report from the *Intergovernmental Panel on Climate Change* (IPCC), the United Nations body charged since 1988 with assessing the science related to climate change, reveals the fragility of global society's condition. On the basis of evidence accumulated from a wide range of scientific disciplines, the panel argues that to achieve stated aims to limit global warming to 1.5°C and to 'ensure a more sustainable and equitable society', 'rapid, far-reaching and unprecedented changes in all aspects of society' are required. This goal demands equally rigorous 'transitions in land, energy, industry, buildings, transport, and cities'. Net human-derived

CO_2 emissions will need to fall by about 45 percent by 2030 from 2010 levels in order to prevent an irreversible tipping point after which physical environmental manifestations, such as melting Antarctic ice sheets and rising sea levels, become inevitable (Maslin, 2014). Already, the report points out, global warming of just 1° C has been linked to 'extreme weather patterns, rising sea levels and diminishing Arctic sea ice' (IPCC, 2018). Fresh studies suggest that up to 630 million people, three times more than previously forecast, are living in coastal regions threatened by sea level rises and flooding by the end of the century, with coastal Asian conurbations most at risk (Kulp and Strauss, 2019). Faced with these threats, action against global heating has become critical, with action specifically required to combat carbonic extraction, whether as coal, lignite, or methane and its deployment to fuel industry, commerce, and air, land or sea transport. In short, current patterns of productive industry, fuelled by political and financial imperatives toward maintaining profitability and economic growth, are threatening the survival of the planet.

Recognising the severity of the crisis, 193 heads of government signed the UN's 'action plan for people, planet and prosperity' in its Transforming the World programme for sustainable development, marking the plan as official UN policy (United Nations, 2015). The programme builds on the Millennium Development goals agreed by the UN in 2000, which provided a framework for sustainable development but without binding requirements made uneven progress with regard to environmental enhancement. The new agenda goes much further in its environmental and human rights ambitions, consisting of 17 broad goals, within which are contained 169 more specific targets. The wide-ranging aspirational goals aim to end global poverty and hunger, ensure equality and healthy lives and provide sustainable energy, economic growth, development, production and consumption within a multilateral framework that seeks urgent action to combat climate change and its effects.

Clearly, achieving all these human rights and environmental ambitions within a planned 15-year time frame would be unlikely but the importance of the document is contained in its global validity, meaning that each country has its own responsibility to undertake the steps necessary to reduce carbon emissions, protect biodiversity and provide environments safe from pollution. In the words of the opening declaration: 'We are determined to protect the planet from degradation, including through sustainable consumption and production, sustainably managing its natural resources and taking urgent action on climate change, so that it can support the needs of the present and future generations' (United Nations, 2015: 3). Nevertheless, the problem of reconciling the broader planetary impacts with localised acts is exposed by the document's affirmation that 'every State has, and shall freely exercise, full permanent sovereignty over all its wealth, natural resources and economic activity' (*ibid.*: 7), a defence vehemently raised by Brazil's President Bolsonaro in August 2019 when confronted by European governments over the planetary environmental implications of his scorched earth policy for the Amazon rain forests. Also, the plan hands initiation and operation of environmental policy to national governments, and as later chapters demonstrate, governments are taking very varied steps and acting at different speeds toward addressing the environmental agenda as well as conforming with the Paris agreement on restricting carbon emissions.

Who creates the pollution?

Despite accelerated accumulation in the present century, the prime historical contributors to GHG warming have been the mature developed economies of Europe and America, many of which are now taking steps, albeit hesitatingly, to reduce their domestic emissions. Newly industrialising countries are now major contributors, but this has been a relatively recent development (CDP, 2017). In a major report, Costello and his colleagues are blunt in their conclusions: 'Climate change will have its greatest effect on those who have least access to the world's resources and who have contributed least to its cause' (Costello *et al.*, 2009: 1694). The top ten percent of emitters generate nearly half of global emissions while the bottom 50 percent generate a mere 13 percent (Raworth, 2017: 58). For the world's most vulnerable communities, dire predictions have already become reality, with food insecurity related to climate change presenting threats to actual survival. Moreover, studies demonstrate associations between land degradation, loss of agricultural diversity and political unrest with countries most at risk of food deprivation. These most vulnerable countries are those with the lowest contributions to carbon emissions (Ware and Kramer, 2019). Burundi has 90 percent of its population working in an agriculture sector that is already being adversely impacted through changing weather patterns. Malnutrition is now endemic in the country, which is calculated to be the least food secure in the world, yet its carbon footprint is virtually zero. Conversely, countries that contrived to block universal adoption of the IPCC (2018) 1.5° limitation report, including the United States, Saudi Arabia and Russia, are those most contributing to high emissions and climate change (Ware and Kramer, 2019). The authors estimate that the average American generates the equivalent CO_2 of 581 Burundians and average Saudi the equivalent of 719 Burundians (*ibid.*: 6). Oxfam (2020) reports that a UK citizen emits more CO_2 in two weeks than individuals in African countries like Malawi and Ethiopia do in a year. The UK was placed 36th in the world for its CO_2 emissions in 2017, with mean per person emissions of 8.34 tonnes annually against a global average of 4.7 tonnes. Eighteen of the 20 lowest emitters were in Africa, with an annual emission average of less than one tonne per person. Other food-insecure countries, notably in the Southern hemisphere, are being affected by combinations of droughts (Malawi, Chad), rise in seawater levels (Yemen), floods (Sierra Leone) and risks to biodiversity (Madagascar) as a consequence of climate change, for which they bear little or no responsibility. Nearly nine out of ten of Chad's population of 11 million live below the poverty line and 40 percent of children under the age of five suffer from malnutrition. In view of these vast imbalances, not surprisingly, there have been differences in opinion as to who should make the biggest contributions toward reducing – and paying for – CO_2 emissions. Further, industrially developing countries can claim that through globalisation and outsourcing, mature economies have simply exported environmentally damaging extraction, transport and production to poorer, less rigorously regulated, countries.

Major agencies like IPCC and USEPA confirm that burning fossil fuels is the primary contributor to CO_2 emissions, with use of coal, natural gas and oil to provide for electricity and heat forming the largest global contributor to GHG emissions (25 percent in 2010). Fossil fuel use also contributes to emissions through deforestation and soil degradation (USEPA, 2017). Industrial use of fossil fuels burnt on site for chemical, metallurgical and mineral transformation as well as for waste management contributed 21 percent to GHG emissions. Agriculture and other land use contribute a further 24 percent, with fossil fuels burnt for transportation by land, sea and air from petroleum-based fuels adding an additional 14 percent. Heating buildings (six percent) and other sources, such as fuel extraction, processing and transportation (ten percent) were other major contributors to CO_2 emissions. In 2014, the highest emitters of CO_2 were, in descending order, China, USA, European Union, India, Russian Federation and Japan (*ibid.*).

The authoritative *Carbon Majors Database* 2017 report by CDP, an independent research agency, identifies the principal emitters of greenhouse gases, finding that just 100 companies world-wide were responsible for 72 percent of global industrial GHG emissions. Since 1988, just 25 corporate and state-owned companies have been responsible for more than half of global industrial emissions. These companies include household names such as BP, Shell, ExxonMobil and Chevron as well as major state-owned assets such as those in China: 'Since the turn of the millennium, growth in Chinese coal production has tripled to nearly 4 billion tons, representing half of global output' (CDP, 2017). Rather than declining, China's dependence on coal is increasing, according to a report by Shearer *et al.* (2019) for the *Global Energy Monitor*. In the 18 months from January 2018, countries outside China decreased total coal-based power capacity by 8.1 gigawatts, mainly through decommissioning. During the same period, China's capacity increased by an enormous 42.9 gigawatts. Forecasts for China's economic plans indicate continued expansion of coal fleets for power stations, despite estimates that the country needs to phase out most of its coal power capacity by 2035 to meet IPCC's 2°C target. An added problem is that China is also financing one-quarter of all new coal projects world-wide. According to the *Climate Accountability Institute*, which has been tracking carbon emissions for decades, just 20 enterprises, representing a mixture of investor-owned, state-owned and government-run companies, have contributed some 480bn tonnes of CO_2 emissions since 1965, representing 35 percent of the global total. Half the emissions of 90 major entities have occurred since 1986, 'demonstrating the increasing speed with which fossil fuels are being burned' (2019: 4).

One glimmer of light from these findings is that this small number of both state-owned entities and investor-owned companies may be sensitive to external and internal pressures to develop transition plans toward more sustainable energy activities. Indeed, there are some signs that this is already happening. Some oil and gas companies are shifting toward new, if limited, sustainable initiatives with carbon capture and storage and investment in diverse renewables projects, though how much companies are investing in these projects compared with continued high levels of investment in oil and gas is disputed (Carbon Tracker, 2019). As oil and gas companies seek out increasingly remote and inaccessible sources of fossil fuels in order to secure a minimum 100 percent 'reserve-replacement ratio' to maintain investor confidence (Klein, 2014: 147), so the dangers of environmental catastrophe increase, as was brutally demonstrated in BP's Deepwater Horizon operations (see Chapter 4). Extending drilling to Arctic regions presents additional environmental risks, with the prospect of melting permafrost releasing methane and other GHGs.

A toxic planet?

The publication of Rachel Carson's *Silent Spring* in 1962 illuminated a number of issues that remain highly relevant today. First was the exposure of widespread irresponsible industrial deployment and abuse of highly toxic substances, in this case organophosphates such as DDT to control pests, in pursuit of large profits for giant chemical companies, notwithstanding their compound disastrous effects on wildlife. Second was the revelation that government, company scientists and corporations could act in concert and be implicated in falsely refuting accumulated solid scientific evidence, portraying it, in contemporary terminology, as 'fake news'. Third was the reaction of the implicated organisations and their media sympathisers, who first attempted to prevent publication of her book and, when this failed, financed campaigns to undermine Carson's personal integrity (she was a woman, after all, interfering in man's work in a man's world), and condemning her findings and arguments as alarmist and even fanatical, a charge that has persisted, like the DDT that Carson investigated, until well into the twenty-first century (Oreskes and Conway, 2012: 223). Finally, though, and on a positive note, DDT was banned in the USA in 1972 and Carson's reach out to the public was instrumental in initiating the modern ecological movement (Lear, 1998). Sixty years later, a similar pattern emerges with credibility-undermining critiques directed at prominent eco-activists, who dare to investigate the dubious activities of global enterprise. Publications by Vandana Shiva, a strong opponent of the use of genetically modified crops in India and Carey Gillam in the USA have led to similar repercussions faced by Carson (Shiva, 2018; Gillam, 2017). Gillam, whose book investigating links between Bayer-owned Monsanto's glyphosate weedkiller and cancer, is alleged to have been targeted in diverse ways by the company, including through social media, with attempts to discredit her and her publications.

While the accumulation of CO_2 emissions threatens to wreak environmental havoc, there have been and continue to be other forms of widespread industrial pollution with the potential to shorten lives and destroy natural habitats. A recent UN-supported report compiled by the Intergovernmental Science Policy Platform on Biodiversity and Ecosystem Services (IPBES) found that a principal threat to global biodiversity is presented by human activities like industrial agriculture (IPBES, 2019). The IPBES chair, Robert Watson, did not mince his words: 'The health of ecosystems on which we and all other species depend is deteriorating more rapidly than ever. We are eroding the very foundations of our economies, livelihoods, food security, health and quality of life worldwide' (IPBES, 2019). And again, familiar patterns of obfuscation and denial by vested interests serve to blur recognition of the impacts of contaminants. Considerable sums are deployed by companies, their opaquely funded think tanks and public relations specialists, to discredit scientific claims and sway public opinion, often through the dubious conduits of social media (Chakrabortty, 2019). Epidemiologist David Michaels, who served in a senior position for the US Occupational Safety and Health Administration (OSHA), has written knowledgably of the egregious – and continuing – efforts of 'product defenders' to 'manufacture uncertainty about the science, then attack the public health agencies trying to protect the public' (Michaels, 2020: 36). This defence, which often translates into aggressive attack, extends to the success of what he terms the climate change denial machine 'in stymying restorative action, even as its case against the science of climate change becomes more desperate and fatuous by the day' (*ibid.*: 181).

In addition to the climate crisis, two areas of particular concern have been micro-plastic pollution of waterways and oceans and deterioration in air quality, especially in newly industrialising countries such as India and China but also in large cities across the globe. It has become established that minute PM2.5 particles (i.e. particles smaller than 2.5 micrometres containing micro-particles of black carbon, sulphates, organics and trace metals (Guan *et al.*, 2014)) and inhaled nitrogen dioxide affect vital organs, reducing longevity and contributing to chronic respiratory, cardiac and other life-threatening illness. Infants and children are at greatest risk of airborne pollution through low birth weight and respiratory ailments like asthma (Gardiner, 2019: 26). Gardiner reports that nearly everyone across the world inhales polluted air; that one out of every nine deaths is attributable to air pollution; and in North America, 'dirty air brings more than 100,000 American lives to a premature end' (*ibid.*: 18). Further, higher Covid-19 death rates have been linked to air pollution, which concentrates in highly populated and less advantaged areas (Friedman, 2020). It has been estimated that over a million premature deaths in China in 2010 were associated with air toxicity (Wang *et al.*, 2012). In 2013, heavy smog covered 15 percent of China's national territory and in Beijing, PM2.5 concentrations reached 40 times the WHO standards for good health (Guan *et al.*, 2014). These authors link the rapid increase in economic growth for the period 1997–2010 with a commensurate rise in energy consumption, and specifically through coal use. Further, between these years, 'China's gasoline and diesel consumption volume has increased almost three-fold ... fuelling the six times increase of vehicle volume' (*ibid.*, 2014). Subsequently, the government announced a number of measures aimed at reducing air pollution, including 'phasing out of inefficient industrial boilers, improving fuel quality, promoting cleaner production, optimizing the energy mix, enhancing regulations, and a marketing stimulus for green energy development' (*ibid.*, 2014).

Throughout industrial history, intense competition has led producers to secure profitability through lowering costs by discounting externalities (see Chapter 4) or even ignoring safety standards, at potential risk to employees and the general public. Historically, at the end of the nineteenth century, we can see the example of Bryant and May's match factory operating with toxic white phosphorous, even though safer red phosphorous alternatives were available. For the 1,500 young women and children forced to apply the phosphorous at the company's East London factory, the health effects were catastrophic in the often-fatal necrosis of the jaw, so-called 'phossie jaw', which many of the workers succumbed to as a consequence of handling the phosphorous. The scandal was revealed by social reformer Annie Besant in 1888 and led to the famous match-girls' strike of the same year. Despite, or possibly because of, the availability of more expensive but safer substitutes, white phosphorous was not formally banned until 1910 in the UK, following protracted agitation by reformers and determined resistance from producers (Best, 2018). Employer responsibility for their impact on the environment can be hard to pin down. In December 1984, a whole community was fatally exposed to toxic fumes when lethal methyl isocyanate gas spilled out from Union Carbide's Indian subsidiary in Bhopal, poisoning thousands of local inhabitants immediately and many more subsequently through untreated environmental contamination. The parent American company passed liability to its Indian subsidiary, contending that it had no responsibility for the disaster, a submission supported by the US courts. Little compensation has been paid to survivors and their families and arguments over liability continue to this day. The authors of a book tracing the Bhopal poisoning and its aftermath conclude that in India

> we continue to see smaller industrial accidents – mini-Bhopals. Hazardous wastes are piling up in many parts, contaminating land and water and endangering lives. But we do have the means or methods to remediate these toxic sites. How do we prevent not just another Bhopal but also the mini-Bhopals from happening?
>
> (Narain and Bhushan, 2015)

Environmental externalities arising from commercial activity help to underline direct links between planetary degradation and social, economic and health inequalities, whether domestic or cross-national. As social commentator Éloi Laurent (2019) argues, 'inequality literally pollutes our planet', with the world's poorest suffering contaminated air and drinking water and high exposure to the disastrous environmental effects of global heating. He points out that 90 percent of deaths related to air pollution occur in poorer countries. Even in rich countries, there can be vast health differences attributable to differences in exposure to air pollution between affluent and less affluent areas. Laurent summarises neatly: 'Inequality is a pollution enabler; pollution is an inequality accelerator'.

Who has responsibility for the environment?

The planet's global health is threatened from different but interconnected directions, of which climate change from GHG emissions is arguably the most serious and urgent. The above analysis indicates that whilst there are numerous sources for GHG emissions, economic and industrial activity is the prime stimulant, specifically through meeting pressures for continuous economic growth in a world of finite resources, thereby demanding ever higher consumption levels and encouraging greater inequality in its wake, which in a vicious spiral, boosts demands for further growth, potentially at the cost of working and community conditions and demonstrably so in less developed and poorly regulated economies. Hence, GHG emissions have been stimulated through globalisation and its impacts on minimum cost production, extraction and transport and made more complex to monitor through outsourcing and offshoring, where responsibility for production processes and environmental safeguards become more diffuse and tenuous to hold to account. Clearly, continuing reliance on extracting and burning fossil fuels in the face of accumulated scientific evidence testifying to its calamitous impact on the planet needs to be confronted. Arguments over the benefits and disadvantages of fracking in Britain demonstrate the sensitivity of the sustainability issue, with environmentalists and their supporters contending that fracking not only contributes to GHG emissions but risks contaminating local habitats and communities (see e.g. Willow and Wylie, 2014), whilst proponents, which have included the UK and other governments, offer measured support to fracking on the questionable grounds that it is an independent domestic source of energy that produces fewer carbon emissions than coal.

Responsibility for protecting the environment can be identified at different levels. There are international coordinating agencies like the IPCC and IPBES, whose recent warnings of irreversible climate damage through carbon emissions have sent a world-wide message of potential life-changing threats to human life, though as Maslin (2014: 136) points out, 'there is no binding international agreement to cut emissions'. Despite the authority of the scientists' analysis, even bodies like the UN, which convened the IPCC, have few if any regulatory sanctions to mitigate emissions. Even when international law prohibits production of toxic chemicals such as ozone-destroying trichlorofluoromethane (CFC-11), used in refrigerators and as insulating material before the ban, countries with weak monitoring and regulatory capabilities may be unable or unwilling to clamp down on rogue CFC-11 production operations like those reported currently functioning in China (Tollefson, 2019).

There are also intergovernmental bodies such as the European Union (EU), which at least has regulatory powers, but implementation of EU policies has been hindered by compromises with national governments (Jordan, 1999) and the weakness of sanctions. Until recently, treating environmental issues has been subsidiary to responding to major economic crises visiting Europe. A recent study was concerned that the 'combined effects of [EU] enlargement and the economic crisis have reduced the EU's appetite for ambitious environmental policy ... In the medium to long term, the EU's environmental policy ambitions and activism will continue to wane' (Burns *et al.*, 2019). The main economic trajectory of the EU has, in contrast, been directed toward support for free-market trading conditions, albeit with some regulatory environmental super-structure. A similar perspective applies to individual governments, who do have legislative powers to enforce environmental safeguards or to engage in policy shifts from fossil fuels toward more sustainable sources of energy, but in many cases, the record is patchy, and as we have seen in the case of North America, supportive of fossil fuels. The British government does have its official scientific advisers, the Committee on Climate Change (CCC) whose 2019 set of proposals advocated a legally binding target to reduce GHG to zero by 2050. Subse-quently, the UK became the first major economy to pass legislation to reduce all GHG emissions to net zero by that year. Heavy responsibility will lie with both government and industry as electric cars and renewable energy sources would make significant contributions to meeting this target. Nevertheless, warning signs are evident. Simon Lewis, Professor of Global Change Science at UCL, argues ominously that the CCC 'shows that it is possible. The question now is if the political will is there to take on the vested interests that will try to stop the UK reaching net zero fast enough to avoid the worst impacts of climate change' (Science Media Centre, 2019).

Conclusions

Recognition of the perilous state of the planet from global heating and climate change is leading to belated action. Authoritative international and national committees of scientific advisers agree on the need for urgent action to curb GHG emissions and protect biodiversity. Proposals have been directed toward consumers to lessen meat consumption, reduce airborne leisure transport, switch to electric cars, reduce domestic heating, recycling and so forth. Impor-tant though such actions are, not least through raising public consciousness, they risk diverting attention from the major climate culprits. Industry has been targeted at different levels, from switching from fossil fuel extraction, transport and burning to providing carbon-neutral energy through renewable sources, such as wind, solar and tidal power. As well as extracting, transporting and burning fossil fuels in processes of production, industry has built and forcefully marketed products that contribute both to GHG emissions and to choking cities through air pollution. In a rational world, international coordinating action, from commissioning adaptability projects through to binding energy use com-mitments are needed to meet temperature rise targets. Nevertheless, pressure to extract, transfer and utilise fossil fuels continues from politicians anxious to sustain economic growth and protect domestic energy supplies; from producers anxious to maintain profits and satisfy shareholder expectations; and from consumers persuaded toward and attached to private car use, frequent and cheap overseas holidays and meat-based diets.

The fulcrum between these opposing objectives is provided by industry, whether primary (extraction), secondary (manufacture) or tertiary (services, such as commercial aviation, leisure cruises). Political action and enforceable legislation at both national and international levels will be needed to ensure that change occurs timeously. There are, however, some glimmers of hope in this direction: we saw above that only 100 companies are responsible for nearly three-quarters of global industrial GHG emissions. Further, most harmful emissions emanate from a small number of countries: already China, one of the biggest contemporary polluters, has taken steps to reduce its dependence on fossil fuels (Qi *et al.*, 2018), though recent developments seem to reverse this trend (Shearer *et al.*, 2019). Conversely, China is a global leader in solar and wind power and is the world's largest market for electric cars. Nevertheless, major questions of investment and compensation would need to be addressed if giant private sector and state-owned corporations, which wield massive political and persuasive as well as economic power, are to abandon current pursuits to invest in innovative, but costly (and possibly commercially risky) sustainability projects. Until now, the record for many of these enterprises has not been encouraging.

An alternative perspective, to which this book is largely directed, is recognition that employing organisations, public and private sector alike, are essentially pluralist in character, meaning that rather than enjoying sovereign status, they form part of an integrated network of actors, or stakeholders, which include equity holders, local communities, suppliers, customers and employees, who all share close interest in organisational directions, activities and outcomes. Foremost among these are employees, who depend on an employer for their livelihood and retirement income and who, for better or for worse, often live in close proximity to its operations. When a large local employer moves or fails, the community itself can become financially and culturally impoverished, and for both employees and community, this can be an ultimate economic and social externality, as shown by the long-term legacy of coal mine closures in Britain following the government's unilateral and heavily contested decision to close loss-making pits in the 1980s with few, if any, transitionary policies to support displaced workers (Waddington *et al.*, 1991; MacIntyre, 2014). Arguably, if companies are to transition to more sustainable operations and more sustainable jobs, a transformative role for all employees and their managers will therefore be key, as their livelihoods are most likely to be affected, for example, if fossil fuel extraction is to cease within a short time frame or heavily carbon-emitting vehicles are to be phased out. With both Paris accords and the UN's action plan heavily dependent on national policy frameworks, strategic operational changes by commercial organisations must also be driven by governmental action. The following chapter explores the political and economic contexts that underlie the environmental tensions that have accumulated over recent years.

Bibliography

Akatsu, M. (2015), 'The problem of air pollution during the Industrial Revolution: A reconsideration of the enactment of the Smoke Nuisance Abatement Act of 1821', in S.

Sugiyama (ed) *Economic History of Energy and Environment*, Tokyo, Springer; 85–109.

Ambrose, J. (2019), 'Report warns of climate threat posed by US oil and gas projects', 25 April, https://www.theguardian.com/environment/2019/sep/06/oil-and-gas-companies-undermining-climate-goals-says-report.

Best, C. (2018), 'The conversation', https://theconversation.com/meet-the-matchstick-women-the-hidden-victims-of-the-industrial-revolution-87453.

Burns, C., Eckersley, P. and Tobin, P. (2019) 'EU environmental policy in times of crisis', *Journal of European Public Policy*, doi:10.1080/13501763.2018.1561741.

Cagle, S. (2019), '"We don't want this" California city confronts the polluter that gave it life', *The Guardian*, 10 October.

Carbon Tracker (2019), 'Oil and gas companies approve $50 billion of major projects', www.carbontracker.org.

Carson, R. (1962), *Silent Spring*, London, Penguin.

CDP (2017), *The Carbon Majors Database: CDP Carbon Majors Report*, London, CDP UK.

Chakrabortty, A (2019), 'In hounding Greta Thunberg the right shows it has no ideas', *The Guardian*, 1 May, https://www.theguardian.com/commentisfree/2019/may/01/greta-thunberg-right-environmental-activist-attacks.

Climate Accountability Institute (2019), *Carbon Majors*, www.climateaccountability.org/carbonmajors.html

Costello, A., Abbas, M., Allen, A. *et al.* (2009), 'Managing the health effects of climate change', *The Lancet*, 373, May 16.

Electricityinfo (2019), www.electricityinfo.org/news/fossil-fuels-446.

FitzRoy, F. and Papyrakis, E. (2016), *An Introduction to Climate Change Economics and Policy*, London, Routledge.

Freeman, J. (2018), *Behemoth: A History of the Factory and the Making of the Modern World*, New York, Norton.

Friedman, L. (2020), 'New research links air pollution to higher coronavirus death rates', *New York Times*, 17 April, www.nytimes.com/2020/04/07/climate/air-pollution-coronavirus-covid.html.

Gardiner, B. (2019), *Choked: The Age of Air Pollution and the Fight for a Cleaner Future*, London, Granta.

Garvey, B., Alves, J. and Morais, M. (2019), 'Amazon in flames: Brazil's Huni Kuin indigenous people count the costs of fire and conflict', *The Conversation*, 2 September, https://theconversation.com/amazon-in-flames-brazils-huni-kuin-indigenous-people-count-the-social-costs-of-fire-and-conflict-122587.

Gillam, C. (2017), *Whitewashed: The Story of a Weedkiller, Cancer and the Corruption of Science*, Washington, Island Press.

Goldstein, J. and Qvist, S. (2019), *A Bright Future: How Some Countries Have Solved Climate Change and the Rest Can Follow*, New York, Public Affairs.

Greenpeace (2018), 'Dangerous pipelines', www.greenpeace.org/usa/reports/dangerous-pipelines.

Guan, D., Xin, S., Qiang, Z. *et al.* (2014), 'The socioeconomic drivers of China's primary PM2.5 emissions', *Environment Research Letters*, 9 (024010).

Harari, Y. N. (2011), *Sapiens: A Brief History of Humankind*, London, Vintage.

Helm, D. (2015), *Natural Capital*, London, Yale University Press.

Hersher, R. (2018), 'After decades of air pollution, a Louisiana town rebels against a chemical giant', *NPR*, 6 March, www.npr.org/sections/health-shots/2018/03/06/583973428/after-decades-of-air-pollution-a-louisiana-town-rebels-against-a-chemical-giant.

IPBES (2019), Intergovernmental Science Policy Platform on Biodiversity and Ecosystem Services, 'Nature's dangerous decline "unprecedented"; Species extinction rates "accelerating"', http://ipbes.net/news/Media-Release-Global-Assessment.

IPCC (2018), Intergovernmental Panel on Climate Change, 'Summary for policymakers of IPCC Special Report on Global Warming of 1.5°C Approved by Governments'.

Jordan A (1999), 'The implementation of EU environmental policy: a policy problem without a political solution?', *Environment and Planning*, 17; 69–90.

Klein, N. (2014), *This Changes Everything*, London, Penguin.

Klein, N. (2019), *On Fire: The Burning Case for a Green New Deal*, London, Allen Lane.

Kulp, S. and Strauss, B. (2019), 'New elevation data triples estimates of global vulnerability to sea-level rise and coastal flooding', *Nature Communications*, 10 (4844), 29 October.

LAIA (undated), Lake Area Industrial Alliance, 'Fighting the cancer alley myth', www.lca.org/resources/chemical-connections/fighting-the-cancer-alley-myth.

Laurent, É. (2019), 'Reimagining a just transition', *Social Europe*, 2 December, www.socialeurope.eu/reimagining-a-just-transition.

Lear, L. (1998), 'Afterword' to R. Carson, *Silent Spring*, London, Penguin.

Leichenko, R. and O'Brien, K. (2019), *Climate and Society: Transforming the Future*, Cambridge, Polity.

Lewis, S. and Maslin, M. (2018), *The Human Planet: How we Created the Anthropocene*, London, Penguin.

Lovelock, J. (2019), *Novacene: The Coming Age of Hyperintelligence*, London, Allen Lane.

MacIntyre, D. (2014), 'How the miners' strike of 1984–85 changed Britain for ever', *New Statesman*, 16 June.

Maslin, M. (2014), *Climate Change: A Very Short Introduction*, 3rd edn., Oxford, Oxford University Press.

McMurtry, J. (2019), 'The Koch brothers and the tar-sands', *Canadian Dimension*, https://canadiandimension.com/articles/view/the-koch-brothers-and-the-tar-sands, July.

Metuktire, R. (2019), 'In the Amazon we are full of fear. Soon you will be too', *The Guardian*, 3 September.

Michaels, D. (2020), *The Triumph of Doubt: Dark Money and the Science of Deception*, Oxford, Oxford University Press.

Monbiot, G. (2017), *How Did We Get into This Mess?*, London, Verso.

Narain, S. and Bhushan, C. (2015), www.downtoearth.org.in/coverage/environment/30-years-of-bhopal-gas-tragedy-acontinuing-disaster-47634.

OECD (2011), *Environmental Impacts of International Shipping: The Role of Ports*, Paris, OECD Publishing.

Oreskes, N. and Conway, E. (2012), *Merchants of Doubt*, London, Bloomsbury.

Oxfam (2020), 'Average Brit will emit more by 12 January than residents of several African countries do in a year', 6 January, http://oxfamapps.org/media/y9rq7.

Parker, A. and Davenport, C. (2016), 'Donald Trump's energy plans: more fossil fuels and fewer rules', *The New York Times*, May 26, https://www.nytimes.com/2016/05/27/us/politics/donald-trump-global-warming- energy-policy.html.

Pilling D. (2018), *The Growth Delusion*, London, Bloomsbury.

Qi, Y., Stern, N., He, J. *et al.* (2018), 'China's Peaking emissions and the future of global climate policy', Beijing, Brookings-Tsinghua Center for Public Policy.

Rainforest Action Network, Palm oil fact sheet, www.ran.org/palm_oil_fact_sheet.

Raworth, K. (2017), *Doughnut Economics: Seven Ways to Think Like a 21ˢᵗ-Century Economist*, London, Penguin/Random House.

Romm, J. (2018), *Climate Change*, 2nd edn, Oxford, Oxford University Press.

Rosen, A. (2015) 'The wrong solution at the right time: the failure of the Kyoto Protocol on climate change', *Politics and Policy*, 43 (1); 30–58.

Science Media Centre (2019), 'Expert reaction to the net zero report', www.sciencemedia centre.org/expert-reaction-to-the-net-zero-report.

Shearer, C., Yu, A. and Nace, E. (2019), 'Out of step: China is driving the continued growth of the global coal fleet', *Global Energy Monitor*, November.

Showstack, R. (2019), 'Global tree cover loss continues but is down from high peaks', *Eos, 100*, https://doi.org/10.1029/2019EO122217.

Shiva, V. (2018), *Oneness vs the 1%*, Oxford, New Internationalist.

Stern, N. (2006), *Stern Review on the Economics of Climate Change*, London, Cabinet Office/HM Treasury.

Thiele, L. (2016), *Sustainability*, 2nd edn, Cambridge, Polity.

Thompson, R., Lassaletta, P., Patra, P. *et al.* (2019), 'Acceleration of global N_2O emissions seen from two decades of atmospheric inversion', *Nature Climate Change*, 9; 993–998.

Tollefson, J. (2019), 'Rogue emissions of ozone-depleting chemical pinned to China', *Nature*, 22 May, www.nature.com/articles/d41586-019-01647-z.

United Nations (2015), 'Transforming our world: The 2030 Agenda for Sustainable Development A/RES/70/1', www.sustainabledevelopment.un.org.

United Nations (2019) 'Sustainable development goals climate action', www.un.org/susta inabledevelopment/climate-change-2.

USEPA (2017), United States Environmental Protection Agency, www.epa.gov/ghgemissions/ global-greenhouse-gas-emissions-data.

Vijay, V., Pimm, S., Jenkins, C. and Smith, S. (2016), 'The impacts of oil palm on recent deforestation and biodiversity loss', *PLoS ONE*, 11 (7).

Waddington, D., Wykes, M. and Critcher, C. (1991), *Split at the Seams: Community, Continuity and Change after the 1984–5 Coal Dispute*, Milton Keynes, Open University Press.

Wang, H., Dwyer-Lindgren, L., Lofgren, K. *et al.* (2012), 'Age-specific and sex-specific mortality in 187 countries, 1970–2010: a systematic analysis for the global burden of disease study 2010', *Lancet*, 380; 2071–2094.

Ware, J. and Kramer, K. (2019), 'Hunger strike: The climate and food vulnerability index', London, Christian Aid, August.

Willow, A. and Wylie, S. (2014), 'Politics, ecology and the new anthropology of energy: exploring the emerging frontiers of hydraulic fracking', *Journal of Political Ecology*, 21; 222–236.

WMO (2019), World Meteorological Organization, *Greenhouse Gas Bulletin, No.15*, 25 November.

Woodward, A. (2019), 'Brazil's new president has started taking steps towards damaging the "lungs of the planet"', *Business Insider*, February 2019.

Wright, S. and Kelly, F. (2017), 'Plastic and human health: a micro issue', *Environmental Health and Technology*, 51 (12); 6634–6647.

4 Neo-liberal or free-market economics
Literally the 'world taken for granted'

The problem with economics

At first glance there would appear to be little to connect increasing deterioration in the quality of working life of so many and the near critical state of the planetary environment. However, the same political and economic imperatives provide the underlying sub-strata to both these phenomena and these can be summed up by the concept of 'neo-liberal' economics.

Economics in general can seem simply to be a system for managing national and international economic affairs involving markets (that is, trading based on the demand and supply of goods and services), prices and the cost of living, investment by firms and individuals, imports and exports and the government's role in regulating some of these. What is not always readily apparent is that any country's economic system represents a series of past and present choices.

One of the most significant of these concerns the appropriate role of government. Immediately following World War II there was general support, by most political parties in the developed nations, for what was called the 'mixed economy', in which governments ran and controlled infrastructure sectors such as transport and power, and the private sector delivered goods and services via the market. In addition, the government was expected to manage the economy, through its policies of taxation and government spending, and maintain full employment. This 'post-war consensus' was partly a legacy of wartime necessity during which the governments of participating nations had to by-pass the market and take control of their countries' productive resources, and partly a reflection of the economic model devised by the economist John Maynard Keynes to avoid a repeat of the severe Depression of the inter-war period.

However, by the early 1970s this economic model was showing signs of strain. The boom fuelled by post-war reconstruction, together with upward pressure on wages by strong trade unions operating in a full labour market, had created the spectre of inflation and rising prices, rather than unemployment, as the main economic enemy. In addition, the rise of the international corporation and increases in international trade made it more difficult for any government to regulate a national economy as a semi-closed system: there were now too many flows in and out of the country for any government to fully control.

However, since the end of the war, some economists, in particular Friedrich Hayek and Milton Friedman, had been urging the rejection of Keynes' model and a return to less interventionist policies and greater reliance on the 'free market' to resolve problems and deliver economic growth. With the coming to power of Margaret Thatcher in the UK and Ronald Reagan in the USA in the early 1980s these policies found ready political acceptance and have, since then, become the new orthodoxy, followed to a greater or lesser degree by parties of all political persuasions in the major developed economies, together with the World Bank, International Monetary Fund and the European Central Bank. Secondly, through the influence these countries and institutions have had internationally, neo-liberal ideology has shaped economic policies of those nations in the global South who are trying to develop their own economies.

The term 'liberal' has different connotations on either side of the Atlantic but as applied to the ideology of market fundamentalism it refers to the philosophical outlook of the early classical economists of the eighteenth and nineteenth centuries. What is different, however, is that while the early economists saw social life as a series of contracts, the terms of which were delivered by the market mechanism, neo-liberalism claims the defining characteristic of society and economy to be competition and the proper role of government to pursue policies to remove competitive barriers.

The major thrust of neo-liberal policies can be summarised in the belief that markets cannot be effective if there are third-party restrictions on the behaviour of market actors. To this end there should be:

- Removal of as many government regulations of market activity as possible – these might include minimum wage setting, gender equality goals, corporate taxation, health and safety requirements, and controls over environmental quality (such as air and water pollution).
- Reduction or elimination of the ability of organised labour to influence labour market terms and conditions: tougher trade union legislation and weakening of collective bargaining and other avenues of worker voice.
- Privatisation of public sector organisations; where there is no visible market for 'public goods' such as education, then artificial markets should be created through league tables of measurable performance criteria.

Many of these goals have been taken up in varying degrees by governments of the advanced economies over the recent past and, where not adopted, there remain significant pressures by powerful economic lobbying groups on governments to pursue these ends (Geoghegan, 2020).

Taken together these features have immediate and severe consequences both for job quality and environmental quality, and begin to explain the economic and political processes behind the developments discussed in previous chapters. They demonstrate that features of contemporary work, such as static low wages, job insecurity, one-way flexibility and the absence of worker ability to influence the terms and conditions of his or her labour are not just some sort of inevitable historical process but the results of policy choices reflecting the steady permeation of market fundamentalism into most sectors of economic and business activity, supported by political and legislative programmes.

Both trade unions and minimum wage legislation are seen as restraints that distort the ability of the true market for labour to fix a 'true' market price (i.e. wage). The economics writer Paul Mason quotes Tidjane Thiam, CEO of Prudential, addressing high-level movers and shakers at Davos in 2012, stating that unions are the 'enemy of young people' and that the minimum wage is 'a machine to destroy jobs' (Mason, 2016: 4). When combined with advocacy for globalisation, the implication is that wages in the mature industrial societies should be in competition with (that is, fall to the level of) those in emerging and newly developing nations. This was undoubtedly one of the ingredients of the ideological offensive behind Brexit: the European Union was seen as 'over-regulated' and 'sclerotic' and once outside its confines, free-market competition would push the British economy on the way to unfettered revival.

It should be stressed that this model is most pronounced in liberal market economies (LMEs) such as the UK and US, where companies are primarily financed by shareholder capital. In coordinated market economies (CMEs) company finance is primarily based on bank lending (as in Germany or Scandinavia) or interconnections between 'families' of companies, as in Japan (Hall and Soskice, 2001). These distinctions are not quite so clear-cut in reality but there is sufficient basis behind the two models for the consequences for jobs and the environment to be significant.

Upholding shareholder value as the prime criterion for economic success, particularly in LMEs, is a principal factor behind the short-termism in management decision-making: shareholders are fickle and their support is instantly withdrawn if better returns are available elsewhere. Thus, their confidence in the organisation has to be constantly boosted and is more easily reassured by company mergers, acquisitions and asset stripping than by long-term investment and innovation, a trend further compounded by the fact that many of the directors in the top 100 UK companies have their pay linked to so-called Long Term Incentive Plans, again based on the performance of the share price. To this end, a 'flexible labour market' is seen as essential for making hiring and firing easier, which in turn allows companies to be bought and sold more easily: as the principle of 'shareholder value maximisation' was eagerly adopted by management in the 1980s, jobs were cut and employees fired and re-hired as non-union agency or contract labour. Also, because the neo-liberal model prioritises the interests of holders of financial assets, leading to the 'financialisation' of the economy (Shafique, 2018: 34), there has been sustained international pressure for greater global mobility of capital (Chang, 2011); this enables assets to be bought, sold and moved around to get the best returns.

Such short-term thinking seems unlikely to prioritise the problems of a destabilised environment: there is even some research that indicates that LMEs have higher per capita carbon emissions (Jackson, 2011: 165). Environmentalist Tony Juniper (2013) points out that, while we measure trends in the natural environment in decades or even centuries, this is unlikely to resonate with the pressure on management to convey good news in every quarterly financial statement. At a deeper level, instead of seeing natural resources such as forests, soils or minerals as capital, to be invested wisely, contemporary economic thinking sees them as a flow or stream of dividends to be liquidated to make a profit. As an example of this thinking Juniper quotes the Vietnamese CEO Doan Nguyen Duc: 'natural resources are limited and I need to take them before they're gone' (*ibid.*: 272). We can see that such a pervasive ideology is a major contributory factor to the continuation of the environmental crisis, decades after the first alarm bells were sounded by the scientific community and that this mindset will view any governmental or inter-governmental environmental initiatives as unwelcome.

Externalities: external to whom?

'The market only values what is priced and only delivers to those who can pay' (Raworth, 2017: 81). Therefore, anything or any human activity that is not bought and sold is given little or no value. This ignores three highly significant areas of human activity and experience. Public goods are those things, such as parks, roads, street lighting and libraries, where their use by one person does not exclude use by others and is at no cost to others. But as these things definitely cost something to provide, there is no market incentive to provide them (for free use) and thus they would never be provided by the market alone, but usually need the help of governments. Such public 'goods' could also include clean air and unpolluted waterways.

The second is all those human activities that create social value but not market value: these include for example domestic labour and voluntary work. As the Swedish feminist writer Katrine Marçal has aptly pointed out, when Adam Smith wrote his famous statement about the self-interest of the butcher and baker bringing food to his table, he ignored the very person whose daily acts of unpaid labour actually prepared, cooked and served his meals – namely his dear old mum (Marçal, 2015). As his mother's domestic work was not purchased in the labour market it had no 'value' and certainly never appeared in the economic output figures for Smith's Scotland. We will return to this when we examine the nature and meaning of work.

The third major failing is that if market transactions, such as for example buying, refining and consuming oil, involve costs to third parties (basically everyone else not directly involved) these costs are not reflected in the market price of oil. Such things as deforestation, loss of biodiversity, air pollution and exhaust fumes are seen by economists as 'externalities' and the costs are usually borne, in the short run, by those living next to the sites of production and consumption and, in the long run, by all of us. When this perspective is combined with the compulsion to externalise costs (Bakan, 2004: 61) resulting in the 'fissured workplace' examined earlier, this can have disastrous consequences for both employees and the environment, in a context of minimal government regulation. This is starkly exemplified in the Deepwater Horizon disaster of 2009–2010.

Oil giant BP were undertaking unprecedented deepwater drilling in the Gulf of Mexico at a depth of 1.5 kilometres when the rig exploded, fountaining thousands of gallons of crude oil into the Gulf for many days in 'the largest accidental marine oil spill in history' (Klein, 2014: 330). Wildlife was devastated, beaches coated in tar, and whole fishing communities were suddenly cut off from their livelihood. The environmental catastrophe was so huge that it is often overlooked that 11 workers were killed when the Macondo wellhead blew. The organisational disaster was compounded by BP's apparent lack of procedures to deal with a blowout at this depth and the fact that government regulators had been happy to accept BP's word that such activity was safe.

The US Chemical Safety Board enquiry concluded that a principal cause of the disaster was lack of coordination between the three organisations involved – BP and its contractors, Haliburton and Transocean – particularly over hazard assessment systems:

> the bridging document that sought to harmonize safety controls between BP and Transocean was a minimal document that focused only on six personal safety issues such as minimum heights for employing fall protection equipment. The document did not address major accident hazards like the potential for loss of well control.
>
> (quoted in Weil, 2014: 19)

One might ask why, given the fact that drilling at such depth had never been done before, was safety relegated to such a minor issue? The other main conclusion from the subsequent enquiries and legal proceedings was that a major priority for the three organisations had been cost saving. The specially convened Presidential Oil Spill Commission found: 'Whether purposeful or not, many of the decisions that BP, Halliburton and Transocean made that increased the risk of the Macondo blowout clearly saved these companies significant time (and money)' (quoted in Klein, 2014: 330).

Joseph Stiglitz argues that, where companies do not take account of the full costs of externalities, we must see this as evidence for market failure, representing a subsidy for polluting activities that may need to be remedied through government intervention (Birch *et al.*, 2017: 197). Here we need to further distinguish between *localised* and *pervasive* externalities. The former can be dealt with by raising prices or specific regulation, the latter are so generalised that they can only be dealt with by major institutional changes. Daly and Cobb (1994) use the example of the coal industry: here the lung disease pneumoconiosis is a localised externality, only directly affecting coalminers and their families. In contrast, the pervasive externalities of the coal industry include acid rain, carbon emissions and environmental degradation left by open cast extraction; here the costs do not fall on a clearly defined group, nor is the cause traceable to a defined industrial location and longer-term impacts are hard to measure or predict. They clearly require much more systemic change.

However, the neo-liberal approach says that *because* there is no market cost, the solution is to create one; the answer, in other words, lies in more market forces, not less (Chang, 2011). It is this that underlies the concepts of 'emissions trading' and 'carbon trading'. The problem here is, if actions are only performed for economic reward, rather than because the actor sees the good sense of the action and is prepared to do it voluntarily, the reward becomes the 'price' of the action. Thus, putting a 'price' on emissions through cap and trade schemes, so that heavily polluting countries can trade their carbon credits with low-polluters, contributes nothing to reducing emissions overall. Under such schemes organisations within a country can similarly trade emissions permits with each other without exceeding the cap for the country as a whole. At the same time so-called carbon 'off-set' activity such as planting trees can qualify for carbon credits, yet the long-term potential of the trees as carbon sinks hardly offsets the direct contribution to climate instability of current emissions. As Naomi Klein (2014: 218) has pointed out, these schemes have come nowhere close to the drastic re-thinking required to meet identified emissions targets and have engendered some very dubious practices in claiming carbon credits.

Many externalities are made truly external by exporting the costs to less developed countries. In 1991, Lawrence Summers, then Chief Economist at the World Bank, sent a memo in which he suggested that it made perfect economic logic for dirty polluting industries and toxic waste to be 'exported' to low-wage developing economies: developed nations would have a cleaner environment and the low-wage countries would gain employment opportunities (Marçal, 2015: 50–51). Although criticised at the time, this is more or less what has happened over the succeeding 20 years through what environmental campaigner George Monbiot has termed 'waste colonialism' (Monbiot, 2019). As an example, he refers to the thousands of tonnes of used tyres that leave the UK every month for India, where they are baked in pyrolysis plants to produce a dirty industrial fuel and, in the many plants that escape government regulation, waste products of heavy metals, benzene, dioxins and other toxic chemicals.

A case that received considerable publicity was the 2006 Trafigura scandal. This started when a cargo of Mexican oil with high levels of impurity was initially shipped to Galveston, Texas to be treated with chemicals to remove the impurities. The cleaned oil was then sold but the problem remained of what to do with the resulting toxic sludge or 'slops'. The global commodity trader Trafigura shipped it to Amsterdam for specialist incineration but it was judged by the Dutch to be too contaminated (and thus expensive) to dispose of at the agreed price. Trafigura then unsuccessfully tried Lagos, Nigeria and finally found a pop-up waste-disposal firm with no previous experience in Abidjan, Ivory Coast. The sludge was pumped from the ship into road tankers, driven away and dumped – poisoning 31,000 inhabitants of Abidjan who suffered headaches, nosebleeds and stomach pains; 12 people died (Lucas, 2015: 13).

Reporters pursuing this story were slapped with heavy injunctions by Trafigura and some were sued for libel for claiming or even suggesting that Trafigura knew that the 'slops' were hazardous waste. It was only when Green Party MP Caroline Lucas used parliamentary privilege to raise the issue in the House of Commons that the media were then allowed to cover it. Lucas demonstrated that Trafigura's traders and their CEO were fully aware of the nature of the cargo, discussed various methods of disposal but were spurred on by the prospect of making 'serious money' (Lucas, *op. cit.*; 15–16).

A major component of neo-liberal policy has been promotion of unfettered international competition through the series of free trade agreements that have marked the last 30 years. This has had major effects both on jobs and environment as both protective social legislation and environmental regulation passed by a country that is party to such an agreement can be seen as a constraint of competition and a violation of World Trade Organization rules. This is the reality of globalisation: national governments locked into undemocratic international legally binding agreements that their citizens have never voted for, and that actively prevent them from undertaking any serious environmental legislation.

The great god of growth

Neo-liberal economics, along with most other economic philosophies, prioritises growth as the goal of national economies and global economic activity in general. Economic growth is usually defined in terms of an increase in size of Gross Domestic Product (GDP), namely the sum total of everything produced within a national economy. For many decades now, the rate of growth has become the arbiter of the health or otherwise of the economy and growth figures are scrutinised on a quarterly basis with all the intensity of a shaman examining the entrails of a goat.

Economic growth is critical to the future of the two main concerns of this book – quality of work and quality of the environment. Supporters of continued growth invariably argue that it is essential in order to maintain or increase levels of employment, and that stagnant growth with a rising population will inevitably lead to an increase in the numbers of those out of work. This leaves many questions unanswered, such as growth of what? If a country's growth in a given time period is, say, down to a surge of investment in new technologies then, while this may lead to increased employment making those technologies, the technologies themselves may be in demand by employers because they *replace* labour, as we examine in a later chapter. It also assumes that patterns and hours of work remain the same, so that a reduction in growth is seen to be bad because it will lead to a loss of jobs for some, rather than everyone in work working fewer hours. Admittedly the former has been the historical pattern during times of rapid change or economic downturn but this does not mean that this has to be the case, as we discuss later. A second unanswered question is growth where? Overall UK GDP figures disguise significant regional disparities in who experiences the supposed benefits of growth and that a large proportion of the UK's economic activity is focused on London and the South-East.

What is not always understood about growth is that it is not a linear but an exponential process, in other words growth in a given period, expressed as a percentage of the whole, feeds on the previous period's growth so that the *amount* added to the whole, even at a constant *rate* of growth, gets ever bigger. This is not a problem if there is an infinite capacity for expansion but if there is a finite limit beyond which growth is impossible, that limit can be reached surprisingly quickly. This is graphically illustrated in a French children's riddle: if water lilies in a pond double in size every day and were allowed to grow unchecked, they would completely cover the pond in 30 days, choking off all other forms of pond life. As the lily coverage seems small for many days you decide not to cut it back until it covers half the pond – on what day would that be? Answer: day 29 (quoted in Meadows *et al.*, 1972: 37).

In the formative years of economic thinking, the eighteenth and nineteenth centuries, it might have seemed that the world and its resources were limitless; new continents were still being 'discovered' by the newly industrialising nations of Europe and North America and the application of science was similarly discovering new raw materials and ingenious uses for them. Human populations, increasingly freed by medical discoveries from high levels of child mortality and fatal epidemics, could expand into the new worlds. Growth seemed the natural order of things.

Today we know better: humans are now squeezed into almost every part of the Earth, much of it in habitats that are only marginally capable of supporting human existence. We are aware that natural resources are not infinite – we can use them up surprisingly quickly. Also, we have seen the images of the 'blue marble' from space and can see it as a finite container for all its life forms, and an extremely small and possibly unique one in the immensity of the universe: as a placard in the school students' demonstrations succinctly states – 'There is no Planet B'. Put simply, continued economic growth will sooner or later come up against the physical limits of the planet and increasingly the signs are that it will be very much sooner than later.

Remarkably, this was first graphically explained over 40 years ago by a team of systems scientists at MIT commissioned by the Club of Rome to examine future prospects for Earth. Taking all the major variables necessary to sustain our existence on the planet they built a World Model and, having run the then trends through the model, were led to sombrely conclude:

> If the present growth trends in world population, industrialisation, pollution, food production and resource depletion continue unchanged, the limits to growth on this planet will be reached sometime within the next one hundred years. The most probable result will be a rather sudden and uncontrollable decline in both population and industrial capacity.
>
> (Meadows *et al.*, 1972: 29)

Their report was written as the full effects of greenhouse gases on planetary heating were only beginning to be understood and so were not included in their model. The tragedy is that few took any notice; economic growth, both as socio-economic activity and as political shibboleth, carried on apace.

As a counter to such warnings, in the 1990s some economists claimed to have discovered an encouraging relationship between growing GDP (measured as income per head) and levels of pollution, which seemed to indicate that, while pollution rose in the early stages of a country's growth trajectory, it peaked and then actually fell with continued growth, creating an inverted U-shaped curve (Cole *et al.*, 1997). This became known as the Environmental Kuznets Curve (EKC), because of its similarity to the economist Simon Kuznets' claim of a similar (but later disproved) relationship between growth and inequality. Stern (2004) argues that 'if the EKC hypothesis were true, then rather than being a threat to the environment ... economic growth would be the means to eventual improvement'. This model provided a significant boost to advocates of both globalisation and economic growth as the means to both alleviate poverty in developing countries and improve the environment, negating the 'race to the bottom' thesis presented by critics of free-market trading: yes, things would get worse but then get better as economic growth cleaned up its own mess.

However, as Stern and others demonstrate, the statistical and empirical evidence for these claims are not robust, casting considerable doubt on the validity of the relationship. The hypothesis was based on figures for water and air pollution only and did not include critical variables such as GHG emissions, oceanic acidification and soil degradation. Also, as we have seen, once countries achieve a certain level of development they no longer have to cope with toxic and polluting industrial processes because they can export them to less developed nations anxious for work.

To overcome the limitations of looking at single or even two or three variables only, Raworth argues that we need to consider the 'global material footprint' of any nation's economy: the sum of all the biomass, fossil fuels, metal ores and minerals used to create the products that a country either makes or imports. And we find that, contrary to the hopes of EKC advocates, from 1990 to 2007 for high-income countries such as the US, UK, New Zealand and Australia, as their GDP grew their global material footprints increased by more than 30 percent (Raworth, 2017: 211).

What the simplistic EKC model fails to address is that there are two kinds of limit to our continued growth. One is what environmental economists call 'sinks' – the shrinking ability of the planet to absorb the waste products of our economic activity, of which the impact of GHGs on global climate is the most acute concern. The other is the finite limit to the 'sources' we rely on for raw materials. Meadows' prediction, 40-plus years ago, that we will see the exhaustion of key planetary resources in the first half of this century has proved remarkably prescient. Jackson (2011) calculates that if every country consumed at the same rate as the US, then copper, tin, zinc, chromium, silver and similar strategic minerals could be exhausted in 20 years. We are already seeing the consequences of dwindling supply in some key resources: for example, as the demand for oil begins to come up against a shrinking supply, it becomes profitable to try to extract it using what have been until now prohibitively expensive means – from the Alberta tar sands, deep water drilling in the Gulf of Mexico or under Arctic ice.

We maintain that it is the unassailable position of the concept of continued growth and its acceptance as holy writ that underpin the twin foci of this book. As we have seen, continued growth on a finite planet cannot be sustained for much longer. However, the other side of the coin of production is consumption and what Tim Jackson calls 'the social logic of consumption' works powerfully against any desires individuals may have to voluntarily reduce their contribution to planetary damage. Consumption is a dominant element in contemporary culture, which repeatedly stresses the themes of consumer choice, the cult of the brand (and the status attached to it) and sees the launch of a new product such as the latest model phone inevitably hailed as progress. But, while ever-increasing consumer choice encourages us to shop more, in our other lives as workers, citizens and family members we experience a diminishing capacity for choice. As summed up by a coalition of NGOs and environmental groups nearly 20 years ago: 'More choices in the shopping malls for those who can afford the price, but more pressure to work long hours, and more tension between the demands of being a worker and being a parent' (Real World Coalition, 2001; 40).

Growth, well-being or degrowth?

Increasing numbers of prominent critics agree that reliance on a single (and distorting) metric such as GDP fails to measure the overall health of a country or the well-being of its citizens (e.g. Stiglitz *et al.*, 2018; Bannerjee and Duflo, 2019; Raworth, 2017; Pilling, 2018). Specific problems are that all marketised productive output is assumed to contribute to GDP, irrespective of its social value (gambling, tobacco, armaments) or environmental impact (air travel; combustion powered vehicles). As we have seen, GDP also fails to take account of how income or wealth is distributed. Economic growth may be high but, as in the USA or UK, its distribution may be heavily skewed to the already wealthy. Hence, the question of alternative targets and measures arises, with a highly significant marker set by the UN's Sustainable Development Goals, which clearly move beyond a GDP focus but leave open ways to measure sustainability (Stiglitz *et al.*, 2018: 92). Heterodox economists like Stiglitz and members of the OECD's catchily titled 2018 High Level Expert Group on the Measurement of Economic Performance and Social Programme point out that though GDP recovered following the 2008 financial crisis, loss of 'social capital' was critically overlooked, with policy failing to identify reduced levels of on-the-job training, declining prospects for young people and growing social inequality. Building on the report of the Stiglitz-Sen-Fitoussi Commission (Stiglitz *et al.*, 2009), they argue that applying different metrics could have led to identification of socially negative impacts of the financial crisis, which had been obscured by GDP recovery. Based on the criterion of what is most valuable to society, Stiglitz *et al.* (2018) argue for a broader set of measures to guide policy toward economic and social stability. These measures take on even greater significance as the Covid-19 induced economic crisis (and the ever-present environmental one) threatens to challenge individual and social well-being.

Assessing well-being relies on welfare concepts associated with 'happiness' or 'life satisfaction' (Layard, 2011), measured through combinations of economic and psychological procedures, which together are based on 'more expansive notions of utility than does conventional economics, highlighting the role of non-income factors that affect well-being' (Graham, 2005: 41). Such ideas help to reduce emphasis on consumption as both source of satisfaction and driver of economic growth. Stiglitz *et al.* (2018: 117–118) present 12 recommendations to identify well-being constituents and proposed means of measurement. These can be summarised as: policies for each country need to be guided by a 'dashboard of indicators' supported by well-resourced statistical measures about people's material conditions, quality of life, levels of income and wealth inequality and sustainability. Appropriate disaggregation of data by gender, age, race, disability and other markers of social status is necessary. Measures should be designed to provide greater understanding of how benefits of GDP growth are shared in society and assess equality of opportunity. Subjective well-being, levels of trust and other social norms should be measured. There should be regular reviews of policies to assess effects of economic insecurity with open access to data for academics and policy specialists. Well-being metrics should be used to inform decisions at all stages of the policy process 'from identifying priorities for action and aligning programme objectives to investigating the benefits and costs of different policy options; from making budgeting and financing decisions to monitoring policies, programme implementation and evaluation' (*ibid.*: 118). A number of countries have introduced well-being indicators to inform policymaking. These include New Zealand (with some 60 indicators), Iceland (39) and Scotland (Exton and Shinwell, 2018). Bids to the New Zealand Treasury will now have to include a well-being impact assessment measured in terms of human capital, social capital, natural capital and financial and physical capital.

As we discuss in more detail in Chapter 10, economists are split between the benefits and effects of growth or degrowth strategies. Optimists insist that growth can be made environmentally sustainable in the form of 'green growth' by decoupling GDP from ecological impacts, i.e. economies continue to grow, but with reduced reliance on unsustainable inputs. However, critics contend that it would be an illusion to suggest that absolute decoupling of GDP growth from environmental damage is feasible (Hickel and Kallis, 2020; FitzRoy, 2019: 62; Raworth, 2017: 259) and that if growth continues to be associated with intolerable environmental and social burdens in advanced (and high polluting) countries, policy should therefore be directed toward progressive degrowth activities.

There is also substantial survey evidence that subjective well-being is not positively associated with long-term economic growth, though in the short term, unemployment and income loss, and widening income differentials are significant factors in promoting negative well-being (FitzRoy, 2019: 63). Major determinants of well-being derive from satisfying work, health, family and social relationships and quality of the environment (Dorling, 2019; Frayne, 2015). In other words, measured or degrowth policies (as expressed in GNDs) are not only environmentally beneficial but, if applied equitably in society and supported by appropriate social welfare support (Kalleberg, 2018), are unlikely to promote widespread dissatisfaction.

What happens when we buy stuff?

Initiatives such as New Zealand's to replace GDP with a more sustainable metric are as yet all too rare and most mature industrial societies display a collective cognitive dissonance: we have been repeatedly told the facts of the increasing difficulty of the planet continuing to support our lifestyle yet our appetite for the shiny products and services on display remains apparently undiminished. And to obtain what are promoted as the expected ingredients of a 'normal' modern lifestyle many of us accept jobs in the gig economy, put up with ever-increasing and stressful workload targets and endure work cultures of bullying and harassment; we also, after over a decade of stagnant real wages, increasingly go into debt. In case the reader is still unclear about links between degraded jobs and a degraded planet, let us consider the socio-economic process of consumption in more detail through examination of a couple of examples of contemporary consumer products. If we construct a composite picture of the current economic and environmental characteristics noted earlier, to look at a couple of hypothetical purchases made in the twenty-first century economy, we see how employment conditions and environmental conditions are all too often inextricably linked.

Electronics

Assume our consumer wishes to buy a new mobile phone and has seen what looks like a good offer from a major online retailer. There is nothing technically wrong with her existing phone but the new model is advertised as being faster, with high definition screen, bigger memory and better camera. As she places her order from the comfort of her own home and arranges for an electronic transfer of funds to cover the price it seems like the very essence of modern frictionless consumption.

The software operating system for the phone she has ordered is designed in California but the phone itself is manufactured and assembled by a 'contractor manufacturer' in a country in South-East Asia where wages are significantly lower than in established industrial nations. As low as they are, they are still higher than can be earned in the traditional rural economy and the phone will be assembled by many young women attracted in from that country's rural hinterland and housed in company dormitories.

The materials that go into the phone's manufacture have had an even more turbulent route before they reach the Asian factory. On average, a mobile phone contains more than 40 chemical elements, including base metals (such as copper and tin), special metals (such as cobalt, indium, and antimony) and precious metals (such as silver, gold, and palladium). One of the core materials in modern electronic devices such as phones and laptops is the rare mineral coltan, which, when refined, produces a heat resistant powder that can hold an electric charge. While the majority of the world's coltan resources are found in Brazil and Australia, a significant percentage comes from the Democratic Republic of Congo (DRC), where control of this resource and Congo's other minerals have underpinned, and it is argued prolonged, many of the armed conflicts suffered by the DRC in recent years, with armed groups using trade in minerals to finance their activities (HCSS, 2013; Wenar, 2016).

Coltan mining in DRC is mainly unstructured and informal, usually controlled by local warlords, and involves miners extracting materials in open pits without benefit of masks, gloves or goggles. Coltan is toxic, a probable carcinogen and miners have been shown to be exposed to high concentrations of dust with consequent respiratory disorders and impaired lung function (Leon-Kabamba *et al.*, 2018). Despite this, at times of a boom in coltan prices mining has sucked in labour from agriculture and even children (and teachers) from schools into the mines. Environmentally, the boom years of coltan mining contributed to significant deforestation in DRC, timber being cleared for mining and habitation, resulting in soil erosion and subsequent river silting, threatening rare plant species and the natural habitat of the gorilla.

The supply chain from mine to export starts with small traders visiting mines to buy coltan and then transporting it using porters to a nearby village. Bigger traders then organise its transportation to main trading centres in Eastern DRC where ore is bought by a mineral trading firm. At every stage in this supply chain the various armed groups operating in Eastern DRC over past decades have used forced labour, pillaging, 'taxes' and obligatory fees, and protection payments, to generate revenue (HCSS, *op. cit.*)

A similar trail of exploitation was recently revealed by human rights groups examining the origins of another scarce mineral, cobalt, used in virtually all rechargeable batteries, from phones to electric cars. Around two thirds of the global supply of cobalt is sourced in DRC with, again, a high incidence of child labour and minimal labour and environmental standards. In December 2019 the International Rights Advocates filed a lawsuit in Washington DC against tech giants Apple, Google, Dell, Microsoft and Tesla on behalf of Congolese families whose children were killed or maimed while working as forced labour mining the cobalt that ended up in the products of these companies (International Rights Advocates, 2019).

Once metals and minerals necessary for manufacture have reached our South-East Asian factory they will be processed into microelectronic components.

Manufacture of semiconductors and other components can involve several known occupational health hazards, particularly from the use of organic solvents and heavy metals. These hazards have been well studied in advanced societies, where there are established (if not always totally effective) occupational health and safety structures and legislation. In manufacturing plants in the global South where pressure from the brand companies on the contractor manufacturers is relentless in forcing down costs and squeezing margins, health and safety is one of the first areas to get squeezed (Chan *et al.*, 2013). A large survey of over 7,500 Chinese female workers in the microelectronics industry found significant levels of occupational health issues; an example was the 137 workers reported to have been seriously injured by heavy exposure to *n*-hexane, a toxic agent used for cleaning the touch-screen glass panels of phones. In addition, it was found that the young women who predominantly make up the electronics industry workforce had a higher incidence of adverse reproductive outcomes (Yu *et al.*, 2013).

For the women working on final assembly of the phones, the working conditions and pressure they are put under has come to light in the past decade due to a series of tragic outcomes. In China's Foxconn factory, assembling iPhones for Apple involves long hours, repetitive single-task operations and extremely tight times allowed for task completion, measured to precise seconds. These are combined with constant surveillance and harsh discipline, with penalties for failing to meet time allowances such as not being allowed the normal half-hour break and various exercises in public humiliation such as being forced to stand to attention after working hours and recite a statement of self-criticism before colleagues. In 2010, this oppressive organisational work culture resulted in a spate of young women workers attempting suicide by jumping from their dormitory roofs: 14 were killed and four left with major physical disabilities. The company's initial response was to deny the suicide rate was anything out of the ordinary and to erect suicide nets around the buildings (Chan, 2013).

Once our phone has been assembled and packaged there then begins the lengthy transport from one side of the world to the other. Global supply chains have undoubtedly increased the demand and tonnage of maritime shipping and, with it, the volume of carbon and other emissions given off by marine vessels. A study a decade ago estimated that shipping was contributing 2–3 percent of total global emissions of carbon dioxide, 10–15 percent of nitrogenous oxides and 40 percent of sulphur oxites (Endreson *et al.*, 2008). In addition, international shipping is a major contributor to global emissions of black carbon from exhaust particulates, the second largest contributor to human-induced climate heating after CO_2. Unlike air pollution in cities where the evidence is visible and breathable, emissions at sea go largely unnoticed and shipping emissions are not yet subject to any international regulation (Azzara *et al.*, 2015).

Once the container ship containing our phone (among thousands of others) arrives in the UK, and is unloaded, the phone makes its way to the warehouse of our online retailer. On receipt of our original consumer's order (forgotten her?), retrieving it becomes just one job in the ten- and half-hour working day of one of the army of 'pickers'. As we have seen in previous chapters, from the accounts of Bloodworth (2018) and others, every movement of these low-wage agency workers is monitored for time completion and 'correct attitude' with a points-based disciplinary system allocating penalty points for days off with illness, low pick-rate or being late. The only way to hit the pick-rate is to run and going to the toilet is logged as 'idle time'.

The phone will then be delivered to our consumer by one of the many courier services used to service online retail. In all probability this will be one of the technically 'self-employed' owner-drivers we examined in Chapter 2 who, while leasing or owning their vans, have their workload controlled by the delivery companies and who face penalties for failing to meet delivery targets or even for going sick (Moore and Newsome, 2018).

But our story does not quite end there. Our satisfied consumer, having unwrapped her shiny new acquisition is then faced with the problem of what to do with her previous, but still functional, phone. She remembers that the local council recycling centre has a bin for small electronic devices and the knowledge that she can at least 'recycle' brings a satisfactory feeling that she is 'doing the right thing'. What she does not know, however, is who is going to do the recycling or where that will take place. The answer is not, as she might imagine, that her phone will be used by a grateful consumer in a less developed part of the world but rather that it will be systematically taken to bits by a small group of 'informal' workers in West Africa.

Agbogbloshie, a slum in the heart of Accra, Ghana, has become one the most polluted urban environments in the world due to its myriad formal and informal enterprises involved in the disassembly of the developed world's unwanted electronic devices. Accra shares with other cities in West Africa a large well-organised recycling sector involved in the recovery and reuse of metals and plastics; this industry extracting re-usable metals from e-waste products is now so large and well established that it has been termed 'urban mining'. Some of the companies involved are large with regular and established links to waste handling companies in the developed world. However, much of the recycling work is undertaken in the informal economy by scrap operators and scavengers who start by dismantling devices with simple tools, such as hammers, chisels and stones, to break them down into parts and/or subcomponents. This might involve incinerating cables and plastic parts to liberate copper and other metal and grinding motherboards into a fine powder containing precious metals. Contemporary electronic devices may contain hazardous metals such as arsenic, beryllium and cadmium, and heavy metals released during the burning and recovery processes contaminate local air, soil and groundwater. Public health assessments show high lead levels in the soil, and blood and urine samples taken from e-waste workers show elevated levels of barium, cobalt, copper, iron and zinc (Grant and Oteng-Ababio, 2016).

In case the above hypothetical example is seen as an atypical outlier, not fully reflective of the mutual benefits of globalisation, let us take an even more common area of consumption, that of clothing.

Clothing

The garment industry is the world's third largest manufacturing sector, after automotive and technology. It is also archetypical 'linear' production, using non-renewable resources and producing goods that are worn for a relatively short time and then disposed of, with waste materials entering the environment at every stage of extraction, production and consumption. The current way in which consumers relate to clothing and fashion has itself been criticised as unsustainable, particularly the recent trend for 'fast fashion' – garments that are so cheap to buy that they are seen as throw-away items: it is estimated that more than half of fast fashion produced is disposed of in under a year (Ellen Macarthur Foundation, 2017).

As garment manufacture is, in virtually all stages of production, an extremely labour-intensive industry, price competition through offering cheaper and cheaper clothing can only mean that labour power is being purchased at ever lower levels of pay and conditions as competition between clothing retailers forces them to 'chase the cheap needle' around the countries of the Majority World. Although fashion retailers are currently to be found marketing clothing with such 'green' messaging as 'Save the Rainforests' and 'Protect our Planet', current methods of garment manufacture, at virtually all stages of production, have huge negative consequences for the environment. The global textile industry creates 1.2bn tonnes of CO_2 a year, more than aviation and shipping combined. The by-products are a huge amount of chemical and plastic pollution.

Contemporary textiles either consist of natural fibres such as wool or cotton, of which cotton is by far the most widely used, or synthetic fibres that basically derive from oil and of which polyester is the most common. One might imagine that in the moves away from oil-derived plastics and synthetics, cotton would be seen as a preferable 'green' alternative but, as the House of Commons Environmental Audit Committee found in their review of the industry's sustainability, intensive cotton farming demands high levels of pesticide and fertiliser use and such huge volumes of water that it has contributed to water supply crises in the cotton growing areas of countries such as India, Turkey and China. To produce one kilo of cotton (the approximate weight of a shirt and pair of jeans) can take between 10,000 and 20,000 litres of water, about a fifth of which is used for diluting chemicals used in production (House of Commons, 2019: 29). In the case of non-organic cotton, the industry uses about 6 percent of global pesticides and around 16 percent of insecticides, despite only covering around 2.4 percent of global cultivated land (Pesticide Action Network, quoted in i newspaper, 2020).

Cotton production has of course historically been closely associated with unfree labour: it was after all one of the drivers of the eighteenth- and nineteenth-century slave trade. Regrettably this association still persists in some cotton growing regions in the global South. The Parliamentary Committee heard reports of autocratic regimes in cotton growing countries such as Turkmenistan and Uzbekistan using forced labour in the cotton fields. Cotton spinning mills in parts of India use the 'sumangli' system of bonded labour in which very young women are bonded to the mill for up to three years. The damp and fibre-laden working environments in cotton spinning in those nations with poor working environment regulations has been reported to contribute to byssinosis or 'black lung' disease, not seen in the UK since the mid-nineteenth century.

The production of polyester and other synthetics demands less water but uses an estimated 342 million barrels of oil every year (Ellen MacArthur Foundation, 2017) and emits up to double the amount of CO_2 than cotton. Synthetics spinning gives rise to the release of micro-fibres into the atmosphere and water tables; atmospheric fibres have been associated with diminished lung function among spinning workers, while it is estimated that 25–35 percent of micro-plastics in the oceans come from synthetic clothing, either during manufacture or from domestic washing (House of Commons, *op. cit.*: 33).

The cutting and stitching of textiles into items of clothing is done almost universally by women in the developing economies of the global South, work often characterised by low wages, long hours, minimal concern for health and safety and low or non-existent levels of unionisation and collective bargaining. Occasionally a spotlight is thrown on the harsh conditions in the clothing factories that supply the big brand names of the high street and online fashion stores. In January 2019 investigative journalism revealed that a Bangladeshi factory was paying 35p an hour to machinists stitching up Girl Power T-shirts in aid of Comic Relief. The workers there reported a regime of verbal abuse and harassment with long 16-hour shifts, and the factory was later forced to pay compensation to a worker who complained about the conditions and was then beaten up on the orders of the HR manager (Murphy, 2019). The event that most alerted the outside world to the conditions behind their clothing was the physical collapse of the Rana Plaza building in Bangladesh in 2013. The building housed five garment factories and its collapse resulted in the deaths of 1,138 workers with 2,500 injured, making it one of the largest industrial disasters in history. While this event led to the establishment of codes of conduct on working conditions and some of the larger brand retailers to establish their own inspectorates, it is widely recognised that it is virtually impossible to effectively check conditions over the vast network of thread and fabric makers, dyeing and stitching companies, all located in countries with poor or undeveloped regulatory regimes.

Nevertheless, the existence of regulatory legislation such as that of the UK can be no guarantee of compliance. Leicester is home to 700 clothing factories, which, because they can supply orders much quicker than factories in Asia, often supply online retailers such as Boohoo, Asos and Missguided. However, research undertaken for the Ethical Trading Initiative found that the majority of garment workers in the Leicester area were paid below the national minimum wage, did not have employment contracts, and were subject to intense and arbitrary work practices. Within this overall picture the researchers noted additional labour market segmentation in which a small segment of workers were paid national minimum wage rates but still did not have an employment contract; beneath them was a larger segment of female workers whose English language limitations limited them to this industry and were paid at levels that forced them to supplement wages with welfare benefits. Below these were migrants on temporary visas or were undocumented and, with no legal right to work in the UK, were on minimal pay, working night shifts and who could be dismissed at will. One consequence of this labour market structure was that those manufacturers who had invested to meet rising demand found themselves undercut by competitors who violated minimum work and employment standards (University of Leicester, 2015).

When we look at the delivery end of online clothing retail, we find similar conditions to those found in our phone example: the distribution centres of online retailer Asos have been criticised by the GMB union for frantic and unregulated pick rates, a culture of on-site bullying and staff 'treated like robots' (TUC, 2018). Once it has been 'picked', clothing is dispatched to the online customer by one of the notionally self-employed delivery services we have already examined.

Finally, what do we do with clothing no longer required? A T-shirt made of a blend of cotton and synthetics is virtually impossible to recycle into new clothing. In the UK we dump 1.3m tonnes of waste clothing a year and despite the significant degree of recycling performed by UK charity shops, 350,000 tonnes of it are either burnt or go to landfill (House of Commons, *op. cit.*).

Inequality: the reality of globalism

This then is the reality of global capitalism. In business schools the above examples would be discussed in terms of value chains, yet we must ask where exactly is the value added and who gets it? We should instead see them as chains that connect degraded and unhealthy jobs on one side of the globe to degraded and unhealthy jobs on the other, connected by unsustainable links that start with the continued extraction of finite resources and end with the toxic pollution of the environment and its inhabitants by waste materials. The resulting characteristic, shouldered by those eking out a living in the world's labour markets and those weathering the impact of climate change alike, is structured and increasing inequality.

As we saw in Chapter 3, it is estimated that developing countries will bear 75 percent of the costs of climate change, despite the global South only contributing around ten percent of CO_2 emissions. In its Transforming the World programme, the UN directly links ambitions for sustainable development with reducing inequality within and among countries, an aim that is also found in many proposed Green New Deals.

As Thomas Piketty (2020) makes clear in his comprehensive analysis of capitalism, inequality is driven largely by ideology, and is, therefore, neither inevitable nor immutable. During the period of the 'post-war consensus' nominally capitalist countries pursued social democratic policies of nationalisation, public education, health and pension reforms and 'progressive taxation of the highest incomes and largest fortunes' to realise widespread and profound egalitarian reform (*ibid.*: 486). Further, Germany and Nordic countries introduced co-management or codetermination initiatives to offer a degree of social ownership to employees in large enterprises. However, the ascendance of neo-liberal ideology served to massively favour concentrations of income and wealth accumulation in the present century (*ibid.*: 966; Piketty, 2014). Growing inequality has in turn led to the growth of social tensions and has been associated with a range of negative and mutually reinforcing indicators of health, well-being, education and societal functionality.

This close relationship between inequality and health and social problems is well established, at least among rich nations, who also comprise the biggest environmental polluters (Wilkinson and Pickett, 2010: 20, 174). As Wilkinson and Pickett demonstrate (*ibid.*: 219), carbon emissions tend to increase as societies get wealthier, so that rich and unequal societies are responsible for disproportionate shares of GHG emissions, while poor countries and communities suffer the worst consequences. Raworth argues that in prosperous, but low equality, countries ecological degradation may be enhanced through status competition and conspicuous consumption by the wealthy, with consequential high carbon emissions (Raworth, 2017: 172). A regular, if ironical, example is the annual sight of hundreds of private aircraft parked at Swiss airports (1,300 in 2018), whilst their executive passengers converge on the World Economic Forum at the Davos luxury ski resort where they pontificate on ways to expand economic growth whilst paying lip service to measures to protect the environment. In contrast, less advantaged people are likely to live and work in areas near emissions and other sources of contamination and often suffer the health and well-being consequences (Michaels, 2020: 33, 78).

Perhaps even more importantly, inequality derives from and emphasises competitive individualism, encapsulated in the *homo economicus* rejection of cooperation, community, and social and regulatory action needed to address the environmental and climate emergency. It is perhaps no surprise that more equal and socially cohesive societies like Japan tend to adopt and practice policies aimed at protecting the natural environment and their business leaders to more willingly endorse international sustainability agreements than in less equal countries (Wilkinson and Pickett, 2010: 233). In short, the environmental crisis is a political crisis underpinned by national and international inequalities and aiming to resolve the former without seriously addressing the latter will have little chance of restoring environmental equilibrium.

The financial implications of cutting carbon

On the face of it one would expect opposition to any meaningful environmental policies to come from the oil and gas industries as what they sell is locked-up carbon, which is then unlocked by purchasers through the processes of production and energy generation. However, the fuel majors are to some extent the 'visible' source of the escalating climate threat; less visible, but of deeper significance, is the role played by the global financial system, which supports and underpins their expanding activities and also contributes to sustaining the inequalities outlined above.

Due to the ever-increasing topicality of the climate crisis we again draw on the excellent work of investigative journalists and campaign groups in probing the complexity of the way in which the resource-exploitative economy is structured and maintained. The US based Rainforest Action Network analysed Bloomberg financial data and company documents to show that, since the 2015 Paris climate agreement, the world's largest investment banks have supported fossil fuel companies with $700bn of funding (including loans, equity issue and debt underwriting) (Greenfield, 2019a), US bank J. P. Morgan Chase alone providing $75bn. The fuel companies are among those most actively expanding in sectors such as fracking and Arctic oil and gas. Of the ten top banks studied, only Barclays seems to have reduced its funding in these areas and has started to seriously consider the advantages of financing renewable energy. Actual investment in the fossil fuel sector continues to be dominated by giant asset management companies, using money from savings and pension funds and such investment has again increased since the Paris agreement. Because asset management firms such as Blackrock and Vanguard can exercise shareholders' voting rights on behalf of their clients, they have frequently opposed motions at fuel companies' shareholder meetings that would have forced directors to include the climate crisis in future policy decisions (Greenfield, 2019b). Black-Rock is the world's largest private equity group and also the largest global investor in coal and among the top three shareholders in all but one of the major oil companies. Its CEO, Larry Fink, is on record as arguing that his company's duty is to make his customers money rather than to combat the climate crisis (Jolly, 2019), though correspondence to clients suggests he may have recanted more recently.

Shareholder pressure is gradually exerting some impact on the fuel companies. For the first time, in 2018, shareholders forced Royal Dutch Shell to set carbon emission targets linked to executive pay. However, when ExxonMobil shareholders tried to get a vote on emission targets, the US Securities and Exchange Commission prevented this on the grounds that it was an attempt to micromanage the company.

The underlying economic reason, which makes the financial world and national governments uneasy at the prospect of public pressure on the fuel companies, is that their stock exchange value is based on their 'assets' and these – approximately 2.8 trillion tonnes of carbon reserves – are primarily still in the ground. UK Green Party MP Caroline Lucas argues that if these assets were extracted and burned it would result in 2,800 gigatonnes of carbon emissions; however, to meet agreed targets for temperature increase we will have to limit carbon emission to just 565 gigatonnes over the next 35 years. In other words, the majority of such assets will have to remain underground (Lucas, 2015: 125). A slightly different calculation concludes that to have at least a 50 percent chance of keeping global warming below 2°C this century then globally a third of oil reserves, half of gas reserves and over 80 percent of current coal reserves should remain unused in the period up to 2050 (McGlade and Ekins, 2015). As such 'high carbon assets' underpin the value of the energy and mining companies that make up about a third of the value of the London Stock Exchange (and presumably make an even larger contribution to traded values on Wall Street) Lucas argues that writing them off in order to meet internationally agreed climate targets could have a significant impact on the world financial system.

Perhaps surprising support for this analysis comes from former Governor of the Bank of England Mark Carney who, speaking to the House of Commons Treasury Committee in October 2019, calculated that the consequences of the current amount of support by the world financial system of carbon producing activities would more than double the target agreed by the 2015 Paris agreement, resulting in a global temperature rise of more than 4°C above pre-industrial levels. This estimate is based on secured funding already agreed rather than on any predicted increase in financial support. Carney has been grittily realistic about the consequences of continuing to ignore the climate crisis, warning that companies and industries that are not moving towards zero-carbon emissions are likely to fail to survive. The only way to avoid a huge shock to the financial system by a sudden drop in asset values is for companies to enable a smoother transition to net zero-carbon by a transparent audit of their risk exposure to climate change. This was also among the recommendations of the House of Commons Environmental Audit Committee's report on 'Greening Finance' (House of Commons, 2019), which criticised the 'structural incentives' in the UK financial system to prioritise short-term profit rather than long-term issues such as the climate crisis. The Committee did not see a voluntary approach as sufficient and felt that unless there was mandatory climate risk reporting by companies and a legal duty on pension fund trustees to consider long-term sustainability, many billions of pounds of savings and investment would be placed in jeopardy.

Conclusions

As we stressed in the previous chapter, an alternative perspective to neo-liberalism must recognise that enterprises, the chief generators of carbon emissions, are essentially pluralist in character. This involves recognition of diverse interests, among which those of employees are prominent. A more transformative and participative role for managers, employees and communities is therefore crucial if processes and jobs are to become more sustainable within the increasingly narrowing time frame.

The following chapters critically examine the role of these major actors and address proposed solutions to the twin problems of work and climate. We also examine the diverse ways in which employees and their representative bodies, notably trade unions, traditionally engage with employers and politically over strategic decisions concerning their livelihood and wider social and economic issues, leading to an examination of the extent to which these approaches can serve to help bring about the transformative but necessary changes outlined above, whilst safeguarding and enhancing jobs and prospects.

Bibliography

Azzara, A., Minjares, R. and Rutherford, D. (2015), 'Needs and opportunities to reduce blackcarbon emissions from maritime shipping', International Council on Clean Transportation, Working Paper 2015–2012, http://admin.indiaenvironmentportal.org.in/files/file/ICCT_black-carbon-maritime-shipping.pdf.

Bakan, J. (2004), *The Corporation: The Pathological Pursuit of Power and Profit*, London, Constable.

Bannerjee, A. and Duflo, E. (2019), *Good Economics for Hard Times*, New York, Hachette.

Birch, K., Peacock, M., Wellen, R. *et al.* (2017), *Business and Society*, London, Zed Books.

Bloodworth, J. (2018), *Hired: Six Months Undercover in Low-Wage Britain*, London, Atlantic Books.

Chan, J. (2013), 'A suicide survivor: the life of a Chinese worker', *New Technology Work and Employment*, 28 (2); 84–99.

Chan, J., Pun, N., and Selden, M., (2013) 'The politics of global production: Apple, Foxconn and China's new working class', *New Technology Work and Employment*, 28 (2); 100–115.

Chang, H-J. (2011), *23 Things They Don't Tell You About Capitalism*, London, Penguin.

Cole, M., Rayner, A. and Bates, M. (1997), 'The environmental Kuznets curve: an empirical analysis', *Environmental and Development Economics*, 2 (4); 401–416.

Daly, H. and Cobb, J. (1994), *For the Common Good*, Boston Mass., Beacon Press.

Dorling, D. (2019), *Inequality and the 1%*, London, Verso.

Ellen MacArthur Foundation (2017), *A New Textiles Economy: Redesigning Fashion's Future*, www.ellenmacarthurfoundation.org/publications

Endreson, Ø. *et al.* (2008), 'The environmental impacts of increased maritime shipping: past trends and future perspectives', Det Norske Veritas, University of Oslo.

Exton, C. and Shinwell, M. (2018), 'Policy use of well-being metrics: Describing countries' experiences', SDD Working Paper No. 94, Paris, OECD.

FitzRoy, F. (2019), 'A green new deal: the economic benefits of energy transition', *Substantia*, 3 (2); 55–67.

Frayne, D. (2015), *The Refusal of Work*, London, Zed Books.

Geoghegan, P. (2020), *Democracy for Sale*, London, Head of Zeus.

Graham, C. (2005), 'The research of happiness', *World Economics*, 6 (3); 41–55.

Grant, R. and Oteng-Ababio, M. (2016), 'The global transformation of materials and the emergence of informal mining in Accra, Ghana', *Africa Today*, 62 (4); 1–19.

Greenfield, P. (2019a), 'Big banks put $700bn into fossil fuel firms despite climate crisis', *The Guardian*, 14 October.

Greenfield, P. (2019b), 'How the top fund managers use your money to block action on climate change', *The Observer*, 13 October.

Hall, P. and Soskice, D. (2001), *Varieties of Capitalism: The Institutional Foundations of Comparative Advantage*, Oxford, OUP.

HCSS (2013), *Coltan, Congo & Conflict*, No. 21, (05/13), The Hague, Hague Centre for Strategic Studies.

Hickel, J. and Kallis, G. (2020), 'Is green growth possible?', *New Political Economy*, 25 (4); 469–486.

House of Commons (2019), 'Environmental Audit Committee: Greening finance: embedding sustainability in financial decision-making', https://publications.parliament.uk/pa/cm201719/cmselect/cmenvaud/1063/106302.htm.

i newspaper (2020), 'High street T-shirts pushing green messages are poison for the planet', 27 February.

International Rights Advocates (2019), www.iradvocates.org/press-release/iradvocates-files-forced-child-labor-case-against-tech-giants-apple-alphabet-dell.

Jackson, T. (2011), *Prosperity Without Growth: Economics for a Finite Planet*, London, Earthscan.

Jolly, J. (2019), 'World's top investor accused of dragging feet on climate crisis', *The Guardian*, 21 May, https://www.theguardian.com/business/2019/may/21/blackrock-investor-climate-crisis-blackrock-assets

Juniper, T. (2013), *What has Nature Ever Done for Us?* London, Profile Books.

Kalleberg, A. (2018), *Precarious Lives: Job Insecurity and Well-Being in Rich Democracies*, Cambridge, Polity.

Kallis, G. (2018), *Degrowth*, Newcastle-upon-Tyne, Agenda Publishing.

Klein, N. (2014), *This Changes Everything*, London, Penguin/Random House.

Layard, R. (2011), *Happiness*, 2nd edn, London, Penguin Books.

Leon-Kabamba, N., Ngatu, N., Kakoma, S. et al. (2018) 'Respiratory health of dust-exposed Congolese coltan miners', *International Archives of Occupational and Environmental Health*https://doi.org/10.1007/s00420-018-1329-0.

Lucas, C. (2015), *Honourable Friends? Parliament and the Fight for Change*, London, Portobello.

Marçal, K. (2015), *Who Cooked Adam Smith's Dinner?*, London, Portobello.

Mason, P. (2016), *Post-Capitalism: A Guide to our Future*, London, Penguin.

McGlade, C. and Ekins, P. (2015), 'The geographical distribution of fossil fuels unused when limiting global warming to 2°C', *Nature*, 517 (8 January); 187–190.

Meadows, D., Meadows, D. L., Randers, J. and Behrens, W. (1972), *The Limits to Growth: A Report for the Club of Rome's Project on the Predicament of Mankind*, New York, Signet.

Michaels, D. (2020), *The Triumph of Doubt: Dark Money and the Science of Deception*, Oxford, Oxford University Press.

Monbiot, G. (2019), 'Britain's dirty secret: the burning tyres choking India', *The Guardian*, 30 January.

Moore, S. and Newsome, K. (2018), 'Paying for free delivery: dependent self-employment as a measure of precarity in parcel delivery', *Work Employment & Society*, 32 (3); 475–492.

Murphy, S. (2019), 'Factory that supplied Tesco compensated abused worker', *The Guardian*, 22 January, https://www.theguardian.com/world/2019/jan/22/bangladeshi-factory-that-supplied-tesco -and-marks-and-spencer-compensates-abused-worker

Piketty, T. (2014), *Capital in the Twenty-First Century*, Cambridge MA, Harvard University Press.

Piketty, T. (2020), *Capital and Ideology*, Cambridge MA, Harvard University Press.

Pilling, D. (2018), *The Growth Delusion*, London, Bloomsbury.

Raworth, K. (2017), *Doughnut Economics: Seven Ways to Think Like a 21st-Century Economist*, London, Penguin/Random House.

Real World Coalition (2001), *From Here to Sustainability: Politics in the Real World*, London, Earthscan.

Shafique, A. (2018), *Addressing Economic Insecurity*, London, Royal Society of Arts/ Nottingham Trent University Nottingham Civic Exchange.

Stern, D. (2004), 'The rise and fall of the environmental Kuznets curve', *World Development*, 32 (8); 1419–1439.

Stiglitz, J., Fitoussi, J-P. and Durand, M. (2018), 'Beyond GDP: measuring what counts for economic and social performance', Paris, OECD.

Stiglitz, J., Sen, A. and Fitoussi, J-P. (2009), *Mismeasuring Our Lives: Why GDP Doesn't Add Up*, New York, The New Press.

TUC (2018), 'Sales season is not a jolly time for Asos workers', *Risks 877*, 1 December.

University of Leicester (2015), *New Industry on a Skewed Playing Field: Supply Chain Relations and Working Conditions in UK Garment Manufacturing*, Leicester, University of Leicester Centre for Sustainable Work and Employment Futures.

Waddington, D., Wykes, M. and Critcher, C. (1991), *Split at the Seams: Community, Continuity and Change after the 1984–5 Coal Dispute*, Milton Keynes, Open University Press.

Weil, D. (2014), *The Fissured Workplace*, Cambridge, Mass., Harvard University Press.

Wenar, L. (2016), *Blood Oil: Tyrants, Violence, and the Rules that Run the World*, Oxford, Oxford University Press.

Wilkinson, R. and Pickett, K. (2010), *The Spirit Level*, London, Penguin.

Yu, W., Lao, X., Pang, S. *et al.* (2013), 'A survey of occupational health hazards among 7,610 female workers in China's electronics industry', *Archives of Environmental & Occupational Health* 68 (4); 190–195.

5 Roles and responsibilities of business

Introduction

Business, alongside government and employees, has a pivotal, complex and multidimensional role in ensuring sustainability. It is pivotal because carbon emissions are highly associated with commercial activity; because business helps to shape and determine consumer behaviour; and is both influenced by and, more often, influences political directions (Michaels, 2020). Powerful transnational companies can even dictate domestic (and usually deregulatory) policy in countries where they enjoy privileged status, as with petrochemicals (Wenar, 2017), overseas mining operations (Klein, 2007) and finance (Tooze, 2018). Business relies on citizens to provide their labour, as well as to consume goods and services that business supplies. Business is complex because it ranges from small enterprises to giant transnational corporations. Multidimensional, as it can take diverse ownership forms, with a principal division between private and public ownership and conglomerate or multi-divisional where the same organisation supplies different markets.

Pressures on business

Fundamentally, the role of a business is to provide goods and services within parameters determined by law and other regulatory bodies. Commercial businesses aim to provide goods and services at a profit for owners, following Friedman's famous dictum (1970) that the sole responsibility of business is to make profits for enterprise owners, who in larger corporations are shareholders, or increasingly, private equity companies. Consequently, in competitive markets, businesses endeavour to operate at low cost and with optimum efficiency. Non-profit-seeking bodies such as public utilities are financed through taxes and are expected to meet both value for money for customers and welfare criteria for employees as well as societal benefit, though they may be expected to operate along equivalent efficiency lines as private sector companies, against which their performance is often measured.

From an environmental and sustainability perspective, the role of business is problematic. Though political and economic policy is gradually shifting in treating the environment as an economic cost rather than an exploitable private resource, organisations facing competitive pressures have tended to navigate ways of maintaining cost minimisation, for instance through lowering employment standards (zero-hours contracts, agency work, etc.), exporting pollution to newly developing countries or by playing fast and loose with regulatory requirements or pressurising governments to relax them. On the other hand, growing political and cultural imperatives cannot be ignored. The UK government has signalled its intention to end the sale of new combustion-engine powered vehicles by 2030. Through the informational and direct action undertakings of pressure groups such as Greenpeace and Friends of the Earth and more confrontational approaches of Extinction Rebellion along with media exposure given to environmental issues by personalities like David Attenborough and Greta Thunberg, citizens are becoming increasingly aware of the impact of consumer behaviour on the environment, for example through taking air travel for granted, excessive meat-eating and driving petrol- and diesel-powered vehicles. A glance at company webpages demonstrates that many commercial organisations are now publicly proclaiming their green credentials and intentions, though a survey of European practice by Vitols *et al.* (2011) found that employers' main concern was to link greening the economy with competitiveness, with employer organisations expressing fears that meeting emissions targets could lead to competitive disadvantage in world markets.

Though many, if not most, large corporations claim sustainability as a distinctive corporate goal, researchers find that managers experience problems in defining and operationalising the concept, leading one study to suggest that it is difficult to assess whether firms are 'merely greenwashing or whitewashing' (Meuer *et al.*, 2020) and for another research group to conclude that 'when it comes to practicing and not just preaching sustainability, many companies struggle, and most flounder in developing and implementing a sustainable business model' (Bhattacharya and Polman, 2017: 71). Peter Dauvergne, of the University of British Columbia, presents some telling mission statements: Nike – pursuing 'sustainable development – a powerful strategy that drives us to dream bigger': Monsanto (really!) – 'leader in innovative and sustainable agriculture'. These, along with others, lead Dauvergne to suggest that 'trusting [big corporations] to lead sustainability efforts is like trusting arsonists to be our firefighters' (2018: 2–5). Nevertheless, former Bank of England Governor Mark Carney, while acknowledging in a recent television interview that ignoring climate change was not an option, still put his faith on companies' ability to respond to market signals and do 'something about it'. According to Carney, the 'system' simply needs to 'manage the risks around climate change, and … move capital from where it is today to where it needs to be tomorrow' (Busby, 2019). This rather ignores the fact that those benefitting most from the capitalist system have so far neither recognised nor responded to the severity of the crisis, and indeed, as we have seen, actively continue to contribute to it.

Credibility gaps between sustainability proclamation and environmental reality are well demonstrated by Ineos, one of the world's largest plastics producers, much of it originating from its petrochemical production facilities in

Scotland. The company is also a major and very combative fracking supporter and licence-holder, notwithstanding concerns that fracking will inhibit the UK from meeting its net zero emissions target by 2050. From its website (ineos.com/business/ineos-enterprises/business/), one of the company's avowed 'key values' is 'excellence in safety, health and environmental performance'. According to the Scottish Environmental Protection Agency, the company was Scotland's biggest emitter of GHGs in 2017, with its Grangemouth refinery emitting 1.6 million tonnes. Four other Ineos facilities were in the country's top 12 emitters (*The National*, 2019). In 2018, it was reported in *The Scotsman* newspaper that 450,000 tiny plastic pellets or 'nurdles', the equivalent of 833 plastic bottles, were found in a single day's sweep on the shores of the Forth estuary, about 12 miles from the Ineos Polymers plant. A local paper reported in 2019 that over 20,000 pieces of plastic were recovered from a 48 square metre site, also on the Fife shores (Henderson, 2019). The company denied responsibility.

Nevertheless, the company was reportedly behind a campaign waged by the British Plastics Federation against the introduction of charging five pence for plastic carrier bags, which the Federation claimed was 'pointless and unnecessary' (since the charge was introduced in England and Wales, plastic bag usage has declined by 85 percent). The world's oceans are being swamped by non-degradable plastic (Eriksen *et al.*, 2014), so it is perhaps fitting that in a classic example of greenwashing, the company applied to enter an Ineos-sponsored team for the 2021 America's Cup, the most prestigious ocean-going event in sailing's calendar.

The Independent newspaper notes that Ineos, whose owner was keen for the UK to leave the European Union, has also been accused of lobbying the government to relax climate change laws. One document, obtained by campaigning group Friends of the Earth, pointed to the alleged commercially competitive benefits of leaving Europe: 'Outside the EU: simplify the UK policy mix and seek a single route to 100% exemption from policy costs and CCL [climate change levy]. Seek a low-cost alternative to EU ETS [the EU's carbon trading scheme]' (Chapman, 2017). Only the previous week, the newspaper reported that

> following a concerted effort by Ineos and other chemical firms the Government … announced a £100 million per year exemption from the carbon trading scheme for the companies that use the most energy, to add to the £250 million in tax breaks it had already granted.

Described in French magazine *L'Obs* (Lacombe, 2020) as 'pollueur en chef', the company's owner is about to manufacture a modified gas-guzzling Land Rover Defender 4x4, though no details are offered on CO_2 emissions expected from the planned 25,000 annual sales.

Responsibilities of business

Commercial activities, from extraction of fossil fuels to promoting consumption, provide the primary source of GHG and global heating as well as contributing to deteriorating air and marine quality. One prime question therefore is to consider business accountability to the society on which it depends for its resources, human and otherwise, and to which it supplies its products and services. How can it identify and best discharge its responsibilities? Clearly, one responsibility is not to poison or otherwise incapacitate the society of which it forms part. This links to the broader question of political and legal responsibility. If industry is the principal contributor to these adverse developments, underpinned and encouraged by neo-liberal economic thinking and political action promoting deregulation and operating in a culture driven by financial hegemony, are there any signs that industry has taken notice of the multiple warnings facing commercial operations and acted to combat these identified problems? Behind these fundamental questions is the extent to which industry should take responsibility for the externalities associated with its activities.

In his excoriating critique of the corporation as a 'psychopathic creature', unable to 'recognize nor act upon moral reasons for harming others', Bakan (2004: 60) points to what are considered 'inevitable and acceptable consequences of corporate activity' found in the 'routine and regular harms caused to *others* – workers, consumers, communities, the environment' by these same psychopathic tendencies. Pilling cites as an example a factory producing plastics, which pollutes rivers and atmosphere by emitting toxic chemicals and dust. The company can save money by not investing in filters and other anti-pollutant devices whilst 'everyone suffers in the form of poorer health and higher taxes to pay for the clean-up' (2019: 175). Whilst a company's profits help to secure its commercial viability and contribute to domestic economic growth, some economists have argued that externalities can be seen as a legitimate constituent of corporate cost-benefit analysis, meaning that 'executives have no authority to consider what harmful effects a decision might have on other people ... unless those effects might have negative consequences for the corporation itself' (Bakan, 2004: 64). And of course, it is very difficult to ascribe specific costs to polluting or exploitative activities, which by their very nature can be extremely difficult to identify or whose spatial and longer-term impacts are hard to measure or predict. Externalities can extend to the treatment of workers and communities alike. As we saw in Chapter 3, the poisoning of Bhopal was an externality, as is the exploitative and often inhumane treatment of outsourced sweatshop workers. As we have seen, work itself, even in mature developed economies, is becoming increasingly externalised through the use of agency, contracted and self-employed labour, for whom the employer takes little or no financial or organisational responsibility for job security, sickness or provision for holidays or retirement (Kalleberg, 2018). Without employee status, there is no agency for labour, individually or collectively, to confront the harmful activities of their employers. If corporations and employers are unwilling, or unable, to constrain or remove cost-minimising externalities, the wider problem of dealing with GHG, pollution and labour resource exploitation will persist in the absence of concerted government action. Kate Raworth (2017: 143) puts the point succinctly: 'Far from remaining a peripheral concern "outside" of economic activity, addressing these effects [i.e. externalities] is of critical concern for creating an economy that enables us all to thrive'. And she could add – to survive.

Today, examples of environmental health and safety being subordinated to the pursuit of market dominance are not difficult to find (Michaels, 2020). Diesel fumes are recognised to constitute a specific public risk to air quality (Brand and Hunt, 2018), leading German carmaker Volkswagen to make deceptive claims about the environmental qualities of their 'clean' diesel cars, which have enjoyed a major market share in Europe. In vigorous promotional advertising in America, 'they had ads with engineers dressed up as white angels because they're ushering in this glorious, green future' (MacDuffie, 2019). Unfortunately, this promised land of a 'green future' was somewhat undermined by the installation of illegal 'defeat device' software in the vehicles, which produced artificially low NO_2 emissions when tested. Since September 2015, the scandal has cost the company upwards of \$30bn in accumulating penalties and early in 2019, the US Securities and Exchange Commission further charged the company and its former chief executive with defrauding US investors, raising billions of dollars through the corporate bond and fixed income markets while making a series of deceptive claims about the environmental impact of the company's '"clean diesel" fleet' (SEC.gov, 2019). Nevertheless, in a saga beginning to take on conspiratorial overtones, VW is by no means alone in its misdemeanours: leading automobile companies like Audi and Nissan have also admitted to adjusting or falsifying emissions data (Lyon, 2018).

Corporate social responsibility

One initiative – or response – from employers has been to turn to demonstrations of corporate social responsibility (CSR) to express their green credentials. CSR has a long and convoluted history, stretching back in the USA at least to the 1930s. Debates were invigorated by concerns about the growing power of enterprise managers in the 1950s and questions about their responsibility toward society (Birch *et al.*, 2017: 74–84). Milton Friedman's (1970) evangelical riposte to CSR emphasised that business has only one responsibility, namely to reward shareholders: diverting profits away from owners without their express permission would be morally reprehensible, doubly so if consumers had to make up the difference in higher prices. Only if CSR activities had a positive impact on the bottom line, could they be justified – a precursor to contemporary debates on 'greenwashing', in which companies detract attention from operating questionable or unsustainable endeavours through presenting attractive green imagery of the enterprise or through association with healthy competitive activities such as sponsoring outdoor sports. Subsequent discussion on CSR has become more multidimensional. Birch *et al.* (2017) point to the work of Archie Carroll (2010) who argued from the late 1970s that as well as dual obligations to pursue profit and act legally, companies do have ethical responsibilities to do good and avoid harm as well as behaving as a good corporate citizen. Clearly these duties would embrace responsibility to the environment, a point emphasised by Raworth, as the doing no harm mantra would involve 'designing products, services, buildings and businesses that aim for zero environmental impact' (2017: 217). Birch *et al.* (2017: 77) align later development of the CSR concept with the open systems model of corporate citizenship 'which comprises a social contract between business and society'. This model also embraces moral and ethical treatment of business's own citizens, namely employees and managers. Finally, Birch *et al.* present the stakeholder concept of CSR, which parallels the pluralist model of employment relations in asserting that the company is in a direct and dynamic relationship with diverse interest groups, including customers, local communities and employees, all of whom have a legitimate stake in what the company does and how it performs.

In contrast to positive interpretations of stakeholdings, Dauvergne (2018) and Dauvergne and Lister (2013) strip CSR of mutuality pretences and like Bakan and Friedman see CSR as a straightforward business preference, extending 'well beyond greenwashing' (Dauvergne, 2018: 42). For these authors, CSR is simply business strategy masquerading as ecology and rather than a response to environmental concerns serves as a means to establish more control over potential risks or 'to gain competitive advantage in a world spiralling into ever-greater environmental crisis' (Dauvergne, 2018: 45). He goes further, asserting that with the massive resources available to geopolitically all-powerful transnational corporations, their control extends to the environmental policy-making process itself, funding supportive think tanks, populating advisory committees and often enjoying close or 'revolving door' proximity to government representatives. Hence, substantive criticisms can be directed at the generic concept and practice of CSR (see e.g. Dans, 2015; Weyzig, 2009), but our perspective is to question the impact that it exerts on sustainability and good ecological practice. Clearly, CSR is a *voluntary* pursuit, which can be adopted by senior managers to portray the enterprise in a positive light and deflect attention from less savoury aspects of a company's activities and though enterprises may present ecologically sound positions, these may simply represent a market or niche opportunity attractive to environmentally aware consumers. Two industries that have vigorously adopted CSR are tobacco and oil and gas. Tobacco provides an example of both externalities (i.e. killing customers) and 'inherent contradictions' between its products and CSR (WHO, 2004). As well as helping to kill off half of its customers, tobacco has a considerable adverse environmental impact in its use of fertilisers and pesticides for plantations, deforestation, water use, transport and packaging. Yet the industry attempts to 'eco-label' its production and products (Houghton *et al.*, 2019) with lavishly funded CSR programmes including small business development, crime prevention, medical treatment, flood prevention in developing countries and even with help to establish an international centre for the study of CSR, maintained by a donation from BAT for £3.8 million. By aiming publicity at adults, companies also claim to deter smoking by young people, a strategy exposed by the WHO as ineffective as young people are happy to be associated with what are seen as attractive adult activities. Conversely, tobacco companies 'vigorously oppose price and tax increases' (WHO 2004: 3), which *have* been shown to act as a deterrent to young people. At the same time, the WHO contends, the tobacco industry continues 'to use a vast array of unethical and irresponsible strategies to promote its products, expand markets and increase profits', including efforts to undermine the health promotion work and tobacco control initiatives of the WHO (*ibid.*: 7). Independent academic researchers, such as those at the Tobacco Control Research Group, are, in the words of the Group's director: 'continuously attacked and abused on social media by those we know are linked to the tobacco industry' (Reisz, 2019). This is an industry that is estimated to kill nearly five million people annually and, in consequence, is continually seeking new markets, often in countries with weak anti-smoking controls. Nevertheless, little support to curtail smoking has been offered by governments, whose 'inaction and public indifference, where it exists, are largely a result of decades of tobacco companies' untoward influence' (WHO, 2004: 10; Oreskes and Conway, 2012: Ch 1).

Oil and gas represent a key sector where CSR initiatives have become commonplace, though corporate objectives are open to question. Frynas (2009) points to oil and gas as prominent champions for CSR, evidenced by high reporting levels and some engagement with community development schemes, though less active involvement with local or domestic governance. In subsequent research he suggests that their CSR agendas barely address governance issues and may seek to divert attention from broader political, economic and social issues (Frynas, 2010). KPMG collects CSR data on 250 of the world's largest companies: a recent report (KPMG, 2016) showed that virtually all major energy companies provided CSR reports. But numerous studies have questioned the intention behind CSR reporting. The website content of six oil companies was analysed by Du and Vieira (2012) who concluded that all had engaged in CSR endeavours in attempts to gain legitimacy for their activities. Companies may also aim to focus on internal moral and strategic management directions rather than impacts on environment, society and health (Cai *et al.*, 2012), echoing Frynas' (2005: 581) earlier conclusion of mounting evidence of a gap between 'stated intentions of business leaders and their actual behaviour and impact in the real world'. Jaworska (2018) examined CSR reports by major oil companies between 2000 and 2013, focusing on frequency of key climate change references. A change of tone was noted: in the mid-2000s, climate change was reported as a phenomenon that could be acted on, though more recent accounts tended to report it as an unpredictable event, distanced from sector activities. There are other signs of corporate intent: the 2016 KPMG survey found that over three-quarters of the oil and gas companies did not report on social and environmental impacts of their supply chains and that the quality of carbon emissions reporting was low. Bearing in mind the recency of the report and the amount of contemporary interest and concern over the activities of these energy companies across the world, it does appear that CSR

activities are not well-embedded in communities impacted by energy exploration and extraction and that reporting, though common, tends toward exercises
in public relations and presenting favourable images to investors and policymakers rather than serious practical engagement with fundamental environmental concerns.

The role of finance

All businesses are reliant on funding for their operations. Small and family
businesses depend on accumulated savings, crowdfunding, loans from banks
and financial support from government. Larger organisations depend for their
financial support on institutions such as investment banks and hedge funds,
whose prime interests are in short-term profit returns rather than long-term
investment in companies (Pettifor, 2019). As we saw in Chapter 4, such
approaches require companies to reward equity holders rather than retain
profits to support research and development, employee training and capital
investment. Cost minimisation policies take priority, potentially inducing productive and service organisations into environmentally impoverishing activities.

Driving these trends and a dominant influential feature of the contemporary
economy is financialisation. The real or productive economy is lubricated by
finance, but it is contended that the two sectors are not sustainably compatible
as finance pursues short-term or immediate returns, hence pressurising the real
economy toward cost minimisation, outsourcing and reducing labour standards.
In a thorough review of the process, Clark and Hermele (2013: 10) demonstrate
that the share of GDP represented by financial interests has grown continuously
in recent years; that financial interests represent an expanding share of
employment; and that its influence has grown over productive investment, over
regulation and over the natural environment. The elevated size and dominance
of finance over the productive (or real) sector is associated with increasing
demands upon it, thereby transferring benefit from the productive to the finance
sector at the expense of worker income, working conditions and equality. Clark
and Hermele argue that high levels of financialisaton are directly associated
with reduced bargaining power by labour, leading to widening social polarisa-
tion (*ibid.*: 25). There are also severe implications for sustainability through
competitive and cost pressures imposed on the real sector but also through
commodification and privatisation of the natural environment. Once commo-
dified, the environment can be treated as a pure and privatised financial asset,
leading to 'displacement of people, livelihoods, knowledge and practices' (*ibid.*:
26). The authors estimate that about ten million hectares of land are being
appropriated annually for commercial purposes by financial interests, including
agro-fuel feedstock, food and forestry, predominantly in Africa and Asia. They
conclude that financialisaton penetrates every aspect of commercial activity,
endangering the fabric of societies as social disadvantage grows. They also
argue that economic sustainability is endangered as financial resources are
increasingly diverted for short-term speculative investments 'to the detriment of

investment in the production of infrastructure, human capital or productive resources in general' (*ibid.*: 70). Basically, there is little evidence that financialisaton has benefitted environmental sustainability. Financial pressures on companies to operate at optimum short-term profitability has inevitably impacted negatively on labour and employment sustainability (Thompson, 2013). Unions have been weakened, terms and conditions for employees undermined and increasingly pressurised and growing numbers of jobs are performed by workers with few employment protections.

The role of banks in funding projects that, though profitable, serve to export emissions and ignore local community interests or welfare, has also been noted by campaigners. One study showed that in 2015 global investment in fossil fuels was three times higher than in renewable sources of energy, notwithstanding the increasing technological efficiency and reduced cost of renewables. Much banking investment has been to support fossil fuel extraction in environmentally fragile South American and African locations. One example of links between funders and owners is shown by the Correjón mine, one of the world's largest open-cast coal mines, owned by Glencore, Anglo-American and BHP Billiton, to whom an estimated $109bn have been committed by Barclays, HSBC, Lloyds and RBS. For a number of years, the mine has gathered a poor reputation for its neglectful treatment of local communities, though this claim has been refuted by the owners. The mine produced in excess of 33 million tonnes of coal in 2015 and is expected to be operative until 2034. Similar cases of bank funding of fossil fuel projects in vulnerable and food insecure countries, often by the same institutions, are evident (Banking on Climate Change, 2020).

As critics of institutional funders have argued, progress can be made by withdrawing from unsustainable investments and transferring to companies engaged in appropriate sustainable activities. There are positive signs: some major pension funds are divesting from fossil fuel companies. Nest, the UK's largest pension fund with nine million members, is banning future investment in all companies with interests in coalmining, tar sands and Arctic drilling. Fund managers are planning to shift £55bn of shares into 'climate aware' strategies, aiming for carbon neutrality across its portfolio by 2050 (FT Adviser, 2020). Similar steps have been taken by Nordic and other major pension funds (Wienberg, 2019; InfluenceMap, 2019). Another positive is the role of 'patient' capital, often provided by state investment banks (SIBs). Patient capital is largely a feature of coordinated market economies like Germany, though bank withdrawal and short-term hedge fund penetration has increased in recent years in Germany (Bessler *et al.*, 2008; Bessler *et al.*, 2013). Patient capital involves firms enjoying long-standing, cooperative and close 'insider' relationships with their investors (Thatcher and Vlandas, 2016). Insider status also provides the opportunity for investors, as well as employees through their unions, to exercise participative 'voice' rather than simply 'exit' from organisational investments that may be slow to mature or whose objectives are not immediately profit-centred (Hirschman, 1970). Often the state supplies the bulk or all of the investment funding. Priority is given by SIBs to environmentally progressive, infrastructural and developmental projects, including support for overseas development. In most cases, long-term systemic growth is a priority (Mazzucato and Macfarlane, 2017). Possibly best known is the German KfW SIB, 80 percent owned by the federal government and 20 percent by German regional states (see Chapter 11). Macfarlane and Mazzucato (2018: 31) identify three stated investment policies prioritised by the bank: (i) support for renewable energies, energy efficiency projects and prevention of environmental pollution (one-third of total budget of $81bn), (ii) support international competitiveness of German industry and (iii) address social and economic consequences of ageing population. KfW has adopted 'exclusion criteria' for environmentally and socially damaging projects. The European Investment Bank, with a budget of some €76.4bn in 2016, has similar exclusions criteria (*ibid.*: 34).

Unfortunately, the situation regarding subsidies to fossil fuel companies is less promising: European energy subsidies have swelled from €148bn in 2008 to 169bn in 2016, most being allocated to the energy sector (€102bn) and fossil fuel subsidies at about €55bn annually have continued. Internationally, subsidies outside of the EU are even higher and largely directed at petroleum products (European Commission, 2019: 10). Based on his research, van Lierop (2019) offered a 'moderate estimate' that in 2013 the 1,801 largest publicly traded oil, gas and coal companies received $700bn in direct subsidies. So why do companies continue to receive subsidies that help to seriously compromise attainment of Paris and SDG targets? Subsidies have often historically been offered to encourage domestic energy production and to provide cheap energy. They are still provided, despite costs of green energy significantly reducing and energy companies remaining profitable. Estimates have been made of direct governmental subsidies in the USA to the fossil fuel industry of at least $20bn yearly (EESI, 2019).

InfluenceMap's (2019) detailed research on the five largest publicly traded oil and gas companies implicates twin subsidy factors involving effective lobbying and constructing a positive climate-effect brand image. Together, these actions provide a narrative that continuing financial support for fossil fuel exploration and extraction is justified in the national interest. The study reveals company aims to maintain public support through 'carefully devised campaigns of positive messaging combined with negative policy lobbying on climate change'. Nearly $200 million is spent by the companies on lobbying 'to control, delay, or block binding climate-motivated policy', an approach that 'has hindered governments globally in their efforts to implement such policies post-Paris'. Coordinated messaging from the heads of the five majors focuses on the need for 'increased fossil fuel production to meet global energy demand'. Social media is used to back up these messages. The InfluenceMap report points out that in the weeks leading to the 2018 US mid-term elections the companies spent $2 million on Facebook and Instagram 'promoting the benefit of increased fossil fuel production and supporting successful opposition to several key climate related initiatives'. Lobbying is accompanied by $195 million spent annually in branding campaigns aimed to convince public and governments that the five companies are aligned with ambitious but voluntary action on climate protection through low carbon investments, though evidence indicates that only about three percent of their capital investments are in these areas. Climate lobbying is prominent in the US and often outsourced to trade groups such as the American Petroleum Institute, which successfully campaigned to deregulate oil and gas development. Clearly, political donations and campaigns are achieving some success with, for example, BP's CEO Bob Dudley thanking the Trump administration in 2018 for 'rolling back the "avalanche of regulation" on the sector' (*ibid.*: 11).

Conclusions: are there alternatives to the profit-maximising model of business?

Seeking short-term profits and remaining competitive are clearly prime roles assumed by business. Though commercial activity is of course subject to legal, regulatory and contractual obligations, the stringency of these demands varies across countries. As we have seen, large companies and their representative and advocatory bodies can exercise considerable political as well as economic power. Economic growth has long been seen as the basic requirement for political success and growth depends upon company performance. Many companies are owned and controlled by transnational corporations, which can exert influence over domestic policies including through implicit or explicit threats of transferring operations to more favourable locations. Further, numerous studies demonstrate uncomfortably close connections and power networks between commercial organisations, interlocking financial providers and sympathetic governments (Klein, 2007; Carroll, 2010; Heemskerk *et al.*, 2016). In consequence, political regulation of the profit-seeking activities of large enterprises is necessarily restricted. In many countries, environmental issues often have had questionable political priority, and companies consequently can feel little pressure, in the absence of binding requirements, to modify their operations.

On the other hand, while the neo-liberal informed shareholder primacy (SHP) doctrine continues to dominate political and economic policy, there are alternative models for corporate behaviour and governance. Cooperative, municipalised and employee-owned enterprises are examined in Chapter 11, but these currently enjoy minority status in advanced economies. Nevertheless, some commentators argue that the standard dominant profit-maximising model is in crisis, whether through executive abuse (e.g. extreme pay differentials; corruption; fraud) or, crucially, through neglecting the (environmental) interests of other parties affected by organisational behaviour (Salazar, 2017). Alternative approaches vary in the extent to which they intervene in SHP or transition to more participative models. Salazar (2017) recognises both insider (i.e. employees) and external interests (e.g. communities, suppliers, customers) affected by corporate activities, noting that stakeholder models are found in both LMEs and CMEs. In LMEs, however, acceptance of wider business responsibilities to employees is largely restricted to providing voluntary channels for consultation and information provision, perhaps underpinned by weak consultative legislation (see Chapter 8) or through minority equity ownership offered by employee share ownership schemes (Baddon *et al.*, 1989; Knyght, 2010). More meaningful engagement with management is undermined by denial of trade union recognition or downplaying collective participative roles.

Stakeholder thought and practice are generally more advanced in CMEs, for example through co-determination in Germany, though recent stock market volatility and the 'toxic combination of financialization, globalization and neoliberalism' (Hyman, 2016: 17) has tended to reduce insider bank involvement and weaken influence of worker directors and works councils (Müller and Stegmaier, 2020). Though broadly sympathetic to co-management principles as practiced in Germany and Scandinavian companies, (where worker directors are eligible in much smaller concerns than the 500 minimum in Germany), Piketty (2020: 498) comments that without allocations of shares to workers or state, the German system is currently loaded against worker representatives, as directors chosen by shareholders control key board appointments and have built-in majority status on the management board that takes operational leadership of co-determination enterprises. This is despite continued efforts by German unions to secure parity on the management board. Nevertheless, Piketty considers that worker directors in the German and Scandinavian systems 'have somewhat shifted the balance of power between shareholders and employees and encouraged more harmonious and ultimately more efficient economic development (at least in comparison with firms in which workers enjoy no board representation)' (*ibid.*: 499). It is hard to dispute that strategic employee and community intervention is vital if companies are to shift towards environmental and employment sustainability, whether to meet national targets for emissions reductions, or potentially, more sustainable organisational development. For this to occur, strategic changes in corporate governance are essential rather than desirable, whether in LMEs or in CMEs: but in the absence of binding legislation, few employers will undertake these changes voluntarily. As indicated in Chapter 11, offering shares to employees is popular with many governments, but problematic as insignificant allocations provided in most government approved schemes offer employee recipients with little or no opportunity to influence enterprise affairs.

So, what can we conclude on the rights and responsibilities of business? First, there is little evidence that employers have voluntarily taken significant sustainable actions. In Chapter 8 we offer evidence that companies are neither doing enough to decarbonise their activities, nor involving employee representatives over decarbonisation plans. The comprehensive study by Vitols *et al.* (2011) pointed to the value of 'well-developed social dialogue structures and good social partnership facilities' to enable organisational sustainable transition, but warned that where these were found, many arrangements were consultative only, tended to focus narrowly on plant level reform of, for example, health and safety or to prioritise efficient work practices in the name of sustainability (see also Renwick *et al.*, 2013). Second, under LME political-economy regimes, companies are pressured by the market to produce short-term returns, potentially mitigating against long-term investment in innovation or transitionary skill development. Though procedures for representative participation are notably more advanced in CMEs and Nordic countries, these too are not immune to financialisaton pressures in a globalised economy. Third, there is little doubt that employees and the wider community tend not to be involved in strategic company decisions. Fourth, the role of trade unions as legitimate social partners has become increasingly marginalised by employers in LMEs. Fifth, the clear lessons are that governments must lead the way in setting terms for corporate sustainability in order to avert an approaching crisis. As business editor Larry Elliott recently observed in *The Guardian*: 'Ministers need to take a tough line. The point about setting a target date is to force companies to innovate – something they are more than capable of doing provided they receive the right signals' (Elliott, 2020). In subsequent chapters, we argue that policies are also needed to wean governments and companies away from prioritising economic growth as a prime measure of performance (Stiglitz *et al.*, 2018; Raworth, 2017: 209). We also maintain that industrial or economic democracy, specifically involving trade unions, must be developed as should alternative organisational forms such as cooperatives. Finally, as Pettifor (2019) has consistently argued, the underlying role of finance must be scrutinised and controlled.

Throughout this book we stress the urgency and profundity of the environmental crisis and its prime source through commercial activity driven by imperatives for continuing economic growth, profit-seeking and by extension, consumerism and cost-minimising employment policies. We have seen that under these conditions, GHGs continue to accumulate and the lived environment to deteriorate, to the extent that even in rich countries, life expectancy, at least for many, has been curtailed through air pollution (Lelieveld *et al.*, 2020). We have seen also that governments, despite the wealth of evidence pointing towards impending calamity, have done too little to influence investment and commercial activity. Under unrestricted conditions, conventional business carries on as usual. Business consultancy Harvey Nash in association with the London Business School conducted a global survey of 640 chairs and non-executive directors, which found that 40 percent of respondents spent zero time reviewing climate-related risk in the boardroom in 2019. Even 35 percent of senior executives of financial services, which are responsible for investment and insurance policies for their commercial clients, reported zero time spent: however, Nordic respondents fared far better than UK ones in time spent discussing environmental impact and in assuming responsibility for their organisations' impact (Alumni/Harvey Nash, 2020).

Bibliography

Alumni/Harvey Nash (2020), 'Predicting the unpredictable', London, Alumni/Harvey Nash Board Report.

Baddon, L., Hunter, L., Hyman, J. *et al.* (1989), *People's Capitalism?*, London, Routledge.

Bakan, J. (2004), *The Corporation: The Pathological Pursuit of Power and Profit*, London, Constable.

Banking on Climate Change (2020), 'Fossil fuel finance report', http://priceofoil.org/con tent/uploads/2020/03/Banking_on_Climate_Change_2020.pdf.

Bessler, W., Drobetz, W. and Holler, J. (2008), 'Capital markets and corporate control: empirical evidence from hedge fund activism in Germany', www.researchgate.net/ publication/228733370_Capital_markets_and_corporate_control_Empirical_evidence_ from_hedge_fund_activism_in_Germany.

Bessler, W., Drobetz, W. and Holler, J. (2013), 'The returns to hedge fund activism in Germany', *European Financial Management*, 21; 106–147.

Bhattacharya, C. and Polman, P. (2017), 'Sustainability lessons from the front line', *MIT-Sloan Management Review*, 58 (2); 71–78.

Birch, K., Peacock, M., Wellen, R. *et al.* (2017), *Business and Society*, London, Zed Books.

Brand, C. and Hunt, A. (2018), 'The health costs of air pollution from cars and vans', www.cleanairday.org.uk.

Busby, M. (2019), 'Capitalism is part of the solution to climate crisis, says Mark Carney', *The Guardian*, 31 July, https://www.theguardian.com/business/2019/jul/31/ capitalism-is-part-of-solution-to-climate-crisis-says-mark-carney

Cai, Y., Jo, H. and Pan, C. (2012), 'Doing well while doing bad? CSR in controversial industry sectors', *Journal of Business Ethics*, 108; 467–480.

Carroll, W. (2010), *Corporate Power in a Globalizing World*, Oxford, Oxford University Press.

Chapman, B. (2017), 'Chemicals giant Ineos "exploiting Brexit to relax climate change laws", documents suggest', *The Independent*, 3 April, www.independent.co.uk/news/business/news/ brexit-ineos-climate-change-laws-lobbying-exploiting-uk-eu-withdrawal-foi-documents- a7665136.html.

Clark, E. and Hermele, K. (2013), 'Financialisation of the environment: a literature review', *Financialization, Economy, Society and Sustainable Development, Working Paper Series*, No. 32.

Dans, E. (2015), 'Volkswagen and the failure of corporate social responsibility', *Forbes Magazine*, www.forbes.com/sites/enriquedans/2015/09/27/volkswagen-and-the-failure- of-corporate-social-responsibility/#39526134405c.

Dauvergne, P. (2018), *Will Big Business Destroy Our Planet?*, Cambridge, Polity Press.

Dauvergne, P. and Lister, J. (2013), *Eco-Business*, Cambridge, Mass., MIT Press.

Du, S. and Vieira, E. (2012), 'Striving for legitimacy through corporate social responsi- bility: insights from oil companies', *Journal of Business Ethics*, 110; 413–427.

EESI (2019), 'Fact sheet: fossil fuel subsidies: a closer look at tax breaks and societal costs', Environmental and Energy Study Institute, Washington DC, July 29.

Elliott, L. (2020), 'Negative UK interest rates were once unthinkable. But tough times lie ahead', *The Guardian*, 6 August, https://www.theguardian.com/business/2020/aug/06/ negative-uk-interest-rates-were-once-unthinkable-but-tough-times-lay-ahead.

Eriksen, M, Lebreton, L, Carson, H. *et al.* (2014), 'Plastic pollution in the world's oceans: more than 5 trillion plastic pieces weighing over 250,000 tons afloat at sea', *PLoS ONE*, 9 (12): e111913. https://doi.org/10.1371/journal.pone.0111913.

European Commission (2019), 'Energy prices and costs in Europe', Brussels, https:// eur-lex.europa.eu/legal-content/EN/TXT/PDF/?uri=COM:2019:1:FIN&from=EN.

Friedman, M. (1970), 'The social responsibility of business is to increase its profits', *The New York Times Magazine*, 13 September.

Frynas, J. (2005), 'The false developmental promise of corporate social responsibility: evidence from multinational oil companies', *International Affairs*, 81 (3); 581–598.

Frynas, J. (2009), 'Corporate social responsibility in the oil and gas sector', *Journal of World Energy Law and Business*, 2 (3); 178–195.

Frynas, J. (2010), 'Corporate social responsibility and societal governance: lessons from transparency in the oil and gas sector', *Journal of Business Ethics*, 93 (2); 163–179.

FT Adviser (2020), 'Nest unveils climate-focused investment strategy', www.ftadviser. com/pensions/2020/07/29/nest-unveils-climate-focused-investment-strategy.

Heemskerk, E., Fennema, M. and Carroll, W. (2016), 'The global corporate elite after the financial crisis: evidence from the transnational network of interlocking directorates', *Global Networks*, 16 (1); 68–88.

Henderson, N. (2019), 'Plastic pollution in the Firth of Forth is much worse than previously thought, new research carried out by Dundee University has revealed', *Fife Today*, 24 May, https://www.fifetoday.co.uk/news/plastic-pollution-forth-worse-expected-new-research-reveals-970589.

Hirschman, A. (1970), *Exit, Voice and Loyalty*, Cambridge MA, Harvard University Press.

Houghton, F., Houghton, S., O'Doherty, D. *et al.* (2019), 'Greenwashing tobacco – attempts to eco-label a killer product', *Journal of Environmental Studies and Sciences*, 9 (1); 82–85.

Hyman, R. (2016), 'The very idea of democracy at work', *Transfer*, 22 (1); 11–24.

InfluenceMap (2019), 'Big oil's real agenda on climate change', London, March.

Jaworska, S. (2018), 'Change but no climate change: discourses of climate change in corporate social responsibility reporting in the oil industry', *International Journal of Business Communication*, 55 (2), 194–219.

Kalleberg, A. (2018), *Precarious Lives: Job Insecurity and Well-Being in Rich Democracies*, Cambridge, Polity.

Klein, N. (2007), *The Shock Doctrine*, London, Penguin.

KPMG (2016), 'Corporate responsibility reporting in the oil and gas sector', June.

Knyght, P. (2010), 'Auditing employee ownership in a neo-liberal world', *Management Decision*, 48 (8), 1304–1323.

Lacombe, C. (2020), 'Sir Jim Ratcliffe, pollueur en chef', *L'Obs*, https://www.nouvelobs. com/economie/20200803.OBS31807/sir-jim-ratcliffe-pollueur-en-chef.html 2909, 30 July.

Lelieveld, J. *et al.* (2020), 'Loss of life expectancy from air pollution compared to other risk factors: a worldwide perspective', *Cardiovascular Research*, https://doi.org/10.1093/cvr/cvaa025, March.

Lyon, P. (2018), 'Nissan admits to falsifying emissions tests', www.forbes.com.

MacDuffie, J. (2019), 'Exhausted by scandal: "Dieselgate" continues to haunt Volkswagen', https://knowledge.wharton.upenn.edu/article/volkswagen-diesel-scandal.

Macfarlane, L. and Mazzucato, M. (2018), 'State investment banks and patient finance: an international comparison', UCL Institute for Innovation and Public Purpose, Working Paper IIPP WP 2018–2011.

Mazzucato, M. and Macfarlane, L. (2017), 'Patient strategic finance: opportunities for state investment banks in the UK', UCL Institute for Innovation and Public Purpose, Working Paper IIPP WP 2017–2005.

Meuer, J., Koelbel, J. and Hoffmann, V. (2020), 'On the nature of corporate sustainability', *Organization and Environment*, 33 (3), 313–341.

Michaels, D. (2020), *The Triumph of Doubt: Dark Money and the Science of Deception*, Oxford, Oxford University Press.

Monbiot, G. (2017), *How Did We Get Into This Mess?*, London, Verso.

Müller, S. and Stegmaier, J. (2020), 'Why is there resistance to works councils in Germany? An economic perspective', *Economic and Industrial Democracy*, 41 (3), 540–561.

The National (2019), 'Government name and shame Scotland's most polluting companies', 2 January, https://www.thenational.scot/news/17337734.government-name-shame-scotlands-polluting-companies/.

Oreskes, N. and Conway, E. (2012), *Merchants of Doubt*, London, Bloomsbury.

Pettifor, A. (2019), *The Case for the Green New Deal*, London, Verso.

Piketty, T. (2020), *Capital and Ideology*, Cambridge, Mass., Harvard University Press.

Pilling, D. (2019), *The Growth Delusion: The Wealth and Well-Being of Nations*, London, Bloomsbury.

Raworth, K. (2017), *Doughnut Economics: Seven Ways to Think Like a 21st-Century Economist*, London, Penguin.

Renwick, D., Redman, T. and Maguire, S. (2013), 'Green human resource management: a review and research agenda', *International Journal of Management Reviews*, 15 (1), 1–14.

Reisz, M. (2019), 'Looking for the smoking gun: the academics taking on Big Tobacco', *Times Higher Education*, 18 September timeshighereducation.com/news/looking-smoking-gun-academics-taking-big-tobacco.

Salazar, A. (2017), 'Corporate governance' in K. Birch *et al.*, *Business and Society*, London, Zed Books, 55–70.

The Scotsman (2018), 'Forth beach is UK hotspot for plastic "nurdle" pollution', 21 May, https://www.scotsman.com/news/environment/forth-beach-uk-hotspot-plastic-nurdle-pollution-289594.

SEC.gov (2019), 'Litigation against VW', www.sec.gov/litigation/complaints/2019/comp24422.pdf.

Stiglitz, J. (2010), 'Government failure vs. market failure: principles of regulation' in E. Balleisen and D. Moss (eds), *Government and Markets: Towards a New Theory of Regulation*, New York, Cambridge University Press, 13–51.

Stiglitz, J., Fitoussi, J-P. and Durand, M. (2018), *Beyond GDP: Measuring what Counts for Economic and Social Performance*, Paris, OECD.

Thatcher, M. and Vlandas, T. (2016), 'Overseas state outsiders as new sources of patient capital: government policies to welcome sovereign wealth investment in France and Germany', *Socio-Economic Review*, 14 (4), 647–668.

Thompson, P. (2013), 'Financialization and the workplace: extending and applying the disconnected capitalism thesis', *Work, Employment & Society*, 27 (3), 472–488.

Tooze, A. (2018), *Crashed: How a Decade of Financial Crises Changed the World*, London, Penguin.

van Lierop, W. (2019), 'Yes, fossil fuel subsidies are real, destructive and protected by lobbying', *Forbes*, 6 December, https://www.forbes.com/sites/walvanlierop/2019/12/06/yes-fossil-fuel-subsidies-are-real-destructive-and-protected-by-lobbying/?sh=4c62d6ec417e.

Vitols, K. *et al.* (2011), 'Industrial relations and sustainability: the role of social partners in the transition to a green economy', Dublin, Eurofound.

Wenar, L. (2017), *Blood Oil*, Oxford, Oxford University Press.

Weyzig, F. (2009), 'Political and economic arguments for corporate social responsibility and a proposition regarding the CSR agenda', *Journal of Business Ethics*, 86 (4), 417–428.

WHO (2004), 'Tobacco industry and corporate social responsibility ... an inherent contradiction', *Tobacco Control Papers*, www.who.int/tobacco.

Wienberg, C. (2019), 'Nordic pension funds invest $700 million in renewable energy', *Bloomberg*, 15 May, www.bloomberg.com/news/articles/2019-05-15/nordic-pension-funds-invest-700-million-in-renewable-energy.

6　The role of government

Introduction

Though the scale and immediacy of the climate crisis has induced increasing numbers of citizens to recycle their waste, change their light bulbs and reconsider their personal means of transport, those citizens would nevertheless normally look to their government as having the only agency capable of taking action on a scale appropriate to the crisis; most would also probably hope for their government to be cooperating with those of other nations.

Several fundamental issues face governments in confronting the climate emergency, and while the extent and intensity of these issues may vary across countries, they all share common features. First, there is dealing with the climate emergency itself. So far, few governments have established consistent procedures to meet Paris accord targets and some, including the worst polluters, are actively campaigning against them. Second, there is the practical problem of lessening dependence on fossil fuels and promoting alternatives. Third, there is the underlying question of inequality and climate justice. Social demands are a growing and hitherto neglected aspect of the environmental crisis, as demonstrated in 2019, when the UN Climate Change Conference (COP25), scheduled to be held in Chile, was moved at short notice to Madrid to avoid escalating street disorder, in part fuelled by the effects of drought and deprivation attributable to years of neo-liberal economic policies.

Governments wishing to take action against each and all of these crises face the power of interlocking vested commercial and financial interests, predicated on pursuit of profit and minimising internal cost, albeit without due regard to external social costs. Governments also need to finance effective transformational programmes, whilst facing competing expenditure demands, such as economic support during the Covid-19 crisis (Naidoo and Fisher, 2020). Further, in an era dominated by globalisation, one governmental policy option is to export polluting processes to less regulated and economically developing countries. These issues prompt the questions: 'who will, and/or *can*, drive global system change?' (Pettifor, 2019: 64).

Meeting emissions targets

Despite decades of increasing emissions and scientifically informed warnings of impending disaster, Naomi Klein points out that between 1990, when serious global deliberations over a climate treaty began, and 2013, CO_2 emissions had risen by 61 percent, prompting the comment that 'the only thing rising faster than our emissions is the output of words pledging to lower them' (Klein, 2014: 11). Neither does the UN's Special Rapporteur on Poverty mince his words: 'more damage has been done in three decades since the United Nations established the Intergovernmental Panel on Climate Change (IPCC) in 1988 than in the whole of human history up until that time' (Alston, 2019: 3). Underlining this book's theme of the close link between the state of employment and the state of the environment, Alston points out that a full 40 percent of global employment relies on a healthy and sustainable environment (*ibid.*: 12). Even though calls are growing in many countries for a Green New Deal (GND), no elected government has yet agreed to formally endorse one, and whilst the incoming European Commission is committed to sanctioning a GND to reduce emissions, guarantees to protect the environment have not been reflected in recent actions of the European Union, which remains wary of tariff threats from the USA (Azmanova, 2019).

The effects of climate change are already being experienced in terms of human misery: harvests are failing and ocean temperatures rising, both affecting the food chain on which millions depend, leading to starvation and mass migration, which has the potential to increase international tensions and exacerbate the growing humanitarian crisis. Countries like Jordan, in the highly volatile Middle East, are facing water shortages through lack of rain and reduced flow from the region's prime water source, the Jordan River. Yet individual countries continue to follow their own economic action agendas, however dysfunctional these may be for the wider environment. This approach was clearly demonstrated by the relative failure of COP25 in Madrid in postponing a treaty on international carbon trading. This setback was in addition to blocking actions by major polluting countries including the USA, Brazil, Saudi Arabia and Australia (which was experiencing its worst-ever bush fires and record high temperatures[1]) and tentative commitment to binding agreements from usually supportive countries like France. Responding to French President Macron's attempt to intervene in the international interest over wildfires ravaging Brazil's Amazon rainforest, a congressman supporter of President Bolsonaro insisted that: 'He [Macron] is not Brazil's president … This forest isn't shared, as he claims. It belongs to a nation which enjoys complete autonomy and authority to decide what happens to the forest and takes every possible care to preserve it' (Phillips, 2019).

Many governments have declared themselves, in varying degrees of commitment, to be phasing out oil-powered automobiles. In the UK, where poor air quality is linked to 40,000 premature deaths annually, debate has centred on the viability of government plans for 'effectively zero emissions' through banning sales on new petrol and diesel cars by 2030. Nevertheless, growing numbers of large cities are taking matters into their own hands by restricting use of cars or applying tariffs for entering city boundaries. Fifteen cities, mainly in Europe, are starting to curb car use in order to reduce emissions and pollution. Ninety major cities, representing a population of 650 million and a quarter of the global economy, are formally committed to making their cities emissions neutral by 2050, in line with the Paris Agreement. The programme involves commitment to make buildings net zero carbon, to reduce waste generation and move to zero-emission transport (C40 Cities, 2018). However, there can be cleavages in policy between central and regional government. In the USA, when confronted with Trump's withdrawal from the Paris Agreement, a number of individual states committed themselves to meeting its demands.

Moving off fossil fuels

Even more worrying is the extent of state support for the same industries that have contributed massively to the environmental emergency. The fossil fuel sector is estimated to be subsidised by \$5.2tn annually by state governments (IMF, 2019). An investigation conducted jointly by Greenpeace and the BBC found that one UK government agency, UK Export Finance (UKEF) spends billions of pounds in support of overseas fossil fuel projects. It is estimated that these projects will eventually emit 69 million tonnes of carbon annually (Clayton, 2020).

The biggest impediment to active regulatory state intervention in economic affairs is the potential impact on the business bottom line. Business often actively lobbies for privatisation of state assets and against legislation which may limit company control or profits. A classic example is provided by the fossil fuel sector, by far the world's biggest contributor to CO_2 emissions. As Alston (2019: 11) points out, in America the industry has been fully aware of the emissions and their impact on the climate for many years, but 'took no action to change its business model' and from 1988 to 2015 produced an equivalent amount of emissions as in the previous 237 years. The American Petroleum Institute, the coordinating body for the sector, actively lobbied to prevent ratification of the Kyoto protocol and spends huge sums fighting against emissions control, obstructing legislation and creating doubts in the public mind over the scientific evidence. If it were not for the influence of the fossil fuel lobby, this emphasis on supporting fossil fuels in America might appear surprising: a recent study by Georgeson and Maslin (2019) estimates that the green economy in America is worth some \$1.3tn and employs more than four percent of the working age population, ten times more than those working in fossil fuels. Moreover, the authors contend that the green economy was a significant contributor to the US recovery after the 2007 financial crisis, and that sustainable economic activity and environmental regulations have the potential to further enhance innovation and economic development.

An additional potential constraint on governments wishing to undertake greener structural change is provided by the question of trust in political processes, acknowledged in many countries to be at a low and declining level. Based on a study of 23 European countries, Fairbrother and his colleagues (2019) argue that trust 'conditions how Europeans' beliefs about climate change translate into their support for [carbon] taxes'. Citizens of Sweden and other Nordic countries express high levels of political trust whilst their support for carbon taxation is high. In the UK, medium political trust was matched by medium tax support, whilst in Poland, both trust and tax support were designated as low. Further, the Nordic model of social partnership helps to underpin governmental efforts to support environmental sustainability, a finding that is also borne out by the high positions occupied by these same countries in the 2020 CCPI climate performance index (Burck *et al.*, 2020).

The breadth and depth of the crisis is reflected in the range of proposed (if indeterminate) actions to combat it, primarily directed at consumer and community behaviour. One effect is undoubtedly to make people more mindful of the scale of the crisis, helped of course by the increasingly visible activities of young people in raising awareness of it. Polls conducted by Ipsos MORI (2019) show that half of respondents are now 'very concerned' about climate change, with three-quarters thinking the UK is already feeling the impact and a majority considering that targets for net zero emissions should be brought forward from 2050.

Many European countries, supported by the EU, are making efforts to transition from high emissions coal-based energy to more sustainable forms of energy. Though progress toward reducing reliance on fossil fuels has been uneven, a number of countries have announced formal timelines for phasing out coal generation, with the UK being the first to do so in 2015. Finland has passed legislation banning coal from 2029. A report from advocacy group Europe Beyond Coal (2019) examines steps taken by European countries in switching to cleaner energy production, arguing that to keep temperature rises within the Paris limits, all European countries need to be coal-free within ten years. Speed of action therefore is crucial, and the report argues that the highest polluting facilities need to be scrapped first. The EU now operates formal and regularly updated pollution standards requiring the dirtiest coal plants to either invest (often substantially) or to close, which many are choosing to do or have already done so. The UK has announced coal phase out by 2025, which coupled with a high carbon price has led to closure of existing dirty plants. The report advocates the full participation of trade unions and community groups to ensure support for those affected by transitional changes, citing the €250 million deal agreed between the Spanish government and coalmining unions as an example of effective good practice. There needs also to be concomitant investment in renewables, the report citing the Netherlands climate agreement for a renewable share of 70 percent in electricity generation by 2030. The European carbon price of around €25 has also helped to make coalmining increasingly uneconomic and thereby allowed for renewables, whose subsidy support has declined in a number of European countries, to compete effectively. These and other complementary positive national developments, alongside the priority offered to a comprehensive European GND by the President of the European Commission, indicates that national and international collaboration over meeting emissions reductions is both feasible and effective.

The unequal burden of the climate crisis

There is growing concern for lack of international action on the third aspect of the environmental crisis: the increase in global social and economic inequality and the erosion of human rights. Raworth points out that: 'higher levels of national inequality ... also tend to go hand in hand with increased ecological degradation' (2017: 172). She identifies two main factors: status competition among consumers, associated with economic policies stressing continuing GDP growth and links between inequality and erosion of social capital. Poorer people are provided with fewer defences against environmental disasters than the better off, who live on higher ground in flood-prone regions or have easier means of escape to, temporarily at least, safer havens. The less equal the community, the greater the degradation and threats to human welfare as well as biodiversity. Inequalities on the basis of gender, race and ethnicity are also associated with vulnerability to hostile climate events (Leichenko and O'Brien, 2019: 144).

It is not generally recognised that both a decent job and a healthy environment come under the rubric of fundamental human rights. Human rights are defined by the UN as a 'globally agreed and universally applicable legal and ethical framework protecting essential freedoms and the minimum requirements of a dignified life' (OHCR-hbs, 2018: 11) and include rights to life, health, education, shelter, water, sanitation and livelihoods. Under international law, states are assumed to be the primary duty-bearers in ensuring human rights but, although climate change has been on the human rights agenda for well over a decade, the UN's Philip Alston (2019: 1) argues that it remains a marginal concern for most actors.

Alston's condemnatory charge sheet, founded on assessment of a wide range of official reports and documents, argues that climate change will exacerbate poverty and inequality with the most impoverished countries bearing 'an estimated 75–80 percent of the costs of climate change', for which they are largely not responsible. Alston estimates that the 'poorest half of the world's population – 3.5 billion people – is responsible for just 10 percent of carbon emissions ... A person in the wealthiest 1 percent uses 175 times more carbon than one in the bottom 10 percent' (*ibid.*: 6). Despite the proliferation of reports and numbers of expert meetings, bodies like the UN and European Court of Human Rights are charged with paying insufficient attention to address environmental human rights and, in turn, state governments are accused of not taking necessary action on climate change and poverty reduction.

The consequences are catastrophic: indigenous peoples are at risk of losing their communities, livelihoods and homes as commercialised deforestation continues virtually unimpeded. Many agricultural pesticides and fertilisers threaten to poison rivers and lakes and are 'known for their ability to cause negative health effects in humans and wildlife and to degrade the natural environment' (Nicolopoulou-Stamati *et al.*, 2018: 8; Mesnage and Séralini, 2018). Human rights are also affected by the competitive race-to-the-bottom nature of globalised employment, from denial of union membership, outsourcing deskilled operations and dehumanised and insecure working regimes in industrialising countries to distribution 'fulfilment' centres across the globe, where human activity is measured, controlled and disciplined through software technology, which was supposed to liberate rather than capture and distort human creativity (Briken and Taylor, 2018). Rights are also being jeopardised in those communities still dependent upon fossil fuel extraction and processing: some countries are affecting efficient transitions to comparable jobs that are secure, decent and environmentally sustainable, while others are left to languish as their once-proud communities decay and political orientations turn to calls of nationalism, isolationism and racism.

The UN's 17 Sustainable Development Goals for 2030 confirm the central role of national governments and their Heads of State in aiming to both protect the environment and advance human rights across the world. Indeed, these two ambitions cannot and should not be separated, a point emphasised in Stroud *et al.*'s (2018: 88) elaboration of the United Nations Environment Programme (UNEP, 2011) definition of a green economy as 'an economic development model "that results in improved human well-being and social equity, whilst significantly reducing environmental risks and ecological scarcities"'. Though the 2030 agenda has been signed by 193 Heads of State or their ministers, each country has the freedom to determine how it aims to meet the objectives and many have now taken some action to address at least some of the 17 goals laid down in the UN programme (Naidoo and Fisher, 2020). This freedom can be of great symbolic as well as practical importance, as most political parties are still locked into supporting environmentally compromising economic growth policies, leading Alston to comment that although the 'Paris Agreement represents the most promising step in addressing climate change … the commitments States have adopted in pursuit of the Agreement are woefully insufficient … and States are failing to meet even their current inadequate commitments' (2019: 10). Hence, downgrading and blocking of environmental protection agencies along with pronouncements from prominent politicians like Trump, Bolsonaro and Bolivia's Evo Morales denying accumulated scientific evidence of environmental despoliation in support of their growth ambitions are likely to influence the determination of emerging and dependent economies to meet agreed targets. Though China has been reducing its internal dependence on coal, there are signs of a reversal in policy and, according to Tooze (2020), China is now opening more coalmines than the rest of the world is closing, worried that worsening economic relations with America requires continuing reliance on internal energy sources. Moreover, China is happy to export coal-fired power plants to other countries and is failing to regulate adequately for domestic methane emissions (Miller *et al.*, 2019).

A fundamental barrier to action is that the people most affected by climate change have little power, experiencing a form of 'climate apartheid', in the words of environmental journalist, Fiona Harvey. A major international report from the *Global Commission on Adaptation* helps to explain why even rich countries are still lagging in efforts to combat the climate emergency:

> Power typically rests with those least affected, most insured, and most able to protect themselves from the impacts of climate change. Major disparities in power also exist between countries. For the rich, climate change can seem like a problem easily overcome, not lifelong sentence to poverty and suffering as it may mean for vulnerable and marginalized people. Those most at risk often have limited ability to shape key decisions that affect them. Without their voice, the urgency of adaptation is muted.
>
> (GCA, 2019: 15)

Further, in their ground-breaking study of the corrosive effects of societal inequality, Wilkinson and Pickett (2010: 233) demonstrate that business leaders in more unequal countries are less likely to support domestic efforts to comply with international environmental agreements.

Why don't governments act?

The question is, therefore, why are states failing so dismally to meet emissions reductions, to protect their citizens and to empower labour to participate in the transitionary process? Part of the answer is a reluctance to regulate the economy: indeed, as we note in Chapter 4, many governments in pursuit of growth ambitions have adopted deregulatory economic policies, which invariably impact upon both environment and most vulnerable workers by seeking to lower production costs and remain economically competitive. Government economic policies basically divide along interventionist and non-interventionist lines. Interventionist policy is closely associated with government support for investment in public ownership and stewardship of productive means, backed by policies such as social housing, minimum wage and active labour market intervention in training and education. Unions are often provided with an established partnership role in industrial policy.

More conservative policies incline toward market remedies, such as carbon offsets and trading, which may have some impact on emissions, though do little to address fundamental issues such as fossil fuel dependence, just transition or climate justice. We have also noted the continued emphasis on maintaining (unsustainable) economic growth by governments, an emphasis that could serve to neutralise or indeed jeopardise market-based efforts to reduce carbon dependence.

While all governments intervene in some aspects of economic policy, extreme non-intervention follows the Friedmanite doctrine of restricting government involvement in economic activity to military and public protection, supporting the judicial system and providing basic services like urgent health care to the most socially deprived. Other free-marketeers go further. Bakan (2004: 113) cites the then Chair of the powerful lobbying group Cato Institute, William Niskanen's insistence that only the military should be under public control. Basically, the neo-liberal ethos is that 'the state has no business in any part of [people's] lives in which the market wants free rein' (O'Toole, 2018: 201). A parallel policy is that, providing business acts lawfully, it should be free from interference in, or regulatory restrictions on, its activities. In their deeply researched history of opposition to environmental regulation, Oreskes and Conway (2012) demonstrate that even prior to the current ecological emergency, Republican policy in the USA was firmly entrenched against regulation in commercial affairs, fearing this as a first step on the road to socialist perdition. Influenced by extreme Friedmanite views, supported by sympathetic and influential cold-war scientists, and confronted by well-funded libertarian think-tanks, backed by a complicit media, it required many years of dedicated independent research for federal action to be taken against tobacco, passive smoking, acid rain, chlorofluorocarbons and global heating. This contest continues unabated, evidenced by recent accounts that technology giants like Google have made substantial contributions to conservative policy groups, many of which are active climate change deniers like the Competitive Enterprise Institute, whose claim to fame includes persuading Trump to withdraw from the Paris accords and continues to press the White House to discard remaining environmental regulations (Bulletin of the Atomic Scientists, 2019). Oreskes and Conway add that 'divergence between the state of science and how it was presented in the media helped to make it easy for our governments to do nothing about global warming' (2012: 215).

The power and resources of big business to manipulate government policy through promotion of the free market are prodigious. Anything, including scientific rationality, that stood in the way of this ideology, from Rachel Carson in the 1960s (and still vilified today) to present-day environmentalists, is seen as a threat to the market system, and thereby personal freedoms, and must therefore be implacably opposed. In his study of the contemporary American economy, Adler (2019) identifies a number of crises attributable to the power of big business and subservience of government to these interests. He points out that the

crisis is due not only to the lack of environmental awareness among corporate leaders, government officials, or the public at large. It is due not only to the campaign of disinformation funded by the fossil fuel industry. It is also a reflection of the basic features of capitalism ... in particular the growth imperative facing individual firms and the subservience of government to that imperative.

(2019: 37–38)

Responding indirectly to Oreskes and Conway's (2012: 255) conclusion that 'environmental deterioration is driven by economic activity so we must consider if there is a fundamental flaw in our economic system', Adler argues that radical transformation of the economy in support of the environment is the *only* remedy, comprising a transformation to democratic socialism, in which exercising democratic control over both enterprises and the national economy is paramount, a significant point to which we shall return in later chapters.

In the UK, deregulation, heralded as the means to promote possessive individualism and enterprise freedom, has been a constant theme of Conservative Party rhetoric and action, as emphasised by previous Prime Minister David Cameron in a letter to government ministers on 7 April 2011 on the launch of the so-called Red Tape Challenge:

> Our starting point is that a regulation should go or its aim achieved in a different, non-government way, unless there is a clear and good justification for government being involved. And even where there is a good case for this, we must sweep away unnecessary bureaucracy and complexity, end gold-plating of EU directives, and challenge overzealous administration and enforcement.
>
> (Prime Minister's Office, 2011)

Unfortunately, deregulating the economy has already had tragic environmental consequences, as Fintan O'Toole perceptively notes with regard to the 1990s BSE disaster, which followed abandonment of food safety regulations designed to protect public health, to allow toxic beef to enter into the human food chain (2018: 48). Excessive deregulation has been cited as a factor in the tragic Grenfell Tower fire in London in 2017, which killed 72 people (FBU, 2019). Unarguably, the spiritual thrust of Brexit has been underlaid by desire to escape from the bureaucracy and alleged stranglehold of EU regulations, early examples being Britain's opposition to European Works Councils, which were considered to be inimical to commercial competitiveness, and opting out from the Working Time Directive, which limits the number of hours that may be worked over a four-week period. Conservatives, pushed by employers and aided by a highly supportive and vocal media, claimed that the Directive ran counter to UK worker preference for the freedom to enjoy flexible work, which in practice operates as freedom to enjoy a regime of low pay buttressed by long working hours.

The use of 'individual freedom' as the ideological justification for non-regulation often has a (hidden) social cost. The costs of the 'freedom to smoke' are borne by socialised health care. (Oreskes and Conway, 2012: 162–164). The 'freedom to drive' one of the 200 million emissions-rich Sports Utility Vehicles (SUVs) in the world (up from 35 million in 2010) has resulted in the second largest contribution to the rise in global CO_2 since 2010 (IEA, 2019). Individual freedom would, of necessity, be restricted through tighter regulation, for example, by imposing severe tariffs on entry to cities by private vehicular traffic and by higher priced or even limitations on air travel, as argued by environmental campaigner George Monbiot (2006) but contemporary political ideology and economic determinism fiercely deny and contest the need for these constraints even as the planet enters its terminal phase. Ironically the current global viral pandemic has very nearly achieved Monbiot's target virtually overnight, although with significant social costs in terms of the thousands of lost jobs, underlining the argument that structural transitions towards a greener economy have to be thought through and strategies put in place to prevent such social damage.

We saw in Chapter 4 that opponents of climate policies still cling to the discredited theory of the Environmental Kuznets Curve, which suggests that preliminary phases of economic growth are associated with initial environmental degradation followed by enhancement as society becomes better off (Raworth, 2017: 209). Not surprisingly, this growth-confirmatory message has been widely broadcast and gleefully accepted among political elites and the business community, despite its theoretical and empirical shortcomings. Raworth notes, however, that environmental quality is directly related to more equal income distribution, high literacy levels and shared civil and political rights: 'it's people power, not economic growth *per se*, that protects local air and water quality' *(ibid.*: 209), arguing that switching to cleaner technologies is influenced by citizen demands and is not simply the effect of higher incomes. Finally, Raworth reminds us that, while mature economies can and do lower their carbon emissions, this is frequently accomplished through outsourcing polluting activities to industrialising countries.

Informed commentators agree that governments must adopt more stringent policies to confront the emergency. Among the many counter-proposals to reliance on market forces we can include the growing interest in potential governmental mitigating policies, including the rise in interest in Green New Deals, founded (at least in more radical interpretations) on just transitions and on climate justice; potential role of policies that subordinate profit to the interests of workers and society generally; raising the profile and political influence of trade unions, which operate on behalf of members and the community at large; enhancing the role of industrial and economic democracy so that workers have a more formal role in decisions that affect them; introducing a universal minimum wage; and shortening the working week. We examine these ideas in later chapters but, of course, in order to enact any or all of these policies in democratic countries requires the consent of the population and radical collective policies can be successfully challenged by a free-market individualistic agenda and undermined by powerful coalitions when their short-term commercial interests are under threat.

Impact of globalisation

Threats to the earth's climate stability have undoubtedly gathered momentum and become more acute during an era of aggressive globalisation. Under globalisation, the world operates as a single interdependent market for exchanging goods, services and labour and, clearly, because of its transnational nature individual territorial action to reduce emissions linked to global trading has proven difficult to accomplish.

Whilst comparative advantage, in which countries trade to mutual advantage on the basis of relative opportunity costs of production (e.g. Britain historically exported textiles and imported wine), has long provided the basis for international trade, globalisation has more zero-sum properties and has developed rapidly under the impact of diverse favourable influences. Prominent among these are neo-liberal economic policies advanced and supported by major economies; the domination of world trade by transnational enterprises; deregulation of finance, product and labour markets accompanied by easy access to transfer costing and tax avoidance by transnational corporations, who are accused of annually shifting some $660 billions of their profits to ultra-low tax jurisdictions (Raworth, 2017: 277); the impact of free trade agreements such as the North American Free Trade Association; readily transferable production technologies, which take advantage of cheap, relatively unskilled and disposable labour; off-shoring of services aided by development of information technology; and foreign direct investment in the form of joint ventures and outsourcing. Globalised markets are typified by continuous uncertainty: if production in one country becomes more expensive through raised resource costs or regulatory barriers, transfer to less regulated countries eager to support inward investment is easily achieved.

Impacts of globalisation on the environment are immediately apparent. First, globalisation is associated with spatial displacement of emissions. This raises significant issues as mature economies have relocated much high-emission manufacturing to emergent economies such as China and India. As national targets for carbon reduction become increasingly a matter for public policy, countries are exporting emissions to help meet these targets, though global emissions could well rise as developing countries may not exercise rigorous emissions controls. Despite renewable sources of energy overtaking fossil fuels for electricity provision in the UK, the Office for National Statistics warns that Britain is now the biggest net importer of carbon emissions among the world's leading economies, with China, the EU and the USA being the biggest contributors to imported emissions to the UK (ONS, 2019). The World Resources Institute estimates that six out of ten top GHG emitters are now developing countries, which as a bloc contribute 60 percent of global annual CO_2 emissions (WRI, 2017). Though products are consumed in mature economies, the carbon emitted in production is not credited to the recipient countries. Through this process, Peters *et al.* (2011) estimate that notional emissions in the EU fell by 6 percent between 1990 and 2008. Leichenko and O'Brien estimate that over a quarter of emissions linked to international trade can be linked to export-linked production (2019: 87).

When we examined our two examples of the purchase of electronics and clothing it was evident that globalisation by its very nature is linked directly to intensified transport of output, whether these are primary products such as coal and oil or manufactured goods. About 90 percent of world trade is carried by sea. Maritime traffic has also increased recently as developing countries like Malaysia and the Philippines demand that wealthy countries take back waste shipped to them for disposal, on the basis that it is often contaminated and cannot be recycled. Other waste, such as used tyres, as described in Chapter 4, is sent to newly industrialising countries like India for reprocessing under dangerous and toxic working conditions, from which fumes emerge to contaminate whole communities. Shipping fuel contains numerous pollutants that would be banned or heavily regulated if deployed for onshore purposes. Studies show that 400,000 premature deaths and 14 million childhood asthma cases annually are associated with emissions from dirty shipping fuel (Sofiev *et al.*, 2018). Maritime transport emits nearly a billion tonnes of GHGs annually, representing about 2.5 percent of total GHG emissions, according to the International Maritime Organization (IMO, 2014). Estimates vary on the carbon emissions of trade attributable to shipping, though some accounts indicate that shipping and air transport together account for six percent of all carbon emissions (Leichenko and O'Brien, 2019: 88). Other reports demonstrate that transportation (including motor vehicles) is responsible for a 'quarter of energy-related CO_2 emissions and half of all oil consumed' (Romm, 2018: 231). Shipping emissions represent about 13 percent of overall EU GHG emissions from the transport sector. Though maritime emissions are forecast to grow significantly, the aim of the IMO is to halve GHGs by 2050. Nevertheless, progress has been slow as shipping and aviation have been excluded from international talks on combatting climate change, leaving emissions reduction to be largely left to self-regulation by individual countries. Tankers have been accused of breaching regulations by discarding fuel at sea and instances of dumping toxic and even radioactive waste materials by ships have been identified. And, as shown in Chapter 3, oil spills are an ever-present danger that have devastating long-lasting effects on the environment and biodiversity. According to Seas at Risk, an umbrella body of European environmental protection NGOs, 20,000 tonnes of waste are dumped each year in the North Sea alone. Ninety percent of plastic found on Netherlands beaches originates from shipping and fisheries. Worldwide, the UN estimates that eight million tonnes of plastic waste are dumped each year in the seas.

A further consequence of globalisation has been significantly diverse outcomes for different countries, with once-prosperous industrialised regions of mature economies facing economic and social decline and neglect in their extractive, steel-making and manufacturing sectors, whilst newly developing countries contribute increasingly to industrial emissions and pollution through their high growth, low regulatory economic policies, relying on low-paid insecure and often exploited workforces. De-industrialisation has led to calls for regeneration of affected areas and governments have adopted different policies for attracting domestic and overseas investment to depleted areas. In addition to the challenge of meeting broad environmental goals through transitional policies, all countries are faced with regionally specific, but differing environmental/human rights challenges. At present, government policies necessary to accomplish any kind of transition to a 'greener' economic structure without incurring social damage to existing jobs and communities are mainly noticeable by their absence.

Government policies for labour market intervention

Differences between coordinated and liberal labour market economies (CMEs and LMEs) over intervention in the labour market provide an indicator for the potential for transitional policy intentions. Active labour market policies would include a range of support provisions with opportunities for retraining, provision of job counselling, support in job searches and so forth (Kalleberg, 2018: 64). All would be relevant contributors to the pursuit of just transition policies. Kalleberg reviewed percentages of GDP expenditure on active labour market intervention in six 'rich democracies', with high proportions allocated in Denmark and Germany and considerably less in the UK and USA. Further, OECD figures show employment protection legislation was highest in Germany, moderate in Denmark and lowest in the UK and USA, where 'most employees ... can be discharged for any (non-discriminatory) reason and without warning' (*ibid.*: 68). Basically, Nordic, CME and LME models provide differential levels of risk, exposure and support for workers faced with labour market fluctuations. Experience from the enforced decline of extractive and manufacturing industry in the UK, in which affected communities were largely abandoned to the operation of market forces as the unsuccessful means to secure revitalisation, indicates the real constraints that transition faces in those advanced liberal market economies wedded to 'hidden hand' market resolutions.

Union ambitions to 'have a seat where key decisions are taken' (TUC, 2019) over environmental transition policies raise a myriad of questions (nature of key decisions; at what level, etc.), and the complex question of *how* needs to be reviewed. Although the crisis is an international one, individual countries are reacting to and contributing to it in different ways owing to different underlying ideologies, models, conventions and traditions of intervention in employment relations that would serve against attempting to introduce, let alone impose, a universal model for reform. As we explore in Chapter 8, one thing is clear – without formal recognition by state or employers, unions will not be able to mobilise 'worker power' (Kalleberg, 2018: 188) to participate in the transformations required to mitigate against carbon emissions or influence employers to transition to products, production and transport policies which are less harmful to the environment.

That government policies can make a difference at central, regional, civic and local levels is demonstrated by research conducted by Stroud and his colleagues (2018: 88) who examine the role of government and institutional frameworks in cultivating 'decent' jobs in transitioning from coal dependency to low carbon production. Their research provides a cross-national comparison of regeneration policies in declining coal regions in Germany and South Wales, supported by an examination of the role of regional development agencies, with the Ruhr region representing the German CME model and the Welsh example the free-market approach associated with the UK. The authors conclude that the Ruhr area is successfully building the foundations for transition 'based on a knowledge-intensive, high skill trajectory, encouraged by effective and robust VET [vocational, education and training] systems, environmental regulation and wider government policy at national and regional levels' (*ibid*.: 93), in which trade unions and other bodies are fully engaged (*ibid*.: 96). The Welsh experience has been less successful, notwithstanding the efforts of local agencies and employers to develop a robust green agenda. The authors conclude that the market-dominated culture is inimical to long-term regeneration with 'a lack of parity for labour as a stakeholder, both reflected in and perpetuated by low levels of employment protection, weak forms of institutional "voice" and "voluntarist" market-based training systems' (*ibid*.: 96). Moreover, there is evidence that investment in technology and skills for transition is insufficient, leading to a workforce unprepared for the demands of a green economy.

A role for public ownership?

In his classic text, Katzarov (1964: 1) defines nationalisation as 'transfer to the community of property and activities … the means of production and exchange – and their utilisation in the interests of the community and no longer in the interests of individuals'. He continues that 'the motivating idea behind nationalisation is the socialist principle of establishing equality between men [*sic*] by enabling goods to be shared out by the community in a socially equitable manner' (*ibid*.: 14). This aim acted as the guiding principle behind the massive programme of nationalisation established by Attlee's 1945 Labour government, which covered electricity, gas, civil aviation, cable and radio communication between 1945 and 1946, followed by transport and coal mines in 1947.

Prior to the accession in 1979 of Margaret Thatcher's government, committed to enterprise and competition, state-owned enterprises contributed more than ten percent of UK GDP and eight percent of overall employment (Vickers and Yarrow, 1988). As Chang (2014) points out, however, Britain was not alone in creating state-owned enterprises in the immediate post-World War II years. Many European countries turned to socialist or social democratic parties following the end of wartime hostilities, many of which followed the aim of establishing public control over key sectors. In recent years, through the neoliberal project to restrict the state's role and inject market discipline into public affairs, privatisation of nationally held assets has become *de rigueur* policy, including in the previously Eastern European Soviet bloc and in still-communist China (Clifton *et al.*, 2006; Xie *et al.*, 2016; Monday, 2017). So, it is important to ask how successful nationalisation has been in accomplishing popular control and whether lessons can be learned for dealing with the environment emergency in which many high-emitting essential sectors are implicated.

Nationalisation has the principal aims of removing private profit from the provision of public utilities and key service sectors of the economy like healthcare and education, where profit-seeking can severely distort social priorities. Often these sectors are natural monopolies, which when nationalised can realise effective economies of scale, arguably making them more efficient and allowing for surpluses to be deployed for investment rather than for allocation to shareholders. There may also be social benefits to nationalisation, including broader public access to essential services (Bayliss, 2002). Opponents to nationalisation take a different perspective: that it is not as efficient as market-driven approaches as, under public ownership, there is little incentive for managers to perform beyond satisficing (i.e. satisfy minimum acceptable requirements to achieve objectives). Second, in the competitive private sector, executives know that poor corporate performance will be punished through lower share price and dividends to shareholders and heightened exposure to takeover: hence, the incentives for profit-seeking outweigh those of meeting political objectives (Letza *et al.* 2004: 163). Further, commercial and competitive demands ensure effective monitoring from direct-interest shareholders (rather than bureaucrats) over senior managers. Third, as these authors point out, the private sector enjoys a straightforward binary principal-agent relationship, unlike the multi-layered distancing and potential differences in objectives between public, officials and senior executives of state-owned enterprises (SOEs). Fourth, as Streeck (2016: 103) emphasises, rising expectations of citizens in their status as customers cannot be satisfied by public sector uniformity of provision. Increasingly sophisticated consumer needs can only be satisfied by the services offered through competitive private markets.

However, Letza and his colleagues review a series of studies that point out that this anticipated private sector superiority is not always, or even often, realised in practice and that the alleged better performance of privatised companies may not be a function of ownership, but structural features such as the overall state of the economy. Moreover, the state, rather than operating as a drag on private enterprise may well be the prime stimulus and risk-taker in technological innovation, from which the private sector draws risk-free benefit (Mazzucato, 2013). A longitudinal study by Martin and Parker (1995: 235) of 11 UK privatised enterprises concluded that 'it is difficult to sustain unequivocally the hypothesis that private ownership is preferable to nationalization on efficiency grounds', a conclusion subsequently reinforced by an extended study of the same 11 firms (Martin and Parker, 1997). The same authors cite several studies that question the positive impact of privatisation on performance (Bishop and Kay, 1988; Hartley *et al.*, 1991); Haskel and Szymanski (1993), commenting that while performance may improve, this can be attributable to factors independent of the privatisation effect. A major review covering different sectors across a range of economically advanced and developing countries concluded unequivocally 'that the evidence shows no significant difference in efficiency between public and privately owned companies in public services', whether through privatisations by sale, outsourcing or through private-public partnerships (PSIRU, 2014: 7).

In their examination of water services privatisation in England, Bayliss and Hall (2017) note that the cost to the private sector was negligible at vestment in 1989 as the government assumed all debts for the sector (valued at £4.9bn) and presented the company with a 'green dowry' of £1.5bn. Subsequently, investors have been the prime beneficiaries. Even when not privatised, internal markets for public sector services such as higher education and healthcare can be established through a process of 'corporatisation' (Zaifer, 2019: 500), in which the state owns and operates services, but these are managed independently. However, studies point to the negative impacts of such policies in terms of managerially imposed cost-cutting, work intensification and stress and consequently deteriorating performance (Carter et al., 2013; Carter and Stevenson, 2012; Hart and Warren, 2015). Moreover, as Zaifer (2019: 496; see also Fleming, 2017: 64; FitzRoy and Papyrakis, 2016: 63) points out, whilst privatisation may primarily benefit stockholders, it can be associated with a range of negative consequences, including rising costs (demonstrated by the costs of competitive healthcare in North America [Garber and Skinner, 2008; PSIRU, 2014: 12–13]), employment loss, 'exclusion from or reduced access to basic services', environmental problems, potentially linked to neglecting externalities and corruption scandals (Hall, 1999). Moreover, privatised concerns often fail to provide efficient or cheaper provision of essential services (Cumbers and Becker, 2018). An investigation by the *Financial Times* (2017) of Thames Water, a monopoly supplier to around one-third of England's water and sewage systems, reveals high dividend payments paid to stockholders, failure to pay corporation tax and the burgeoning pay of senior directors. The environmental consequences of privatisation to local communities have been disastrous, with the company confronted by a massive £20.3m fine for river pollution, in what the judge termed 'borderline deliberate' behaviour. As mentioned in Chapter 2, suicides at France Télécom also demonstrate the human cost of privatisation programmes.

Questions that emerge from assessing the potential of publicly owned enterprises to benefit the environment are: first, that if nationalised and state-owned enterprises are responsible to and financed by the public, is there any evidence that these entities benefit the environment better than conventional profit-driven concerns? Second, if it is acknowledged that employees in conventional private sector organisations have little control or voice over strategic decisions, is the case different in cases of state-owned companies? Empirical studies of environmental impacts are not conclusive. One study of 44 developing countries between 1987 and 1995 indicate that higher degrees of private sector involvement were linked to lower levels of environmental degradation, an effect possibly enhanced under conditions of an effective domestic capital market and when supported through inward investment by developed economies (Talukdar and Meisner, 2001). Other commentators suggest that as state-owned enterprises act, or are expected to act, in the same profit-driven directions as those of the private sector, little difference in appreciation or application of community or environmental priorities may be anticipated.

The second question is, of course, of considerable environmental concern. In private sector companies, the opportunities for employees to challenge organisational directions or management decisions are constrained by combinations of restrictive government policies, organisational culture and reduced collective strength of employees. While union density in the UK public sector has remained relatively stable, government policies for their own employees have also served to weaken collective influence. Employment legislation has been restrictive and privatisation a common element of the underlying neo-liberal agenda. Even where direct privatisation has not been undertaken, governments have encouraged or required competitive tendering and outsourcing to increase cost-efficiency and introduced target-based internal markets for increased competitiveness. Under these conditions, the influence of labour on organisational decision-making, even in the public sector, is heavily circumscribed.

Nationalisation may serve different purposes. For newly emerging economies, supporters can claim that it helps to stimulate economic growth and provide wide community access for essential services like water and energy provision, though other commentators doubt whether these putative benefits are actually realised. Further, centralised management under direct or indirect government control may offer opportunities for corruption and self-serving. For mature economies, rationales for nationalisation vary. The massive post-war programme initiated in the UK was based on several premises. One was of universal provision of essential services like a national health service 'free at the point of use' (Bew, 2016: 403), whereas for coal mines, concern was over dangerous working conditions and with coal as primary energy source for manufacturing industry, the need for efficient production. Another objective was to promote full employment, coupled with greater involvement by labour in the planning and supervision of nationalised concerns (MacInnes, 1987: 18).

Many remain unconvinced that this latter ambition was realised. Middlemas (1979: 412) points out that: 'instead of a home for full, sustained collaboration between unions, management and government, freed by public ownership from impediments of wage and class struggle, the nationalised industries became an element apart, subjected to arbitrary government interference'. From the workers' and union perspective little seemed to change under public ownership and, as the UK nationalisation programme was heavily influenced by wartime 'command and control' practice, its centralised, bureaucratic and top-down organisation formed a central focus of the subsequent free-market critique. Despite the availability under public ownership of extensive consultation machinery, the scope of joint consultation was all too often typified by the 'tea, towels, and toilets syndrome' (Ramsay, 1977: 482). Despite national negotiating procedures with recognised trade unions backed by high levels of union membership, the 'two sides' of industry remained entrenched in their oppositional stances with publicly funded management control periodically confronted by well-organised unions when government policy dictated wage restraint or sectoral restructuring (Coates, 1989).

Richard Hyman notes that while in some European countries 'nationalisation provided a favourable environment for trade union activity, its bureaucratic character did little to enhance democratic control. Most social democrats soon abandoned explicitly or implicitly, the idea of comprehensive public ownership' (2015: 14). In essence, this lack of strategic labour input continues in remaining

SOEs and public services to this day. As Allyson Pollock comments in her pioneering review of the UK's health service shift to privatisation: 'The main qualification for running the NHS now is knowledge of business methods ... Doctors, nurses, public health specialists and other health care professionals do not feature except at the front line' (2004: 1), the front line being a location, of course, where most casualties may be anticipated. In contemporary economically developing countries, similar estrangement from decision-making processes have been noted (Zaifer, 2019: 499). Clearly, if neither nationalised industries nor privately owned strategic industries, both of which in many parts of the world tend to operate in high carbon-emitting extractive and productive sectors, do not offer stringent opportunities for employees or their representatives to influence environmental policies, it would appear that converting private companies to 'traditional' centralised nationalisation is unlikely to allow those most affected, whether through their work or their communities, opportunities to take direct action for alternative policies.

Finally, the case for nationalisation has been undermined by the dominance of globalisation, a process in which states have become 'located in markets, rather than markets in states' (Streeck, 2016: 22), leading Dauvergne (2018: 33) to comment that 'ruling parties in democracies have become indebted to the owners and CEOs of big business', suggesting that national governments not only see benefits in reducing state intervention in domestic economies (Birch *et al.*, 2017: 104) but 'this global dominance of business is supported *and* enforced by national governments, especially those in the Global North' (*ibid.*: 110). This argument is supported by Fleming (2017: 44) in his depiction of 'wreckage economics' as governmental policies partly based on 'capture or enclosure of community-based resources and economic activities', with profits syphoned off to overseas shell companies. Globalisation permits free movements of assets between countries and one option for high-emitting transnational companies faced with nationalisation is to threaten to shift to more accommodating economic regimes, an approach guaranteed to focus the minds of both politicians and affected trade unions, and in the extreme phase of unfettered capitalism, state subservience to or endorsement of the 'market' serves to undermine the democratic process itself. In the words of the former chair of the Federal Reserve, Alan Greenspan, 'thanks to globalisation, policy decisions in the US have been largely replaced by global market forces ... The world is governed by market forces' (cited in Pettifor, 2019: 33).

Faced with these ideological and practical pressures, few political parties in mature economies – or voters – are likely to give much support to centralised nationalisation proposals these days, and when they do, opposition from money markets, free-marketeers and media are likely to condemn such suggestions to oblivion. Nevertheless, as Cumbers (2020) points out, any democratically grounded GND may well require publicly owned and coordinated central energy systems to ensure equitable supply. Also, as we have seen, privately owned companies are failing to take the increasingly urgent action required to curb emissions and other forms of pollution, are likely to oppose or resist regulation and only governments are in a position, albeit compromised through globalisation, to take remedial action. So, as the environmental emergency becomes increasingly urgent, the question is raised, what other alternatives are available for governments to take?

Increased democratic control

The above analysis confirms that in many areas, democratic control over state actions has been seriously diminished with the ascendancy and dominance of globalised, privatised and corporatised public service and commercial institutions. In the private sector, collective labour influence has been progressively weakened, especially as *national* unions face *international* enterprises often controlled far from the sites of physical operation. Even the praised German model of single enterprise co-determination is now being 'forced to accommodate to the externally imposed requirements of intensified global competition' (Hyman, 2015: 17). Under these conditions, calls for closer labour participation with potential for establishing some democratic control over organisational environmental issues are increasingly being expressed. A number of different approaches have been advocated, and in some cases, established. Zaifer (2019: 503) identifies four practical alternatives to privatisation, which can help to make 'public assets more responsive to the needs of the general populace', namely, remunicipalisation; various dimensions of worker control; user cooperatives; and hybrid initiatives.

Remunicipalisation is a process whereby previously privatised essential services are returned to the public domain, usually by city councils and local authorities, often following civic or popular demands or referenda. The growth in such initiatives is demonstrated in the water sector, where there were two reported cases in 2000 and 210 in 2015. By 2017 the Transnational Institute recorded 267 remunicipalisation cases in the water sector, 304 in energy and 835 in total across seven essential services in both mature and developing economies, with 347 of these initiatives in Germany (TNI, 2018). Rationales for adopting remunicipalisation vary and there may be combinations of reasons: growing civic opposition to privatisation can provide the impetus, especially in cities with histories of local public provision, for example for energy services in Berlin and Hamburg (Cumbers and Becker, 2018). Others serve in response to the failure of privatisation to deliver effective or equitable services or conversely, through recognition that while centralised public ownership faces too many objections, locally owned and controlled services can benefit the communities to which they are directed. Others may be influenced directly by environmental concerns or through demands for greater local democratic governance and accountability. In the UK a number of local authorities have established municipal energy companies, although several have found survival difficult in the cut-throat energy market. Though not specifically directed at environmental enhancement, growing civic initiatives are compatible with GND aspirations for national localisation offering practical alternatives to globalisation through combining local interests with self-sufficiency (Hines, 2000; Livesey, 2017).

Diversity of intent means that it is difficult to measure the performance of these new public institutions as they may serve different or multiple objectives, but more public involvement and a concern to serve all members of a community through effective delivery of services and close attention to environmental needs may offer genuine benefit, especially as the environmental and social failures of privatisation and neo-liberalism are exposed to critical scrutiny. Sympathisers do serve warnings, though. Pettifor cautions that to safeguard the eco-system, society needs to shift from fossil fuel dependence and governments need to regain control over key sectors like water and energy but opposition from private finance will be formidable: 'Their aim is to wear down states and privatise *all* investment, especially risk-free investment backed by taxpayers and governments' (2019: 58). Cumbers and Becker warn that finance and established political interests always hold the whip hand to frustrate – or control – ambitious social programmes. They also warn of the transfer of risk that can occur when private sector providers face challenges to profitability.

> Many remunicipalisations, both in Germany and more globally are driven by either pragmatic cost considerations by local public officials or the withdrawal of private entities. This would imply that now, at a time when there is need for substantial investment – for example to meet the massive bill in adapting energy infrastructure to tackling climate change – costs are socialised, whereas in the more profitable past, rent-seeking gains could flow to private – or state – shareholders of commercially operating utilities.
>
> (2018: 514)

Also, of course, while these initiatives appeal to local civic needs and are less abstract than visualising global meltdown, they operate as small islands confronting often unpopular privatised projects and, though offering environmental enhancement in these designated areas, the overall dominant frameworks of environmental despoliation and social inequality identified by many prospective GNDs have yet to be confronted.

The state also needs to ensure that workers are provided with the institutional means to agree and protect formal emissions targets, based on the road maps of international requirements. Compliance by employers will benefit workers and society as a whole, rather than serve minority interests represented by international capital. Unions and employers both understand and appreciate the validity of agreements negotiated between parties and securing agreements forms the essence of union activity. We see in Chapter 11 that broader calls for economic and industrial democracy as a means to combat the triple crisis of the environment, inequality and democratic deficit are being expressed from a range of directions (e.g. Adler, 2019; Pettifor, 2019; Cumbers, 2020; Klein, 2019). As asserted by these and other authorities, economic democracy does not assume revolutionary transformation; social democracy providing the framework for expression of social interests within a capitalist society is feasible. It simply requires the willingness of all parties, political and commercial, to recognise that the best interests of the majority can be secured through institutional reform.

Collective labour activism during the past 30 years of free-market dominance has been seriously undermined, and even countries with enduring social democratic and pluralistic traditions have responded to globalised competitive forces with increased labour market liberalisation (Kalleberg, 2018). Yet it is recognised that the 'business as usual' emphasis built into these policies is seriously undermining efforts at GHG mitigation. This hands-off approach has been graphically demonstrated by the Australian government's claimed mandate 'to prioritise the needs of business' rather than reduce carbon emissions at a time when the country has been devastated by wildfires linked to rising temperatures and GHG emissions (Murphy, 2020). When national governments appear helpless in confronting the global power of transnational enterprises, what realistically can be proposed to offer the labour movement and the communities it represents the means to establish social-need policies which meet at least some principal issues identified in GND proposals?

It will be recalled that GHG emissions arise largely from commercial activity, especially from extractive and manufacturing sectors, backed by government ambitions to maintain economic growth. We have also demonstrated that private and public enterprises have largely failed to heed the warnings of scientists to mitigate polluting activities and different models of shared management must be considered. Markey *et al.* (2019) point out that employees have a direct interest in emissions reduction programmes as, in addition to potential environmental benefits, these will require changes in their working regimes, training, etc. History shows that employees and their representatives possess the skills and motivation to contribute to the development of mitigation and transitionary projects. At times of national emergency unions have successfully collaborated with governments and employers and there is every reason to expect that confronted by global emergency, similar positive responses can be anticipated.

Conclusions

In this chapter we pose the critical questions of who can and will drive global system change given the irrefutable scientific evidence for anthropogenic environmental destruction and associated climate change, the effects of which are already being experienced through more extreme and unpredictable weather patterns, leading to droughts, wildfires, melting glaciers and ice caps and floods (Wallace-Wells, 2019).

The role of government in combatting the twin challenges of environmental despoliation and human rights is a complex one. Though the lead has been taken by a number of international bodies, treaties and targets are not binding on individual nations, and recent experience demonstrates that the most powerful economies are often least likely to commit to targets set by scientists and adopted by international policy agencies. This was the case with Kyoto and continues by country abstentions and, incredibly, through uninformed disputation of the evidence by the USA. Even countries that are setting targets tend to aim for long-term ambitions, such as to be carbon-neutral by 2050 (UK, France) but eschew binding regulation and are less clear on setting short and medium-term intermediary objectives to reduce emissions and pollution. One concern is that states take short-term adaptation measures in the form of reinforcing sea walls, levees and firebreaks rather than comply with scientific recommendations for more coordinated, rigorous and binding environmental strategies while countries and companies also continue to support and invest in fossil fuels which are the cause of climate and environmental devastation.

We also know that the carbon footprint of rich mature economies and rapidly industrialising countries like China and India is far greater than that of poorer less developed countries but nevertheless, the impacts of carbon and other forms of pollution are not bounded by frontiers. The poorer, lower carbon-emitting countries are facing the worst effects of environmental degradation from ecological disasters over which they have had no responsibility and for which they can offer few remedies for their endangered populations, though rich countries too are now beginning to experience the catastrophic impacts of climate heating. Notwithstanding all the warnings, governments are implicit in the continued rise in GHG emissions through their growth and business as usual policies and, more insidiously, abject surrender to the imperatives of global financial capitalism. Broad questions of growing inequality and democratic deficits are implicated with this loss of state sovereignty. In pursuit of growth, some governments, as in Brazil, even deliberately flout all the warnings and physically constrain or silence oppositional environmental movements. Others, such as Australia, adopt short-sighted economic policies, irrespective of longer-term domestic and global consequences. Some countries, like the UK, adopt a 'holier than thou' attitude, pointing with satisfaction to reduced domestic emissions, while ignoring those caused through exporting their emissions to less-favoured countries. Other, predominantly Nordic countries, though not blameless, take a wider, more corporatist approach to dealing with climate emergency and its consequences. As recent international climate conventions have demonstrated, differences between countries over origins of GHGs, national economic rivalries and domestic governments' desires not to falter in pursuit of economic growth have conspired to make binding international cooperation over an international emergency difficult to uphold.

However, propelled by growing awareness of the disaster, there has been an increasing range of proposals for dealing with the emergency. Most of these rely on state actions, though many states are fuelling the crisis, whether by abstinence or by design, whilst being charged with resolving it. Proposals are either interventionist and regulatory, arising largely from more radical or social democratic traditions or market-based, involving minimal regulation, carbon taxes, cap and trade policies. We argue that the continuum from extreme LME non-intervention (represented by a number of South and Central American governments) through moderate LMEs (UK, USA), CMEs (Germany, France), to highly interventionist corporatist economies (Scandinavia) suggest greater inequality, oppression and relative poverty in the far end of the LME spectrum with less attention to environmental concerns by government and employers who have easy access to influence policy through lobbying, financial pressure, threats to relocate, etc. Non-intervention is also associated with greater resource depletion (rain forests, fracking) and physical degradation of air, sea and land. CMEs have a history of effective partnership between employers, unions and governmental bodies that sanction at least some shared means of addressing the environmental crisis, whilst this partnership is further developed in the corporatist economic model of the Nordic countries. In these circumstances, it is difficult to know whether LMEs, wedded to individualistic neoliberalism economic doctrine, can be shifted or supported into adopting more collective approaches to address the climate emergency. It is inarguable that industrial and agricultural society is now facing unprecedented challenges and that equally unprecedented political action is needed to meet the accumulating emergency.

Protecting the environment is going to require dramatic and universal switches of policy towards safeguarding human rights to a healthy life through abandoning short-sighted growth policies by advancing regulation, including binding provisions to encourage union membership and participation. Corporative or social democratic models can be effective in distributing influence beyond minority capital ownership through mobilising efforts for mutual benefit. However, for reasons outlined above, states are hesitant about implementing more radical programmes, whether in whole or in part. Governments tend to work to short-term agendas in order to gain public support and endorsement and 2050 seems more than a political lifetime away.

Unions, in addition to working for their members' benefits, have a positive record of adopting a progressive and authoritative role in corporatist exercises in response to national emergencies. Establishing legislation that requires enterprises to recognise unions for the specific purpose of collaborating over emissions controls and the workplace changes required to help bring this about would be an essential (and urgent) step to take. The experiences of Nordic countries and, to a lesser extent, Germany suggest that government policies to support union revitalisation and active labour market interventions, for example, in providing retraining or job counselling, and which offer strategic collective labour input, can be implemented. A tripartite national level forum could set the broad terms of mitigation to which industry needs to respond. Worker directors and co-determination have a substantial and generally successful history in monitoring and working with senior managers and in those countries receptive to the idea (which, bearing in mind the gravity of the situation, should include many economically developed ones), governments could promptly introduce legislation and funding to provide a reasonably straightforward and timely intervention in ensuring that large companies agree to follow appropriate nationally coordinated emissions reduction programmes whilst adopting effective just transition work regimes. In the European arena a similar monitoring or supervisory approach could be adopted for the currently under-performing European Works Councils, along the lines of health and safety committees, for which European legislation provides statutory safeguards for worker health and welfare.

Governments must take prime responsibilities for the actions of corporations acting in their countries and recent judicial decisions in Europe, based on a state's duty of legal care for its citizens, are potentially helping to change the tone of the political debate. The Dutch Court of Appeal in 2018 required the government to reinforce efforts to cut GHGs to at least 25 percent below 1990 levels by the end of 2020. In response to the finding, the Chief Executive of an international group of environmental lawyers considered that 'climate litigation has become a powerful tool in holding decision-makers accountable for climate inaction. This is the climate case that started it all, inspiring similar lawsuits worldwide' (Schiermeier, 2018). Early in 2020, the UK Court of Appeal dismissed proposals to extend the number of runways at London's Heathrow airport with a ruling that expansion would conflict with the 2015 Paris Agreement. *Nature* (Viglione, 2020) also reports that the Supreme Court in Colombia has ruled that the government must introduce appropriate steps to protect the Amazon rainforest. Welcome as they are, these judicial acts are, however, largely prohibitive in nature, designed to prevent additional emissions rather than promote the remedial sustainable actions needed to positively impact on corporate behaviour toward the environment or on employment practice. However, although international governmental cooperation and agreement has appeared elusive, a climate law proposal by the European Commission in March 2020 could effectively influence action by the 27 EU states. The proposal is based on support for the European GND announced in December 2019 (see Chapter 10) based on cutting emissions by 50–55 percent by 2030, compared with 1990 levels. The new law would allow the Commission to establish binding short-term targets for member states. The move is controversial (and likely to be opposed by some states) as short-term targets would be legally binding and not require unanimous member state approval.

Further movement is underway with international bodies like the UN and its ILO labour standards agency recognising the failures of profit-seeking economies to combat global harm, environmental degradation or to ensure decent and secure employment. The imperative now is for these and other authoritative institutions to insist on radical change in corporate governance. There is no longer any doubt that a state of emergency exists, which requires that governments stop subsidising fossil fuels; that finance institutions withdraw promptly from investing in fossil fuels; that the public interest be recognised in the democratic oversight of industry; that different models of workplace citizenship be explored through worker directors, trade union participation in line with national traditions and development of cooperatives; that liberalising union membership and recognition be treated as established governmental policy; that company boards expanded with worker director members be required to adopt ecological agendas that must meet with internationally-agreed targets. The potential contribution of the much-maligned state has been vividly demonstrated during the coronavirus epidemic emergency and by those interventionist European governments embracing corporatist and coordinated economic policies in combatting GHG emissions. As Greta Thunberg has starkly warned, the planet is on fire, and governments have just one final opportunity to reject the failed neo-liberal model of self-governing markets and do what governments are expected to do – to govern on behalf of their citizens.

Note

[1] Australia also occupies 56th out of 61 places for the 2020 CCPI aggregate results for GHG emissions, renewable energy, energy use and climate policy. Bottom place was occupied by the USA (Burck *et al.*, 2020).

Bibliography

Adler, P. (2019), *The 99 Percent Economy: How Democratic Socialism Can Overcome the Crises of Capitalism*, New York, Oxford University Press.

Alston, P. (2019), 'Climate change and poverty: Report of the Special Rapporteur on extreme poverty and human rights', UN, Human Rights Council, A/HRC/41/39, June.

Azmanova, A. (2019), 'The big Green New Deal and its little red social question', *Social Europe*, www.socialeurope.eu/the-big-green-new-deal-and-its-little-red-social-question.

Bakan, J. (2004), *The Corporation*, London, Constable.

Bayliss, K. (2002), 'Privatization and poverty: the distributional impact of utility privatization', *Annals of Public and Cooperative Economics*, 73 (4); 603–625.

Bayliss, K. and Hall, D. (2017), 'Bringing water into public ownership: costs and benefits', London, Public Services International Research Unit (PSIRU).

Bew, J. (2016), *Citizen Clem: A Biography of Attlee*, London, Riverrun.

Birch, K. *et al.* (2017), *Business and Society*, London, Zed Books.

Bishop, M. and Kay, J. (1988), *Does Privatisation Work?*, London, London Business School.

Briken, K. and Taylor, P. (2018), 'Fulfilling the "British way": beyond constrained choice – Amazon workers' lived experiences of workfare', *Industrial Relations Journal*, 49 (5–6); 438–458.

Bulletin of the Atomic Scientists (2019), 'Google funds climate deniers', thebulletin.org/2019/10/google-funds-climate-deniers.

Burck, J. *et al.* (2020), *Climate Change and Performance Index 2020*, Bonn.

C40 Cities (2018), *Climate Action Planning Framework*, New York, C40 Cities Climate Leadership Group.

Carter, B. and Stevenson, H. (2012), 'Teachers, workforce remodelling and the challenge to labour process analysis', *Work, Employment & Society*, 26 (3); 481–496.

Carter, B., Danford, A., Howcroft, D. *et al.* (2013), '"Stressed out of my box": employee experience of lean working and occupational ill-health in clerical work in the UK public sector', *Work, Employment & Society*, 27 (5); 746–767.

Chang, H. (2014), *Economics: The Users' Guide*, London, Penguin.

Clayton, J. (2020), 'Carbon emissions: Scale of UK fossil fuel support "staggering"', *BBC Newsnight*, www.bbc.co.uk/news/science-environment-51216084.

Clifton, J., Comín, F. and Díaz-Fuentes, D. (2006), 'Privatisation in the European Union: Pragmatic, ideological, inevitable?', *Journal of European Public Policy*, 13; 736–756.

Coates, D. (1989), *The Crisis of Labour: Industrial Relations and the State in Contemporary Britain*, Oxford, Philip Allan.

Cumbers, A. (2020), *The Case for Economic Democracy*, Cambridge, Polity.

Cumbers, A. and Becker, S. (2018), 'Making sense of remunicipalisation: theoretical reflections on and possible political possibilities from Germany's *Rekommumalisierung* process', *Cambridge Journal of Regions, Economy and Society*, 11 (3); 503–517.

Dauvergne, P. (2018), *Will Big Business Destroy Our Planet?*, Cambridge, Polity.

Europe Beyond Coal/Sandbag (2019), 'Solving the coal puzzle: lessons from four years of coal phase-out policy in Europe', Brussels, December.

Fairbrother, M., Sevä, I. and Kulin, J. (2019), 'Political trust and the relationship between climate change beliefs and support for fossil fuel taxes: Evidence from a survey of 23 European countries', *Global Environmental Change*, 59, 102003, November.

FBU (2020), 'FBU response to the Grenfell Tower Inquiry Phase 1 Report', Fire Brigades Union, Kingston-upon-Thames, January.

Financial Times (2017), 'Thames Water: the murky structure of a utility company', 4 May.

FitzRoy, F. and Papyrakis, E. (2016), *An Introduction to Climate Change Economics and Policy*, London, Routledge.

Fleming, P. (2017), *The Death of Homo Economicus*, London, Pluto Press.

Garber, A. and Skinner, J. (2008), 'Is American health care uniquely inefficient?', *Journal of Economic Perspectives*, 22 (4); 27–50.

GCA (2019), *Adapt Now: A Global Call for Leadership on Climate Resilience*, Rotterdam, Global Commission on Adaptation.

Georgeson, L. and Maslin, M. (2019), 'Estimating the scale of the US green economy within the global context', *Palgrave Communications*, 5 (121).

Hall, D. (1999), 'Privatisation, multinationals, and corruption', *Development in Practice*, 9 (5); 539–556.

Hart, S. and Warren, A. (2015), 'Understanding nurses' work: exploring the links between changing work, labour relations, workload, stress, retention and recruitment', *Economic and Industrial Democracy*, 36 (2); 305–329.

Hartley, K., Parker, D. and Martin, S. (1991), 'Organisational status, ownership and productivity', *Fiscal Studies*, 12 (2); 46–60.

Haskel, J. and Szymanski, S. (1993), 'The effects of privatization, restructuring and competition on productivity growth in UK public corporations', *Working Paper 286*, Department of Economics, Queen Mary and Westfield College, University of London.

Hines, C. (2000), *Localization: A Global Manifesto*, London, Routledge.

Hyman, R. (2015), 'The very idea of democracy at work', *Transfer: European Review of Labour and Research*, 22 (1); 11–24.

IEA (2019), *World Energy Outlook 2019*, International Energy Agency, Paris, November.

IMF (2019), 'Global fossil fuel subsidies remain large: an update based on country-level estimates', https://t.co/2n6IgtrD61.

IMO (2014), *3rd IMO Green House Gases Study*, London, International Maritime Organization.

Ipsos MORI (2019), 'Concern about climate change reaches record levels with half now "very concerned"', www.ipsos.com/ipsos-mori/en-uk/concern-about-climate-change-reaches-record-levels-half-now-very-concerned.

Johnson, E. (1952), 'Joint consultation in Britain's nationalized industries', *Public Administration Review*, 12 (3); 181–189.

Kalleberg, A. (2018), *Precarious Lives: Job Insecurity and Well-Being in Rich Democracies*, Cambridge, Polity.

Katzarov, K. (1964), *Theory of Nationalisation*, The Hague, Martinus Nijhoff.

Klein, N. (2014), *This Changes Everything*, London, Penguin/Random House.

Klein, N. (2019), *On Fire: The Burning Case for a Green New Deal*, London, Allen Lane.

Leichenko, R. and O'Brien, K. (2019), *Climate and Society*, Cambridge, Polity Press.

Letza, S., Smallman, C. and Sun, X. (2004), 'Reframing privatisation: deconstructing the myth of efficiency', *Policy Sciences*, 37; 159–183.

Livesey, F. (2017), *From Global to Local*, London, Profile Books.

MacInnes, J. (1987), *Thatcherism at Work*, Milton Keynes, Open University Press.

Markey, R., McIvor, J., O'Brien, M. and Wright, C. (2019), 'Reducing carbon emissions through employee participation: evidence from Australia', *Industrial Relations Journal*, 50 (1), https://onlinelibrary.wiley.com/doi/full/10.1111/irj.12238.

Martin, S. and Parker, D. (1995), 'Privatization and economic performance throughout the UK business cycle', *Managerial and Decision Economics*, 16 (3); 225–237.

Martin, S. and Parker, D. (1997), *The Impact of Privatisation: Ownership and Corporate Performance in the UK*, London, Routledge.

Mazzucato, M. (2013), *The Entrepreneurial State: Debunking Public vs Private Sector Myths*, London, Anthem Press.

Mesnage, R. and Séralini, G-E. (2018), 'Editorial', in R. Mesnage and G-E. Séralini (eds), *Toxicity of Pesticides on Health and Environment*, Lausanne, Frontiers, 4–5.

Middlemas, K. (1979), *Politics in Industrial Society*, London, André Deutsch.

Miller, S., Michalak, A., Detmers, R. and Hasekamp, O. *et al.* (2019), 'China's coal mine methane regulations have not curbed growing emissions', *Nature Communications*, 10, Article number: 303 (2019), January 29.

Monbiot, G. (2006), *Heat: How we can Stop the Planet Burning*, London, Penguin.

Monday, C. (2017), 'Privatization to putinization: the genesis of Russia's hobbled oligarchy', *Communist and Post-Communist Studies*, 50 (4); 303–317.

Murphy, K. (2020), 'Some political leaders find their natural authority in a crisis - not Scott Morrison', *The Guardian*, 2 January, https://www.theguardian.com/global/comm entisfree/2020/jan/02/some-political-leaders-find-their-natural-authority-in-a-crisis-not-scott-morrison

Naidoo, R. and Fisher, B. (2020), 'Reset Sustainable Development Goals for a pandemic world', *Nature*, 583; 198–201, www.nature.com/articles/d41586-020-01999-x?utm_source=Nature+Briefing&utm_campaign=13e014bf84-briefing-dy-20200707&utm_m edium=email&utm_term=0_c9dfd39373-13e014bf84-44223933

Nicolopoulou-Stamati, P., Maipas, S. *et al.* (2018), 'Chemical pesticides and human health: the urgent need for a new concept in agriculture', in R. Mesnage and G-E. Séralini (eds), *Toxicity of Pesticides on Health and Environment*, Lausanne, Frontiers; 6–13.

OHCR-hbs (2018), *The Other Infrastructure Gap: Sustainability*, Geneva, UN Human Rights: Office of the High Commissioner with Heinrich Böll Foundation.

ONS (2019), *UK National Capital Accounts: 2019*, www.ons.gov.uk/economy/environm entalaccounts/bulletins/uknaturalcapitalaccounts/2019.

Oreskes, N. and Conway, E. (2012), *Merchants of Doubt*, London, Bloomsbury.

O'Toole, F. (2018), *Heroic Failure: Brexit and the Politics of Pain*, London, Head of Zeus.

Peters, G., Minx, J., Weber, L. and Edenhoffer, O. (2011), 'Growth in emission transfers via international trade from 1990 to 2008', *Proceedings of the National Academy of Sciences of the United States of America*, 108(21); 8903–8908.

Pettifor, A. (2019), *The Case for the Green New Deal*, London, Verso.

Phillips, T. (2019), '"Chaos, chaos, chaos", a journey through Bolsanaro's Amazon inferno', *The Guardian*, 9 September, theguardian.com/environment/2019/sep/09/amazon-fires-brazil-rainforest.

Pollock, A. (2004), *NHS plc: The Privatisation of Our Health Care*, London, Verso.

Prime Minister's Office (2011), 'Letter from the Prime Minister cutting red tape', 7 April, www.gov.uk/government/news/letter-from-the-prime-minister-on-cutting-red-tape.

PSIRU (2014), 'Public and private sector efficiency: a briefing for the EPSU Congress', Public Services International Research Unit, University of Greenwich, May.

Ramsay, H. (1977), 'Cycles of control: worker participation in sociological and historical perspective', *Sociology*, 11 (3); 481–506.

Raworth, K. (2017), *Doughnut Economics: Seven Ways to Think Like a 21st Century Economist*, London, Penguin/Random House.

Romm, J. (2018), *Climate Change: What Everyone Needs to Know*, Oxford, Oxford University Press.

Schiermeier, Q. (2018), 'Dutch court rules that government must help stop climate change', *Nature,* www.nature.com/articles/d41586-018-07007-7, October.

Sofiev, M., Winebrake, J., Johansson, L. *et al.* (2018), 'Cleaner fuels for ships provide public health benefits with climate tradeoffs', *Nature Communications*, 9 (406), 6 February.

Streeck, W. (2016), *How Will Capitalism End?: Essays on a Failing System*, London, Verso.

Stroud, D., Fairbrother, P., Evans, C. *et al.* (2018), 'Governments matter for capitalist economies: regeneration and transition to green and decent jobs', *Economic and Industrial Democracy*, 39 (1); 87–108.

Talukdar, D. and Meisner, C. (2001), 'Does the private sector help or hurt the environment? Evidence from carbon dioxide pollution in developing countries', *World Development*, 29 (5); 827–840.

TNI (2018), 'The future for democratic water: resistance and alternatives', www.tni.org/en/article/the-future-for-democratic-public-water-resistance-and-alternatives, 5 April.

Tooze, A. (2020), 'The fierce urgency of COP26', 20 January, www.socialeurope.eu/the-fierce-urgency-of-cop26.

TUC (2019), 'A joint transition to a greener, fairer, economy', London, Trades Union Congress.

UNEP (2011), *Towards a Green Economy: Pathways to Sustainable Development and Poverty Eradication*, Geneva, United Nations Environment Programme.

Vickers, J. and Yarrow, G. (1988), *Privatisation: An Economic Analysis*, Cambridge, Cambridge University Press.

Viglione, G. (2020), 'Climate lawsuits are breaking new legal ground to protect the planet', *Nature*, 28 February, https://nature.com/articles/d41586-020-00175-5.

Wallace-Wells, D. (2019), *The Uninhabitable Earth*, London, Penguin Books.

Wilkinson, R. and Pickett, K. (2010), *The Spirit Level*, London, Penguin.

WRI (2017), 'This interactive chart explains world's Top 10 emitters, and how they've changed', Washington DC, World Resources Institute, 11 April.

Xie, F., Chi, J. and Liao, J. (2016), 'From share issue privatisation to non-tradable share reform: a review of privatisation in China', *Asia Pacific Economic Literature*, 30 (2); 90–104.

Zaifer, A. (2019), 'Constructing a substantial alternative to privatization: promoting democratic control and social commitment', *Science & Society*, 83 (4); 495–521.

7 International bodies and non-governmental organisations

Introduction

Previous chapters have shown that nation states have been slow to react to the climate emergency and many continue to subsidise fossil fuels rather than invest in renewable energy. Stances taken to sustainability by states whose economies have been severely disrupted by the coronavirus pandemic will be instructive: will governments heed and act upon the warnings and objective evidence of scientist specialists with regard to the vulnerability of the environment and to society? Will governments recognise and act upon the interconnectedness of nations? Or, once the viral emergency subsides, will it again be a competitive pursuit for economic gain, predicated on consumerism, depletion of resources, disregard for biodiversity and loss of natural habitats?

An early warning of governmental inclinations toward economic nationalism has been posted by Japan, one of the world's largest economies and a major emitter of GHGs. The Japanese government published its updated plans to meet Paris Agreement targets well in advance of the COP26 climate talks originally planned for November 2020. Basically, Japan's original post-Paris targets of reducing CO_2 output by 26 percent by 2030 based on 2013 levels remain unchanged, a target then as now assessed as 'highly insufficient' by the authoritative Climate Action Tracker. Almost alone among the world's major economies, coal remains a major provider of Japan's electricity, in addition to which, the country funds development of coal-fired power plants overseas. The reaction to the government's announcement was severe, both domestically and internationally. Christian Aid's global climate lead described the failure to act as 'feeble' and 'shameful', whilst Laurence Tubiana, a key force in securing the Paris Agreement, reflected that 'at one of the most challenging times of recent memory, we need bolder, mutually reinforcing plans that protect our societies from the global risks we all face' (Lovett, 2020). An even more ominous sign was given, perhaps not surprisingly, by the Trump administration, which over four years severely reduced Environmental Protection Agency staffing, weakened its protective functions and dismantled over 70 environmental protections and regulations including loosening standards for CO_2 emissions from power plants, cars and trucks and launched plans to open up land for oil and gas projects (Popovitch, Albeck-Ripka, and Pierre-Louis, 2020).

So, if two of the planet's biggest economies have been prepared to take either negligible or retrograde steps to control emissions and with insufficient progress having been achieved since Paris, change may need to be driven from outside the nation state. Diverse authoritative international organisations concerned with promoting and protecting personal and community health, social justice and the environment have been long established. The major contributor to the environmental and societal agenda and for setting targets has been the UN, both directly and acting through principal specialised agencies.

United Nations (UN)

The UN has membership from 193 states. It is composed of several different bodies, with its main policy-making body being the General Assembly, comprising representatives from all member states. The Security Council is composed of 15 members, whose main role is to maintain international peace and security, alongside the International Court of Justice, which is responsible for adjudicating in legal disputes submitted by states. The Economic and Social Council makes policies and recommendations concerning economic, social and environmental matters. Many issues concerning international relations are linked, with climate change affecting the number and duration of droughts, melting sea-ice and flooding, forest fires and even locust plagues, which jeopardise food security and livelihoods in East Africa. These different phenomena lead in turn to changing migration patterns and growing flows of internal and cross-border refugees, often putting pressure on neighbouring states and threats of resource disputes, for example over access to water (Sanderson, 2009).

In the 1960s, the UN's concern for the environment and sustainability was sharpened by a series of industrial catastrophes amidst growing agitation by concerned environmental groups. Action was also stimulated by influential cautionary publications like the widely read *Limits to Growth*. With growing scientific and popular concern for sustainable development, the UN held a conference in 1972 on the human environment in Stockholm, followed by the establishment of the United Nations Environment Programme (UNEP), which had ambitious objectives to act as a 'central coordinating mechanism' to 'provide political and conceptual leadership' for global environmental issues (Ivanova, 2005: 14). The role of UNEP is largely to collect and analyse data from UN agencies and other sources and present broad environmental assessments (*ibid.*, 2005: 14) and whilst acknowledging partial success in monitoring, coordinating and assessment of initiatives, Ivanova argues that UNEP has faced a number of constraints that have restricted its effectiveness (*ibid.*, 2005: 9).

The UN has established numerous agencies to deal with environmental and human rights issues. One of the most important is the Intergovernmental Panel on Climate Change (IPCC), a forum aimed to bring together research from leading scientists from around the world. Based in Geneva, the IPCC was a joint creation of UNEP and the World Meteorological Organization in 1988 and tasked with collecting and assessing scientific evidence on climate change, its environmental and socio-economic impacts and to formulate appropriate response strategies. Its first Assessment Report in 1990 provided the platform for establishing the UN Framework Convention on Climate Change (UNFCCC) in 1992, the principal international treaty on stabilising atmospheric GHG concentrations. The Fourth Assessment Report published in 2007 received virtually universal governmental support and paved the way for the IPCC's Fifth Assessment Report, which provided the authoritative scientific basis for the UNFCCC's 2015 Paris Agreement. The UNFCCC invited the IPCC to draft a special report on how global temperature rises could be kept to within 1.5°C of pre-industrial levels. The resulting report argued that restricting temperature rise to below this level was possible but would demand 'far reaching and unprecedented changes in all aspects of society' and required carbon emissions to attain net zero by 2050 (IPCC, 2018).

As we saw in Chapter 3, the UN Agenda on Sustainable Development made an important holistic attempt to integrate approaches to resolve global inequality, access to education, poverty and climate damage with 17 sustainable development goals (SDGs), which have been accepted by all UN member states. An early problem was how these different targets could genuinely interact to represent 'an indivisible whole', as at least some targets could be incompatible with others, such as boosting consumption and its impact on climate mitigation (Nilsson *et al.*, 2016). In a subsequent empirical study of interactions, Nilsson *et al.* (2018) consider that on balance, positive interactions have outweighed negative ones, but warn that problems in target measurement and interpretation will vary according to context, with aspects of governance, geography, technology and available resources being especially relevant. Because of these significant differences, decisions to achieve targets have been left to individual countries and there are no enforcement mechanisms or binding requirements. This has led the UN to acknowledge that actions to meet the goals are falling behind on many objectives and that ambitious steps are needed to deliver the goals by 2030. In a 2019 address to an international forum on sustainable development made before the coronavirus pandemic, António Guteres, UN Secretary-General, made clear the challenges to be faced and the persistent distance between aspiration and achievement. In his view, attainment of the goals was being seriously undermined by a mix of continuing conflict, trade tensions and climate crisis with the result that global poverty and inequality were increasing. Fair globalisation was only possible through following the Agenda and by transitioning economies toward net zero emissions by 2050 (UN, 2019).

Explanations for this lack of progress are not too difficult to find and have been examined at numerous points in this book. In summary, nation states are signatories to UN protocols but are less willing to restrict business activity by introducing procedural regulation to provide the necessary backing for environmental reforms. Further, many countries have been wedded to economic growth and public austerity, affecting both environment and employment security and working conditions. Key sectors have been privatised with primary allegiance shifted to satisfying shareholder expectations rather than provide a community service. Globalisation has enhanced the power of private corporations, whilst weakening the control of states where transnational companies operate and suppressing the regulatory influence of collective labour. These same corporations also exert considerable and often covert political influence, for example, through offering 'revolving-door' executive positions to high-ranking public officials and politicians, funding and associating with business-sympathetic political parties and supporting free-market think tanks.

The International Labour Organization (ILO): 'well-intentioned but with questionable effectiveness'?

Founded in 1919 at the conclusion of the First World War, the ILO was inspired by the conviction that peace and social justice were directly linked and that a widely recognised international institution bringing together governments, labour and employers would have the legitimacy and authority to oversee and monitor initiatives and policies directed toward attaining social justice with economic progress.

The ILO is distinctive in being the only tripartite UN agency, bringing together governments, employers and workers from member states to establish labour standards, develop policies and promote 'decent work' for men and women worldwide. The main aims of the ILO are to 'promote rights at work, encourage decent employment opportunities, enhance social protection and strengthen dialogue on work-related issues' (ILO, 2019). According to the current Director-General, the ILO was established to 'bring together representatives of governments, of employers and of workers as joint decision-makers' (Ryder, 2014). He also argues that rising levels of inequality make 'economies more volatile and crisis-prone'. A main accomplishment of the ILO, Ryder argues, is demonstrated by its international labour standards comprising conventions and recommendations adopted through tri-partite discussion at annual conferences and subsequently ratified by member states. However, not all states do ratify and recommendations are non-binding, but if a country ratifies a convention, it commits itself to meeting its obligations. Conventions and recommendations are seen as guidelines and are often loosely worded, allowing considerable scope for interpretation. Moreover, as Peter Rossman, director of an international union federation, points out, core 'conventions are all routinely breached – not only in poor countries, but also in some of the richest countries of the world' (2013: 62). And if a country does not like a convention it can easily 'deratify' from it (Standing, 2008). In 1998, core labour standards were bundled into a *Declaration of Fundamental Principles and Rights at Work* covering rights to organise, collective bargaining and freedom from discrimination. Standing, who knows the ILO well, was highly dubious about the choice and impact of the core conventions, which were arbitrarily chosen and questionably worded, citing as an example 'the abolition of the worst forms of child labour', leaving scope for wide interpretation. He doubts whether the Declaration had a definitive effect in reducing social injustice and inequality (2008: 367).

One problem is that the conventions have made little impact on the worldwide growth of the informal economy and precarious forms of working that underlie dominant business models, which in turn support increasingly unsustainable productive and service economies. As we saw earlier, union organisation is difficult to establish and maintain when agency working, zero-hours contracts, temporary and seasonal work and denial of employee status dominate large areas of labour recruitment and work organisation. Reduced access to unions and to collective bargaining also serve to enhance both domestic and international inequality (Brown, 2017). These market vulnerabilities persist despite conventions being formal treaties binding on ratifying states. As recently as 2013 it was acknowledged by a senior ILO official that the 'unstated premise behind most ILO conventions is a direct and stable employment relationship' (Cunniah, 2013: 5) and while the Organization has undoubtedly extensively researched and reported on the decline of this model relationship (*International Journal of Labour Research*, 2019), its practical impact on labour market behaviour has been questionable (Demaret, 2013). The British TUC considers that core labour standards underpinning the conventions are 'more often ignored than honoured' (TUC, undated) and argues that governments and transnational enterprises need to ensure more effective implementation and enforcement. Moreover, the ILO has been charged with increasing reliance on non-legally binding 'soft governance' instruments, requiring relatively lower levels of rigour and offering more flexibility in compliance (Jakovleski *et al.*, 2019). In contrast, Luc Demaret, an ILO director, advocates adoption of a new convention to strengthen the rights of all workers, recognising that existing labour standards 'fail to sufficiently protect workers against precariousness as such' (2013: 20).

Standing also has harsh words about the notion of 'decent work', a concept vigorously taken up by the ILO, criticising its 'inherent vagueness', 'timidity' and 'lack of coherence'. He castigates its failure to attempt any meaningful measurement so that 'decent work' simply became a mantra used by the Organization rather than an empirically validated set of values or practices. Interestingly, he questions whether the ILO itself adheres to notions of decent work, pointing out systematic internal casualisation, politicisation of governance and management, with key appointments made through 'contacts' and 'factional allegiances' (2008: 373). Nevertheless, meeting the SDG goal of 'decent work and economic growth' (two goals that may not be necessarily compatible) is a centrepiece of the ILO's commitment to the UN's 2030 development agenda. The Director-General of the ILO summarises this commitment as: 'promoting jobs and enterprise; guaranteeing rights at work; extending social protection and promoting social dialogues ... the four pillars of the ILO Decent Work Agenda, with gender as a cross-cutting theme' (ILO, 2020). In view of the questions raised about the ILO's capabilities to identify and deliver these individual pillars, it is difficult to see how, in an economic world still dominated by liberal, globalised and financialised economic forces, the institution will find the intellectual and policy resources to deliver these diverse commitments.

From these accounts, from commentators with considerable experience of working in and investigating the organisation, it is clear that the ILO would face problems in acting as a force in adopting and leading an international strategic green agenda. Its contribution to protecting labour during the past years of supply-side economics or confronting the worst effects of globalisation has been seriously questioned. One consequence of its inability to adapt to changing conditions is that the ILO has 'drifted into the margins of international influence' (Standing, 2010). Other commentators recognise the institute's shortcomings but tend to be more sympathetic to the reasons. Jakovleski *et al.* (2019) recognise that progress is impeded by the need to compromise between the stances of fairly inflexible parties, so that change, when it comes, tends to be incremental rather than decisive in a period of rapid macroeconomic and labour market developments. Indeed, the main ILO role has been in providing technical assistance, publishing information and offering moral support or persuasion (Bolle, 2014).

Moreover, the ILO has always been at the mercy of more powerful members, notably the USA, which withdrew membership in 1977 in response to perceptions of the Organization's selectivity in treatment of convention violations and its shift to an overtly and divisive political role. The ILO was confronted with financial collapse following the loss of the USA's contribution, which amounted to a quarter of its funding. In response to this loss policy changes followed, notably hardening its stance toward the then USSR and softening its critical approach toward Israel, following which the USA re-joined in 1980. Referring to the contemporary era defined largely by globalisation, the Director-General suggests that the organisation is 'well-intentioned but with questionable effectiveness'. For the future Ryder acknowledges a major challenge: 'transition of production systems to a sustainable development path is a necessity. The greening of the ILO will need to distinguish the second century of ILO work from its first' (Ryder, *op. cit.*: 15).

Notwithstanding these reservations, the ILO has contributed to the development of sustainable development recognition and thinking, though perhaps less to action. In their review of the ILO's role in advancing environmental values in the UN and in wider society, ILO insiders Olsen and Kemter (2013) point out that since the 1972 Stockholm conference on the human environment the ILO contribution has been based on three main areas: (i) collaboration within the UN system and other institutions, (ii) capacity building of the constituents and (iii) linking working conditions to the external environment. Most importantly Stockholm recognised that 'the working environment is an important and integral part of the human environment as a whole, since those factors that harm the working environment are also among the major pollutants of nature and of people's living environment' (*ibid.*: 42).

Several reports in the 1970s and 80s emphasised that the 'improvement of working conditions and environment and the wellbeing of workers remains the first and permanent mission of the ILO' (*ibid.*: 43) and recognising rising CO_2 levels, a joint report with UNEP made early reference for the need for just transition for workers affected by shifts to environmentally supportive activities. Lack of national action was also noted: in 1993 an ILO report argued for the 'irreplaceable role' of unions in promoting social and environmental change but notes that 'this role has to be accepted and implemented by governments and employers in practice' (*ibid*.: 47). A report in 2008 on green jobs noted 'the chronic and worsening levels of inequality, both within and between countries as the major impediments for the expansion of green jobs' (*ibid*.: 50). Olsen and Kemter conclude that though 'the search for a fairer, greener and more sustainable development model is clearly gaining momentum' (*ibid*.: 54), the talk is still of potential rather than achievement and interventions by workers and their representatives, a fundamental ILO demand, as well as actions by states and employers, are rarely being achieved.

The observed shortcomings of the ILO in presenting a counter to the widespread economic and labour market developments of the past few decades lead to the question of its role in developing a strategic position with regard to the environment. As we saw above, the Organization has committed itself to the UN's SDGs. It has also demonstrated its support for the 2015 Paris Agreement in a major summary report and call for action published in 2018 (ILO, 2018). The ILO report follows current thinking in arguing that transition to sustainability is urgent as economic activity continues to exert an adverse effect on both environment and climate, which in turn exerts adverse effects on workers through, for example, heat stress. Hence environmental sustainability is crucial for decent jobs. The report emphasises the importance of social dialogue to ensure just transitions and the value of social protection policies to support workers during transition to a greener economy. It points out that whilst skills development programmes are an essential element of the greening process, these have not yet been sufficiently 'mainstreamed'. Finally, the need for closer integration of policies is emphasised. Though few would argue against these analyses or steps, the question remains as to how many of them may be accomplished while economies fail to collaborate and where worker and community interests are subordinated to the demands of profit maximisation. By failing to recognise and act upon changing labour market patterns and pressures the ILO is in danger of falling into insignificance and there will be no shortage of populist politicians ready to exploit its weaknesses in return for short term gain and in support of vested interests.

World Health Organization (WHO)

The WHO was established in 1948 as a specialised health agency of the UN with an original objective to promote international cooperation to improve healthcare for all peoples. The Organization's recent stated priorities include addressing the role of social, economic and environmental factors in public health and promoting 'public health and well-being in keeping with sustainable development goals' established by the UN. Hence, the institution has a direct role in promoting a healthy work and public environment.

Since its establishment, the WHO has been an institution in more or less perpetual crisis, mainly as a consequence of political and ideological tensions among which the Organization is often at the centre (Brown *et al.*, 2006). Early tensions emerged with the admission of decolonised nations to the UN and with them the spread of nationalist and socialist influence as well as more radical approaches to public health and social development. Perhaps in response, by the 1980s, the World Bank had become a serious rival to the WHO in promoting public health, though World Bank projects were based on support for a market platform for health care provision. With its influence diminishing, in 1982 the WHO budget was frozen by its decision-making body, the World Health Assembly. In 1985, the USA decided to withhold its contribution to the WHO's regular budget, a move supported by American profit-seeking pharmaceutical companies. By the late 1990s, World Bank influence over health care had grown, whilst the prestige and influence of the WHO declined. The Organization was also confronted by the growing ascendancy of the neo-liberal political economy and associated promotion of private sector health care by the World Bank and its supporters. However, in 1998, the highly respected Gro Harlem Brundtland, an authority on associations between health and the environment, was appointed Director-General and during her five-year tenure the fortunes of the WHO improved as she shifted it towards more of an independent position promoting public health, rather than being subordinate to the interests of economically powerful nations.

Political, funding and role problems have increased rather than diminished in recent years. The financial crisis from 2008 led to a drop in its funding. The WHO operates as a global institution at a time when nationalism and nationalist politicians question its relevance, added to which, it has little actual power. It can offer guidelines, assistance and advice, but no formal means to sanction member states or bind them to its action plans. For a huge international organisation, it enjoys comparatively meagre funding, around $2bn in 2019. Main funding providers are the USA (in excess of $400 million in 2019) and the UK (about $195 million) as well as big and influential private donors like the Bill and Melinda Gates Foundation. All the main providers want – and expect – the WHO to operate according to their beliefs and priorities, which can put the Organization in a delicate position: demonstrating its independent professional authority in possible defiance of influential members may well lead to retaliation. Alternatively, it can face accusations of being impotent in the face of threat. There have been criticisms that it reacts too slowly to crises (for example, failing to declare a public health emergency after the Ebola outbreak in Western Africa in 2014) whilst others accuse it of over-reacting and creating unnecessary panic when forecasted epidemics remain limited in spread or casualties (Wibulpolprasert and Chowdhury, 2016). Critics argue that the Organization is too bureaucratic, spread too thinly relative to its dwindling resources, that it is now competing with other, possibly more effective, bodies, for example in gathering global health data, and that it could operate better by focusing on a limited range of core activities and act as a coordinating intermediary for those agencies that might be more suited to implementing these activities. In other words, critics say that it should contract out some of its current functions (Negin and Dhillon, 2016).

Notwithstanding these almost constant difficulties and pressures, the contribution of the WHO to understanding and confronting issues of environment and public health cannot be underestimated, as demonstrated by the widely publicised Air Quality Guidelines, which advocate a limit of 25 micrograms of particulate per cubic metre (WHO, 2018a) and that now defines air pollution as a significant contributor to lung cancer. The WHO also publishes lists of the world's most heavily-polluted countries and cities (WHO, 2018b). In China, where toxic air has become a major factor in premature death and childhood respiratory disease, there has been a response: schools and nurseries in Shanghai close down and sporting events are cancelled when dangerously high levels are reached (Klein, 2014: 351). The publicity given to poor air quality in parts of India has also led to some initial, if inconsistent action (Gardiner, 2019: 55). Nevertheless, the WHO estimated that in 2016, nine out of ten of the world's population lived in areas where poor air quality levels exceeded guideline limits. As a consequence of WHO reports, member states adopted a 2015 resolution and 2016 road map to encourage a wider global response to the adverse health effects of air pollution as well as acting as custodial agency for three air pollution related SDG indicators, namely mortality from air pollution; access to clean fuels and technologies; and air quality in cities (WHO, 2018b). It also assists member states by sharing information on successful approaches to pollution reduction and in monitoring initiatives.

The problems facing the WHO and other international organisations are highly relevant to identifying difficulties in confronting the environmental emergency and its associated ramifications. These are organisations that are attempting to chart a coordinated response to ecological and societal calamity. Nationalist politicians such as Brazil's Bolsanaro, India's Narendra Modi and of course America's Trump have been keen to manipulate these institutions in pursuit of their own narrow economic and political interests and when differences arise, financial and political pressures are rapidly exerted to undermine multilateral international efforts. These actions were most evident in Trump's dismissive and threatening treatment of the WHO and its leadership during the coronavirus pandemic (Politico, 2020).

World Meteorological Organization (WMO)

The work of the WMO is international in scope: weather patterns, climate change and quality and quantity of water resources may impact differentially on specific regions but are not defined by boundaries or by borders. Hence, international cooperation is vital for the study of meteorological phenomena. As well as analysing changing climate patterns, the WMO undertakes a range of projects such as assessing flood risk and flood management, especially in low-income countries; providing weather and climate services to support farmers, herders and fishing communities; and support for agricultural planning and sustainable agricultural development. At its Congress in 2019, the WMO adopted its 2020–2030 strategic plan, containing three 'overarching priorities': (i) to enhance preparedness for hydrometeorological extremes, (ii) to support climate-smart decisions and enhance the socio-economic benefits of related services and (iii) to contribute to societal needs reflected in the global agenda to realise sustainable development (WMO, 2019a).

The WMO presents valuable scientific data on global temperatures, weather patterns and their socio-economic consequences. In 2019 the UN General Assembly President and the WMO Secretary-General presented the WMO's *Statement on the State of the Global Climate in 2018* (WMO, 2019b). The report established that the impact of record GHG emissions had driven global temperatures to ever more dangerous levels. It also confirmed that 2018 was the fourth warmest on record; that ocean heat content had reached a record high and that global sea levels were continuing to rise, imperilling low-lying coastal communities, especially those in developing countries. The report also recorded that the extent of Arctic and Antarctic sea-ice was well below previous levels. Extreme weather was impacting negatively on lives and sustainable development on every continent and average global temperatures had reached about 1°C above pre-industrial levels. Finally, and ominously, the report concluded that the world was not on track to meet climate change targets or to rein in rises in temperature. The Statement was subsequently consolidated into a summary report based on the most recent climate science information, with contributions from the IPCC and other authoritative bodies (WMO, 2019c). In introducing the Report, in a possible riposte to climate-denying national leaders, António Guteres, UN Secretary-General, pointed out that 'science informs governments in their decision-making and commitments' and continued by urging 'leaders to heed these facts, unite behind the science and take ambitious, urgent action to halt global heating and set a path towards a safer, more sustainable future for all'.

Non-governmental organisations (NGOs)

NGOs aim to support societal security and range from local groups, which focus on protecting specific communities, to well-established international organisations whose interests cover the planet. Select groups of these non-state agencies have been granted observer status at some UN based meetings. UNESCO, for example, cooperates with diverse NGOs, forming official partnerships with nearly 400 of them. Thiele (2016) points out that NGOs are neither political parties nor government agencies. They can be defined as non-profit, citizen based, voluntary organisations, often cooperative in nature, which formally act independently of government. They have considerable variety of form and function (Cousins, 1991), designed to serve specific or general societal or political purposes. The World Bank defines NGOs as 'private organizations that pursue activities to relieve suffering, promote the interests of the poor, protect the environment, provide basic social services or undertake community development'.

Growth of NGOs has been relatively recent, with over 90 percent established since 1975. NGO growth accelerated in tandem with the consolidation of neo-liberal economic thinking, which advocated a diminishing role and expenditure by government, which in turn encouraged NGOs to initiate services from which government increasingly abstained. NGOs can offer their services at all levels of civil society, but our prime interest is in the activities of those operating across national borders in support of the environment, social justice and public health, often popularly referred to as BINGOs (big international NGOs). There are calculated to be over 80,000 BINGOs worldwide. These NGOs are often single-purpose charities such as Oxfam (dealing with global hunger) or Reprieve, which campaigns against capital punishment. Aiming to protect the environment is of course a multi-dimensional activity so, for example, Greenpeace conducts diverse and often highly active campaigns against whaling, despoliation of the planet and confronts fossil fuel extraction. In all these circumstances, it faces hostility from major corporations and their financial and political supporters as well as from countries who feel disadvantaged by the NGO's vigorous and sometimes provocative campaigning. Workers whose livelihoods may be jeopardised by these campaigns may also, and understandably, be less than impressed by an environmental NGO's activities. However, NGOs sometimes act as unofficial sponsors for workers who would otherwise be constrained by law or by employers from taking industrial action. Thus, in some cases NGOs have helped to raise labour standards in developing countries where production has been outsourced by well-known brand names. Campaigns by NGOs have targeted consumers of easily recognised clothes and sportswear companies, whilst also pressing the companies to improve their practices though requiring companies and their suppliers to conform to codes of good conduct. The problem is that supply lines can be long and sometimes obscure, and codes of conduct may have little relevance to or impact upon the working lives of the workers (Hasmath and Hsu, 2007).

Nevertheless, rather than challenging and confronting corporate and political interests, NGOs have been charged with operating complicitly with them by acting in partnership with governments, intergovernmental agencies and corporate donors, who can provide the biggest source of their funding (Godrej, 2014), an important factor when post-2008 austerity led to reduced donations for many NGOs (Shahin *et al.*, 2013). Also, many heads of BINGOs are drawn from the corporate world of technocratic management, accustomed to looking for measurable outcomes from easily quantifiable projects. The head of a major BINGO confirmed that by becoming 'mired in bureaucracy' and avoiding 'approaches or issues that might threaten our brand or upset our donors', NGOs reinforce the 'social, economic and political systems we once set out to transform'. He concludes that 'primary accountability must be not to donors but to all those struggling for social justice' (Sriskandarajah, 2014). LeBaron (2013) also identifies the 'corporatisation' of the established environmental movement, which has become increasingly compromised through commercial associations with major producers. Thus, respected NGO Conservation International has entered into partnership with Starbucks and Walmart. Even the venerable Sierra Club is reported to have accepted significant donations from the gas sector. LeBaron argues that these shifts – and she details many more – represent a change in NGO strategy in treating major corporations as allies rather than adversaries. At the same time, she notes a move towards 'market-friendly and consumer driven activism' by providing eco-certification and eco-labelling to legitimise big business commercial activity. As an example, she cites the tie-up between WWF and Coca-Cola to protect polar bears, which led to a massive increase in sales of the drink, whose can featured a picture of a polar bear. Whilst the major NGOs seek financial protection and publicity, LeBaron argues that such acts 'reinforce unsustainable patterns of production and consumption world-wide'. Dauvergne and LeBaron (2014) note in some detail how the respectability afforded these NGOs contrasts with the somewhat harsher treatment meted out to grass-roots protesters and activists by corporations, police and the courts when they attempt to confront the damaging activities of these same organisations.

Naomi Klein also spells out the potentially complex entanglements of what she calls 'Big Green' environmental NGOs with both corporates and politicians. Reviewing the history of NAFTA, the North American Free Trade Agreement, concluded in 1993 by then President, Bill Clinton, an agreement that sought to bring about deregulated free trade between the USA, Canada and Mexico, she points out that many major NGOs were partially or even enthusiastically supportive, whilst others, notably Friends of the Earth and Greenpeace were implacably opposed (2014: 84) for fear of driving down environmental standards and undermining labour protections. Later campaigns in support of migrants to the USA were also supported or led by major NGOs, including the Sierra Club, the BlueGreen Alliance and again, Greenpeace (*ibid.*: 156). Conversely, Klein also notes that fossil fuel and other polluting companies have, in some cases, helped to fund NGOs including a 'long-term, close strategic relationship' between the World Resources Institute and the Shell Foundation (*ibid.*: 197). She also notes that some NGOs with corporate affiliations tend to set their campaigning sights low with regard to actions against fossil fuel operations, supporting, for example, natural gas and carbon trading as effective means to reduce carbon emissions (*ibid.*: 199).

Conclusions

The central question concerning international bodies is to consider how effective they have been in achieving their climate, environmental and social justice objectives and also to review reasons for their success or lack of it. The fundamental purposes of the UN are laid down in its Charter, namely: to maintain international peace and security; to develop friendly relations between nations; to achieve international cooperation in resolving international problems; and to act as a harmonising centre in attainment of these objectives. From its inception the UN has often served as a political football, reflecting the varied and contesting interests between factions from East and West, developed and emerging economies and latterly between America and China. Basically, the United Nations has rarely been united and undoubtedly this has hampered pursuit of its fundamental aims. Also, political divisions within countries manifest themselves in different attitudes to international bodies, sometimes linked to perceived unfair treatment of client countries and sometimes to ideological opposition to outside intervention in a country's internal or economic affairs. A recent case was the government's response to the UN Special Rapporteur's 2019 report on the prevalence of extreme poverty and human rights shortcomings in the UK, prompting denial and caustic attacks on the rapporteur from government ministers and threats to make an official complaint to the UN, though little action to interrogate the evidence or address the issues was subsequently taken.

A broader critique has been the failure of the UN to recognise or act upon the claimed negative effects of globalisation on the world's poor, whilst other commentators have commended globalisation as serving and benefitting the same poor (Nayyar, 2002). There is no doubt, however, that the UN has enjoyed some success in its humanitarian missions, though its record in peacekeeping intervention has been mixed, with routine criticism coming from strongly conservative or isolationist commentators (Jacobson, 2012). More balanced appraisals have been offered by analysts such as Howard (2008), who recognises the practical difficulties faced by many UN peacekeeping units placed into situations over which they have little control.

Some broader UN initiatives, though tinged with contemporary management-speak, have met with mixed results. For example, in 1997, past Secretary-General Kofi Annan proposed that human rights should be fully integrated across the full range of UN activities in a process of 'mainstreaming'. The intention was to adopt mainstreaming as a cross-cutting issue to guide all UN activities. This is possibly a surprising proposal, bearing in mind that one of the fundamental aims of the UN and its agencies is to ensure that member states uphold human rights as a fundamental, and presumably mainstream, value. More broadly, ideological and power-play factors have undoubtedly hampered the UN in achieving its objectives. The most recent development has been the escalating trade war conducted between China and the USA since 2018, initiated by the USA under Trump, who in defence of domestic production implemented a series of measures to cut imports including steel. Universal tariff increases were followed by specific targeting of China imports. Throughout 2019, both countries then successively proceeded to raise import trade tariffs, which resulted in mutual economic detriment, with Chinese exports to the USA substantially diminished and American consumers facing price increases (UNCTAD, 2019). During 2020, mutual hostility escalated following the outbreak of the Covid-19 pandemic. These tensions were compounded by growing doubts about North American commitment to multilateralism under the Trump presidential regime and also by China's increasing presence and assertiveness in high-level positions in the UN and its agencies. The decision by the Trump administration to withdraw funding from the WHO over the coronavirus outbreak coupled with criticism of China's purported non-disclosure role, in addition to the UN's refusal to recognise Taiwan, together cast doubt on the directions of America's recent global leadership as its political orientation became more internally focused.

The ILO also faces problems of legitimacy founded in its failure to keep pace with labour market developments, including extended supply chains and increasing vulnerability and commodification of labour, as globalisation extends its reach to ever-growing numbers of national communities. During the past few years there have been few matching developments in its legislative machinery, sharp internal differences over the right to strike and no new measures for defining or operationalising 'good work' (Van der Heijden, 2018). However, the ILO's basic model remains located in the tri- or bi-party negotiating frameworks once common to Western Europe, Japan and the USA despite the model's growing fragmentation since the 1980s (La Hovary, 2015). Moreover, as we saw above, ILO conventions can be breached without too much difficulty, and many are not ratified if they conflict with nationally identified priorities. And, as with other agencies, influential countries can threaten to withdraw, or, as was the case with the USA, did withdraw in 1977, leaving the Organization in financial jeopardy, to which it responded by adapting its policies to be in closer alignment with American political sympathies.

NGOs, and especially those dealing with global issues, often act in sympathy with, though independently of, UN agencies or in tandem with them. Their objectives are directed to support disadvantaged communities and people and in protection of the environment. They also face numerous problems. Their field activists who defend human rights or the environment can face obstruction, harassment or worse when operating in sensitive areas. More generally, NGOs are dependent on funding, which is volatile, for example, during economic recessions. They may also depend on large corporate donors for whom attachment to an NGO may confer legitimacy to their commercial activities. Tighter finances have also forced NGOs to operate with greater caution in terms of specifying their objectives and to appoint senior personnel whose skills and interests in managing budgets may be greater than their experience with social organisations, leading to possible internal conflicts or with supporters.

We can see, therefore, that international organisations have faced numerous attacks, some justified, and many of them contradictory. They are variously accused of being authoritarian, bureaucratic, slow to adapt, afraid to lead and submissive to political pressures. Yet recognition of their authority as informed international bodies is essential as nationalism, and even isolationism, grows. As we have seen, the environment, and the range of human issues associated with its despoliation, does not respect borders, with especial consequences for less developed countries, but the influence of international agencies, stripped of finances, serving as a political football and no binding authority in countering these developments is too weak. Nevertheless, the UN and its agencies have established, through multilateral agreements with the international community, ways of measuring, monitoring and proceeding which should serve to avert climatic, environmental and social calamity. Whatever the ideological, political and economic differences between nations, these are secondary to the need for global cooperation and solidarity when faced with an emergency of these dimensions. The coronavirus pandemic has demonstrated the horrendous consequences of disregarding the interconnectedness of nations and the importance of ensuring that the common links that bind communities to one another neither weaken nor break.

Bibliography

Bolle, M. (2014), 'Bangladesh apparel factory collapse: the background in brief', Congressional Research Service, CRS Report, Prepared for members and committees of Congress.

Brown, R. (2017), *The Inequality Crisis*, Bristol, Policy Press.

Brown, T., Cueto, M. and Fee, E. (2006), 'The World Health Organization and the transition from "international" to "global" public health', *American Journal of Public Health*, published online 2011, https://ajph.aphapublications.org/doi/full/10.2105/AJPH.2004.050831.

Cousins, W. (1991), '"Non-governmental initiatives" in Asian Development Bank', *The Urban Poor and Basic Infrastructure Services in Asia and the Pacific*, Manila, ADB.

Cunniah, D. (2013), 'Foreword: Meeting the challenge of precarious work: a workers' agenda', *International Journal of Labour Research*, 5 (1); 5–8.

Dauvergne, P. and LeBaron, G. (2014), *Protest Inc: The Corporatization of Activism*, Cambridge, Polity Press.

Demaret, L. (2013). 'ILO standards and precarious work: strengths, weaknesses and potential', *International Journal of Labour Research*, 5 (1); 9–22.

Gardiner, D. (2019), *Choked*, London, Granta.

Godrej, D. (2014), 'NGOs – do they help?', *New Internationalist*, 1 December, https://newint.org/features/2014/12/01/ngos-keynote.

Hasmath, R. and Hsu, J. (2007), 'Big business, NGOs and labor standards in developing countries', *Asian Journal of Social Policy*, 3 (1); 1–15.

Howard, L. (2008), *UN Peacekeeping in Civil Wars*, Cambridge, Cambridge University Press.

ILO (2018), 'Greening with jobs: world employment social outlook', Geneva, International Labour Organization.

ILO (2019), http://ilo.org/global/about-the-ilo/lang–en/index.htm.

ILO (2020), ilo.org/global/topics/sdg-2030/targets/lang–en/index.htm.

International Journal of Labour Research (2019), 'The future of work: trade unions in transformation', *International Journal of Labour Research*, 9 (1).

IPCC (2018), *Special Report: Global Warming of 1.5°: Summary for Policy-Makers*, Geneva, IPCC.

Ivanova, M. (2005), 'Can the anchor hold? Rethinking the United Nations Environment Programme for the 21st Century', *Forestry and Environmental Studies Publications Services* (27), New Haven, Yale Center for Environmental Law and Policy.

Jacobson, T. (2012), 'UN peacekeeping: few successes, many failures, inherent flaws', International Diplomacy and Public Policy Center, LLC, March-April.

Jakovleski, V., Jerbi, S. and Biersteker, T. (2019), 'The ILO's role in global governance: Limits and potential', *International Development Policy*, Geneva, Graduate Institute, https://doi.org/10.4000/poldev.3026.

Klein, N. (2014), *This Changes Everything*, London, Penguin.

La Hovary, C. (2015), 'A challenging "*ménage à trois*": tripartism in the International Labour Organization', *International Organizations Law Review*, 12 (1); 204–236.

LeBaron, G. (2013), 'Green NGOs cannot take big business cash and save planet', *The Conversation*, 30 September, https://theconversation.com/green-ngos-cannot-take-big-business-cash-and-save-planet-18770.

Lovett, S. (2020), 'Japan's latest climate change plans criticised as "feeble" and "shameful"', *The Independent*, 30 March, www.independent.co.uk/environment/climate-change/japan-climate-change-plan-emissions-paris-agreement-greenhouse-gas-a9434916.html.

Nayyar, D. (ed.) (2002), *Governing Globalization: Issues and Institutions*, Oxford, Oxford University Press.

Negin, J. and Dhillon, R. (2016), 'Outsourcing: how to reform WHO for the 21st century', *BMJ Global Health*, 1 (2), https://gh.bmj.com/content/1/2/e000047.full.

Nilsson, M., Chisholm, E., Griggs, D. *et al.* (2018), 'Mapping interactions between the sustainable development goals: lessons learned and ways forward', *Sustainability Science*, 13, 1489–1503 (2018). https://doi.org/10.1007/s11625-018-0604-z.

Nilsson, M., Griggs, D. and Visbeck, M. (2016), 'Policy: map the interactions between Sustainable Development Goals', *Nature*, 534; 320–322.

Olsen, L. and Kemter, D. (2013), 'The International Labour Organization and the environment' in N. Räthzel and D. Uzzell (eds), *Trade Unions in the Green Economy*, London, Routledge; 41–57.

Politico (2020), 'Trump aides consider creating alternative to WHO', www.politico.com/news/2020/04/10/trump-aides-debate-demands-who-179291.

Popovitch, N., Albeck-Ripka, L., and Pierre-Louis, K. (2020), 'The Trump Administration is reversing nearly 100 environmental rules', *The New York Times*, https://nytimes.com/interactive/2020/climate/trump-environment-rollbacks-list.html.

Rossman, P. (2013), 'Food workers' rights as a path to a low carbon agriculture', in N. Räthzel and D. Uzzell (eds), *Trade Unions in the Green Economy*, London, Routledge; 58–63.

Ryder, G. (2014), 'Relevance of the ILO in the twenty-first century', *Warwick Papers in Industrial Relations*, 98, June, Industrial Relations Research Unit, University of Warwick.

Sanderson, M. (2009), 'Globalization and the environment: implications for human migration', *Human Ecology Review*, 16 (1); 93–102.

Shahin, J., Woodward, A. and Terzis, G. (2013), 'Study on the impact of the crisis on civil society organizations in the EU – risks and opportunities', Brussels, European Economic and Social Committee.

Sriskandarajah, D. (2014), 'NGOs losing the war against poverty and climate change', *The Guardian*, 11 August, https://www.theguardian.com/global-development-professionals-network/2014/aug/11/civicus-open-letter-civil-society-professionalisation.

Standing, G. (2008), 'The ILO: an agency for globalization?', *Development and Change*, 39 (3); 355–384.

Standing, G. (2010), 'The International Labour Organization', *New Political Economy*, 15 (2); 307–318.

Thiele, L. P. (2016), *Sustainability*, 2nd edn, Cambridge, Polity Press.

TUC (undated), 'Core labour standards explained', www.tuc.org.uk/sites/default/files/gettingtothecore_0.pdf.

UN (2019), 'Remarks to high-level political forum on sustainable development', www.un.org/sg/en/content/sg/speeches/2019-09-24/remarks-high-level-political-sustainable-development-forum, United Nations, September.

UNCTAD (2019), 'Trade and trade diversion effects of United States tariffs on China', *UNCTAD Research Paper*, 37, Geneva.

Van der Heijden, P. (2018), 'The ILO stumbling towards its centenary anniversary', *International Organizations Law Review*, 15 (1); 203–220.

WHO (2018a), 'Ambient (outdoor) air pollution', Geneva, World Health Organization, 2 May, www.who.int/news-room/fact-sheets/detail/ambient-(outdoor)-air-quality-and-health.

WHO (2018b), 'WHO global ambient air quality database', Geneva, *World Health Organization*, www.who.int/airpollution/data/cities/en.

Wibulpolprasert, S. and Chowdhury, M. (2016), 'World Health Organization: overhaul or dismantle?', *American Journal of Public Health*, 106 (11), 1910–1911.

WMO (2019a), 'WMO Strategic Plan 2020–2030', World Meteorological Organization, WMO-No. 1225, Geneva.

WMO (2019b), 'WMO Statement on the state of the global climate in 2018', World Meteorological Organization, WMO-No. 1233, Geneva.

WMO (2019c), 'United in Science', World Meteorological Organization, Geneva.

8 The role of workers and their unions

Introduction

Major challenges face workers and the labour movement in confronting the twin crises of environmental despoliation and the destruction of meaningful work. The first challenge, as noted earlier, is that increasing numbers of workers do not enjoy the status of employees and hence are distanced both from organisational decisions and the supportive activities of trade unions. Secondly, as part of management's emphasis on treating humans as resources to be deployed to optimum organisational benefit, employee involvement programmes such as engagement and empowerment are initiated and driven by management to contribute to organisational profit and efficiency objectives. As union influence has waned, these individual programmes have become increasingly ubiquitous (van Wanrooy et al., 2013) and channelled into narrow areas of task discretion and top-down communication described by Pateman (1970) as pseudo-participation. Through indirect employment control techniques including metrics, target setting, key performance indicators and 'bureaucratic rules [which] present alternative logics of control to professional autonomy' (Hodson, 2001), higher discretion work and professional autonomy are also increasingly subordinated to profits and managerially defined objectives (Mather and Seifert, 2014). A weakened or divided professional voice may fail to advocate alternative approaches to production or sound warnings over sustainability deficiencies. Moreover, degradation of professional authority may serve to undermine specialists' voices when offering evidence-based scientific arguments over global heating and its attendant effects (Oreskes and Conway, 2012).

A third challenge is that collective influence at work has been progressively undermined through the dominance of financial institutions in establishing criteria for productive objectives in both newly industrialising economies where global competition pressurises labour costs and in the West where individualism, precarious work and reduced union consciousness among young people impact negatively on collective voice for employees. Whilst transnational corporations can develop strategies and oversee operations from a central command centre, unions are locally based and are often forced by employers into multilateral competition rather than cooperation; competition that can have adverse environmental and employment implications. Further, when employers do recognise unions, they negotiate only over a restricted range of bargaining topics, which are unlikely to include major corporate direction changes. In consequence, in recent years established unions have had little impact on the organisation of work and have been unable to stem the rise in precarious work, though newer union entrants are attempting to mobilise membership and confront the worst excesses of non-conventional employment. These developments have an added negative dimension. Political parties that support working people tend to be funded through trade unions, and the lower the union membership, the fewer resources flow into sympathetic political parties. These are the same parties that are likely to identify with transitions into a greener and more sustainable economy, give support to healthy and secure employment and to offer policies that encourage union membership, recognition and supportive legislation. Despite these trends, unions across the world are expressing growing interest in protecting both their environments and their jobs, though, as noted earlier, the two campaigns are not always immediately compatible.

In previous chapters we outlined links between consumption, production and distribution and the adverse impacts upon both environment and work resulting from the international management mantra of 'act local, think global' (Bartlett and Ghoshal, 2002), which only too often translates into treating both workers and environment as local externalities, whose costs can be absorbed by impoverished host countries, desperate for internal domestic 'investment'.

Yet, what unites forests and mines, factories and distribution centres is that the work environment is central to broader ecological issues. This presents both challenges and opportunities. The challenge is that the frontier of control over work between employers and workers has shifted significantly toward the former. The opportunity is that workers and their communities are the ones most affected by the growing environmental crisis and whilst employers have taken marginal steps at best to offset their carbon emissions or polluting activities, unions and supportive groups are subscribing to JT and GND programmes aimed at averting the climate emergency. Some political parties are recognising the potential contribution that workers collectively can make to mitigating the impending crisis. Though the scale and intensity of challenges differ across sectors and countries, workers and, by association, unions face similar problems of employer intransigence, often helped through state and regional government complicity.

The weakening of trade unions

For worker interests to be converted into potential action, their common interests through collectivism need to be recognised, organised and mobilised. We have indicated the positive contribution that unions can make towards defending and progressing worker (and community) safety, health and welfare, whether in mature economies like those of Europe and North America, in the rainforests of South America and textile industries in Asia. In general, the extent and depth of union involvement in climate-change strategies will be linked to the social and economic role ascribed to them at the political level. At the extreme end of economic liberalism, authoritarian regimes deny legitimacy to independent unions, while free-market deregulatory policies in mature liberal democratic economies treat unions as an obstacle to economic progress. In these circumstances, unions may be tolerated, but their activities are heavily circumscribed. In Britain, since the Thatcher governments of the 1980s a succession of policies have been implemented to weaken trade union influence and action, including restrictive legislation, privatisation of strongly unified sectors (van Wanrooy *et al.*, 2013: 193; Arrowsmith, 2003) and outsourcing and contracting out public services (Huws and Podro, 2012) accompanied by encouragement and enforcement of workplace individualism.

Throughout the world, restrictive policies against unions predominate, union membership continues to decline and collective influence lessens. The most usual measure of union power is by union density (the percentage of a given workforce that are union members), and though corporatist economies like Denmark retain high levels of density, even here decline is evident, from about 80 percent in 1980 to around two-thirds in 2013. Japan shows a similar rate of density decline, from 31 to 18 percent between 1980 and 2012 (Kalleberg, 2018: 47). The overall adverse pattern is most marked in those countries identified as free-market orientated. In Australia, 2.5 million workers were in a union in 1976, and 1.5 million in 2016 and, over the same period, density plummeted from 51 percent to 14 percent. Density in the USA halved from 20.1 percent in 1983 to 10.5 percent in 2018, with most of the decline being accounted for by the private sector, where only 6.4 percent of employees were in a union in 2018, compared with 34 percent in the public sector (Bureau of Labor Statistics, 2019). Decline in UK membership is equally precipitous: from half the workforce organised in unions in 1980 to 23.5 percent by 2019 (DBEIS, 2020). Just over three-quarters of all members were aged 35 years and over, whilst just 4.4 percent of 16- to 24-year-olds were in a union (*ibid.*, 2020). However, union membership in the public sector has stabilised, an important finding, for as we discuss later, one approach to confronting the climate emergency is to expand public and cooperative ownership of key commercial sectors, while concomitantly ensuring that unions are provided with a central role in the strategic management of public and employee-owned enterprises. Surveys indicate that when offered the choice, as in the public sector, significant proportions of employees do join unions (Kalleberg, 2018: 189). Many young people may not be in unions because of the precarious nature of their employment, possibly exacerbated through employer opposition. Prior to the coronavirus pandemic, food and hospitality were among the fastest growing sectors in the UK, but union density was a mere 2.5 percent (Tait, 2017), attributable to a combination of small employer opposition, insecure work and reliance upon young workers (see Chapter 2). In Bloodworth's personal account of working undercover in Britain, a zero-hours care-worker colleague considered the options:

> People either love them or hate them [unions]. They either love them
> because they think they're going to stick up against bullying, or they're just

stirring little bastards that ruin companies ... I remember when I were seventeen, I was just like, 'Yeah, I'm not interested. What's the point of giving them an extra £2 a week?' Ten, fifteen years later it's like, they're worth their weight in gold, they really are.

(Bloodworth, 2018: 131)

So, do workers actually want to join trade unions, and if they do, are they being prevented from doing so? In other words, is there a gap between the influence workers have and that which collective representation could offer, what is often referred to as a (union) representation gap (Towers, 1997)? A number of studies have been conducted to examine worker views regarding union membership. A celebrated early series of coordinated studies of workers in six Anglo-American countries confirms (i) that unionisation decline is a common feature – outside of the public sector (Boxall *et al.*, 2007: 207), (ii) appreciable proportions of people working in non-union locations would want to be organised, though the relative strength of this ambition is difficult to gauge across different countries, sectors and organisational contexts, (iii) some workers want greater workplace influence whether through unions or through management-directed 'voice' initiatives, though modern HRM engagement techniques do not appear to have narrowed the desire for representation, (iv) management opposition to unions can undermine worker aspirations for more voice, (v) the role of government in supporting or suppressing collective ambitions is a crucial determinant of union membership (*ibid.*: 210–211), (vi) the union representation gap was wide among young workers in all six countries, which disputes popular assumptions of youth indifference or hostility to unions and (vii) most high skill, high pay private sector workers were fairly indifferent to unions, a finding replicated by studies of software engineers in Scotland (Hyman *et al.*, 2004) and skilled IT workers in India (Noronha and D'Cruz, 2009). Conversely, many professional workers (such as teachers) in Anglo-American public sectors are unionised (Boxall *et al.*, 2007: 207). A survey of 30,000 respondents covering 15 European countries by D'Art and Turner (2008) found that nearly three-quarters of respondents considered that employees need union protection, including 69 percent of non-unionised respondents and high proportions of professionalised and women workers. Four-fifths of under-25s expressed a need for unions. Even private sector workers express continuing support for trade unions, with 59 percent in a recent UK survey thinking they are necessary to protect working conditions and nearly half disagreeing with the statement that unions have no future in modern Britain (Tait, 2017).

These findings suggest that union membership has declined not so much through disinterest or opposition but rather through the structural conditions imposed by a combination of hostile legislation, globalisation, rapidly changing industrial landscape linked to decline in manufacturing and shifts to smaller, service orientated establishments, increased use of individualistic HR practices and the rise of the gig economy. Together, these factors serve to repress union membership and growth but not necessarily the values that underpin collectivism.

Further, the much-vaunted rise in individualistic attitudes among employees is also challenged by research (Peetz, 2010).

Union recognition in the contemporary economy

One thing is clear – without formal recognition by state or employers, unions will not be able to mobilise 'worker power' (Kalleberg, 2018: 188) to participate in the transformations required to mitigate against carbon emissions or influence employers to transition to products, production and transport policies that are less harmful to the environment. Securing recognition rights is therefore a priority, which under strong market-facing economic policies has proven not to be easy.

It is often overlooked that collective representation is recognised, albeit somewhat tenuously, as a fundamental human right both by the UN and the ILO. This poses an important question: if union rights are a human right, why is union recognition and membership under threat? One explanation has been that labour rights have been associated with ambiguous philosophies centred around freedom of association. In this respect, Hilgert's (2019: 517) warning is entirely apposite:

> the risk that now presents itself is the risk to the legitimacy of the idea of human rights for workers. Given the aggressive cultural opposition to unions, the United Nations failure to articulate a strong human rights-based philosophy for trade union rights could lead to workers viewing human rights as a largely irrelevant philosophy.

The equally significant problem is that many employers and economically liberal governments are taking precisely this oppositional view. In India, for example, liberalisation of the economy has led to significant expansion of contract work, with access to union membership blocked and accompanied by growth in anti-union strategies by employers and often given support by local government (International Commission for Labor Rights, 2013; Singh and Saini, 2016). In a speech to the ILO in 2010, the General Secretary of the International Confederation of Arab Trade Unions commented on the poor state of government-union relations in the region: 'The Decent Work Agenda still faces major challenges, especially in the form of serious violations of the fundamental Conventions. Trade union rights and freedoms in many Arab countries are still subject to government interference and persecution of trade union activists' (TUC, 2010), a situation that has not improved in the succeeding decade.

According to the ITUC (2020) report on worker rights, conditions for organised labour are deteriorating across the world with the number of countries excluding workers from union membership jumping from 92 in 2018 to 106 in 2020. Moreover, 85 percent of countries have violated the right to strike. Four out of five countries deny some or all workers the right to bargain collectively. Workers had no or restricted access to justice in nearly three-quarters of the 144 countries analysed in the Report. Moreover, 61 of these countries arrested and detained workers in 2020. Authorities impeded the registration of unions in well over half of the countries examined. Equally worrying is the violation of workers' rights in Europe: 38 percent of countries excluded workers from establishing a union, over half violated the right to bargain collectively and nearly three-quarters violated the right to strike (ITUC, 2020: 23).

The perils of organising: emerging economies

Many of the countries that restrict union recognition and bargaining rights are also active in depleting the ostensibly unproductive natural environment in pursuit of profitable agricultural products. Local actions such as deforestation contribute to global effects, specifically through reduced levels of carbon absorption, and union organisers and community campaigners often put themselves physically at risk in confronting the activities of mining companies, oil extraction, large-scale beef farms and agricultural projects. Governments dependent on incoming corporate investment, embracing economic growth or in protecting the property rights of major landowners may well act to suppress union activity, to the extent of putting campaigners in justifiable fear for their lives (ITUC, 2020).

Though problems facing activists and unions are similar in many countries whose economies are exposed to exploitation by internal or external forces, the case of Brazil is highly instructive because the Amazon rain forests, comprising some five million square kilometres, and that act as one of the world's largest absorbers of carbon, are under risk from two related phenomena: one is that diminished rainfall associated with climate change increases the risks of catastrophic forest fires, which in turn, release huge quantities of carbon into the atmosphere (Maslin, 2014: 111). The second, and possibly more immediate, threat is posed by aggressive intrusion into the rain forest's natural environment. As we discussed in Chapter 3, between 1990 and 2005, 80 percent of the forests cleared were linked to beef production, compounded by more clearance for soy plantations, whose products are fed in turn to livestock (Gustin, 2016). One study of America's major beef purchasers found nine of 13 companies, including giants like Burger King and Pizza Hut, lacked commitment to purchasing zero-deforestation beef. Even companies like McDonald's, which have adopted limited deforestation-free policies, had policy gaps that helped obscure sourcing, meaning that not a single company was able to guarantee deforestation-free beef (Union of Concerned Scientists, 2016).

Again, the non-virtuous patterns explored in Chapters 3 and 4 between cheap, insecure and McDonald-ised labour, meat processed for fast-food consumption, environmental – but profitable – vandalism and global climatic consequences become clear. But there is more: indigenous people, described by the Brazilian President as little more than 'zoo animals', are offered little protection from displacement and indeed, opponents of forestry and other incursive redevelopments face threats, assaults and murder from gangsters, often hired by landowners (Araújo, 2019). In addition, Bolsonaro has continued the attacks on trade unions and their capacity to defend workers (Fox, 2019); even before Bolsonaro, in 2017, at least 57 land and environmental defenders were killed and many more injured following protests against industrial scale agriculture, mining, plantations and cattle-ranching. Four-fifths of these casualties occurred whilst attempting to defend the natural Amazon environment (Global Witness, 2018). Few if any of the perpetrators have been identified, though government agents and security forces acting on behalf of farm, plantation and logging interests are suspected (Purdy, 2017). This is not an issue peculiar to Brazil: in Colombia, one of the most hazardous countries for unions, 34 unionists were killed in 2018 (ITUC, 2019), again followed by no action from the authorities.

Brazil is no stranger to direct action by workers. Mass demonstrations helped to remove the military dictatorship in 1985. Strikes by public sector and urban workers increased from 2010 as Brazil confronted the economic problems of being a dependent economy with austerity programmes following the 2008 financial crisis, peaking with over 2,000 strikes in 2013, the year of mass protests. However, from the point of view of defending the rainforests, numerous problems persist, exacerbated by the difficulties faced by unions and activists combatting neo-liberal economic policies under increasingly authoritarian governance. First, there has been little unity between dispossessed rural and urban citizens; second, virtual incorporation of the main union grouping into the then ruling Workers' Party helped to weaken union resistance and solidarity during the Lula periods of government (Purdy, 2017); third, the rights of indigenous communities have rarely been treated as a priority by politicians, established unions or the bulk of the urban populace. Finally, the importance of the agribusiness complex to Brazil's economy cannot be overstated and toleration, by the authorities, to any disruption is low.

Deforestation is reckoned to be responsible for about ten percent of global carbon emissions and therefore taking action to save the rainforests would have positive global consequences. However, indigenous communities, agricultural workers and their unions in countries like Brazil, Colombia and Indonesia are excluded by intransigent governments from decisions affecting their civic rights and in consequence the world faces a prospective global calamity, over which local agencies have little control. However, the decline in deforestation and carbon emissions experienced in Brazil in the ten or so years from 2000 demonstrated that environmental controls are compatible with socioeconomic progress. Initially, following a series of international negotiations, Brazil adopted a National Climate Change Plan, with ambitious and subsequently legislated-for commitments to significantly reduce emissions from deforestation, whilst protecting both communities and jobs. The integrated programme included closing illegal sawmills and providing designated protected areas for indigenous communities. At the international level, overseas aid was provided through the Norwegian International Climate and Forest Initiative, which offered compensation for reductions in carbon emissions (Boucher et al., 2013). Finally, Boucher et al. stressed the essential role of organised grassroots social movements, including environmental NGOs, unions and indigenous representatives in shifting governmental and business thinking and actions toward emissions reductions (*ibid.*: 443).

The lessons from the Brazil experience are complex. Clearly, unions can be, and increasingly, are marginalised under authoritarian regimes, as has been demonstrated beyond agriculture in countries as politically diverse as China and India and labour repression in developing countries is spreading, especially in the wide-ranging informal sectors (Wood, 2010; ITUC, 2020). Neo-liberal governments reject regulation of business activities to the benefit of inward-investing multi-national corporations and privileged national subjects. They also reject intervention by bodies supportive of workers, whose precarity is growing in emerging economies (Kalleberg and Hewison, 2013) and preclude progressive policies that threaten these privileges. As Rossman (2013: 62) observed: 'workers remain excluded from virtually all policy analysis of the crisis of the global food system'. Now, though, global circumstances have changed. Continued non-intervention in polluting activities at national levels can only contribute to subsequent global contamination. The lessons from Brazil show that unions acting alongside community interests can contribute to solutions at local and national level, which in turn may help to mitigate the planetary crisis presented by carbon emissions and global heating (Veiga and Martin, 2013). And, accumulating evidence indicates that protecting and restoring natural rainforests is a highly effective way of removing atmospheric carbon (Lewis *et al.*, 2019; Bastin *et al.*, 2019).

European initiatives

The challenges faced by labour movements in mature economies in moderating pollution are less perilous but possibly more entrenched than in industrialising countries. In many countries, two centuries of relatively uninhibited capitalist production have led to gradual accommodation and closer legal and societal acceptance of unions as legitimate bargaining and social partners. Nevertheless, as outlined above, the strength and vitality of unions across the developed world have diminished in recent decades, resulting in substantially reduced bargaining power, and critically, lowered political influence. Faced with the very real challenge of preserving legitimacy, union attention has been focused on attracting and retaining members, combatting a growing unitarist HRM-influenced managerial agenda, and in maintaining positive links with sympathetic political parties. As recently as the early 1990s, awareness of environmental issues was relatively low by both employers and unions (Eurofound, 1994) and was seen, if at all, as an establishment-level health and safety matter. Yet, across the world, union awareness and concern for environmental issues have undoubtedly grown, and supported by environmental voluntary bodies and NGOs, unions are developing constructive plans to combat environmental degradation, as well as preserve jobs.

The Paris Agreement potentially offers opportunities to trade unions for more strategic interventions in establishing, monitoring and maintaining decarbonisation efforts. Under the Agreement, countries are expected to submit periodic accounts detailing their success in meeting energy targets and the policies and measures adopted to accomplish these. At EU level, climate objectives are specified in the EU 2030 Framework, which identifies three principal objectives: a minimum 40 percent reduction of GHGs relative to 1990; a minimum 27 percent share of renewable energy consumption; and similar improvement in energy efficiency. Meeting these objectives 'requires a deep and rapid change of the way we produce, move and consume' (ETUC, 2018: 6). Equally, the policies adopted are expected to advance and protect job quality and security. The ETUC is insistent that there can be 'no just transition without workers' participation' and that workers must be involved in any policy process. The ETUC also adopts ILO guidelines for a just transition: 'consultation and the association of trade unions in the elaboration and implementation of low-cost policies at all possible levels and stages' (*ibid.*: 8). As ever, though, the thorny question of how unions are to participate is unresolved. From a Europe-wide survey of union representatives, the ETUC identifies a number of potential obstacles. First, trade union involvement generally and specifically in sustainability matters differs widely across the EU, though involvement in tri-partite bodies is common in those countries with established partnership arrangements, such as France and Germany, with sector level involvement deemed the most productive. Conversely, union representatives from the United Kingdom, Greece and Malta point out that they have not been involved in discussions over national long-term decarbonisation plans (*ibid.*;14). Their omission is vital as 'implementation of the Paris Agreement requires ambitious climate policy planning and design of mid and long-term decarbonisation strategies ... [and] is crucial in order to ensure a just transition for workers' (*ibid.*: 28). The ETUC recognises the obstacles facing unions. When asked about barriers to their involvement in the design of long-term decarbonisation strategies, respondents to the ETUC survey identified the lack of priority given by their respective employer organisations to green transition issues. Moreover, in many cases, respondents considered that their organisation was not well equipped to participate in discussions linked to decarbonisation strategies. Other surveys are indicating that organisations are simply not doing enough to involve employees in reducing carbon emissions (Carbon Credentials, 2018).

More generally, with the decline in union influence, instruments available to workers and their unions to use their voice to influence employer and governmental policies are restricted (Gumbrell-McCormick and Hyman, 2013). Further, there is little evidence of cross-national union collaboration: cooperation among European unions has largely been limited to exchange of information on collective agreements and developing common training programmes (Furåker and Bengtsson, 2013). A European Commission directive gave a boost to consultation and information provision in 2004 with its requirement for establishments with over 150 employees (50 from 2008) to establish a consultative committee (ICE) on request by a minimum of ten percent of employees. The committee is expected to deal with significant organisational issues, including the economic situation of the enterprise, employment prospects and major changes with implications for work organisation or contractual relations, though evidence to date suggests that a more passive and understated role has been experienced in the UK (van Wanrooy *et al.*, 2013) where two-thirds of consultative committees were composed of non-union members and the impact of the ICE regulations have 'proved peripheral, leaving wide scope for management inaction, of unilateralism, and for unenforceable and sub-standard consultation arrangements' (Hall *et al.*, 2015).

Whilst prospects for consultation appear to be questionable in Britain, experience in continental Europe has been more varied: in the more strongly defined systems associated with Northern Europe, formalised works councils operate alongside models of co-determination, in which informative and advisory processes are overlaid with elements of decision-making. Some 80 percent of companies in France have established works councils and Conchon (2015) points out that representatives can submit resolutions to AGMs of company boards, requiring shareholder consideration. Council members can also be delegated to attend board meetings. Works councils cover about 40 percent of employees in Germany and 80 percent in Sweden. As with consultation in Britain, larger enterprises are most likely to establish a works council. Notwithstanding their original three aims of regulating conflict between capital and labour, providing a societal stake and offering a voice for employees, tensions with employer performance priorities undoubtedly compromise these ambitions (Gumbrell-McCormick and Hyman, 2010).

It is also difficult to isolate the impact of works councils as they tend to operate systematically alongside employee representation on supervisory boards of directors in a co-determination procedure. The German model of worker directors has been the forerunner of European schemes and it is the German system that has been most comprehensively investigated. The research has mainly assessed contributions of worker directors to company performance and productivity, rather than to corporate governance (Scholz and Vitols, 2019: 3). Also, there are indications that their main inputs have focused on personnel issues (Carley, 2005). Recognising that 'the relationship between codetermination and company policies with respect to the environment and society ... is almost completely unexplored in the literature' and that research has tended to ignore unions and works councils as stakeholders in shaping enterprise policies (Scholz and Vitols, 2019), the authors developed a co-determination index designed to measure the strength of worker influence over this issue. Their quantitative study of sustainability policies of 96 German firms between 2006–2014 indicated that employee representatives tend to support policies that involve 'real changes to operations' and 'commitment of organizational resources', including targets for emissions reduction, publication of sustainability reports and commitment to employee security. Their findings also indicate that worker representatives are 'an important factor in explaining the spread' of sustainability policies in coordinated market economies like Germany. The authors recognise that these findings are indicative only and would benefit from deeper case study insights; however, it would seem that, in contradistinction to other forms of employee voice, there are genuine opportunities presented through the co-determination system for representatives to influence company policy.

While the ICE regulations and their European equivalents are directed at domestic establishments, since 1996 European Works Councils (EWCs) have aimed to provide an integrated system of representative participation in international enterprises with operations in Europe. Under EU regulations, companies with a minimum 1000 employees and 150 in two or more member states are entitled by agreement to establish information and consultation procedures covering all of a company's European operations. In August 2016, there were over 1,000 EWCs, covering about half of eligible companies. Most of the continent's largest employers, including those headquartered in America and Japan, have negotiated EWCs. Initial expectations for EWC influence were high, but research suggests that reality has failed to match the early promise. In a wide-ranging study, Vitols and co-workers (2011: 2) were charged by the European Foundation to study cases in which social partners successfully 'contributed and accompanied the transition to more energy efficient, low carbon-emitting ways of production or more environmentally friendly ways of workplace organisation'. The research examined case studies in different sectors across Europe, exploring whether social dialogue can contribute to environmental innovation or environmentally sustainable ways of work organisation; whether unions and employer organisations contribute to environmentally friendly ways of production and working; and whether companies can remain competitive with environmental innovation whilst maintaining employment. Finally, the enquiry explored the processes through which the social partners engaged in their efforts to become more environmentally sustainable.

The findings confirmed that 'above all, well-developed social dialogue structures and good social partnership facilitate projects that seek to green the economy', with positive examples drawn from both Germany and France, where these arrangements were strongly embedded (Vitols *et al.*, 2011: 50). Some of the findings demonstrated the uncertain state and even contradictions of environmental partnership. Though they found examples of active involvement by union representatives to develop good practice, many interventions were consultative rather than transitionary and aimed at plant level reform of, for example, health and safety practice or to use sustainability as a project to drive through efficient work practices. The authors acknowledge that it was too early to measure the actual impact of the various initiatives. Though most of these were voluntary, the research confirmed the importance of governmental support; however, at the time of the study, the authors point out that both sides to the employment relationship demonstrated joint opposition toward state-imposed policies. Even today, when the existential threat to the environment is well-recognised, positive concrete governmental actions, as opposed to exhortation and dubious target-setting exercises (Rankin, 2019), are still under-developed. In its report to Parliament, the UK Committee on Climate Change (2019: 8) pulled few punches:

> tougher targets do not themselves reduce emissions. New plans must be drawn up to deliver them ... Climate change adaptation is a defining

challenge for every government, yet there is only limited evidence of the present UK Government taking it sufficiently seriously.

Recommendations deriving from Vitols *et al.*'s (2011) sectoral case studies are instructive. Primarily, the study shows that change can take place within the parameters of effective and cooperative social partnership (a mechanism that is comprehensively under-developed within the Anglo-Saxon neo-liberal political economy). While state support was acknowledged as important, the partners downplayed legal enforcement, instead suggesting that while a participative framework can be state-endorsed, responsibility for dialogue rests with the social partners at national, sector and plant levels, though the contemporary weakness of the labour movement would clearly put them at a disadvantage when environmental action threatens to impact upon company profitability and potentially union members' jobs. Indeed, reservations have been drawn from a recent UK study, which concluded that organised workers can play 'a vital role in efforts to mitigate climate change and adapt to its effects', but with the caveat – if given the opportunity (Hampton, 2018). His study was limited to examination of union and TUC sources relating to environmental concerns and mainly to early union material up to about 2010, when only relatively minor workplace interventions were made, often with the focus on collectively determined savings or health and safety concerns as a means to legitimise union involvement. Further, many of these early interventions, though securing some environmental improvements in organisational practice, did not comprise the 'fundamental transformation' demanded by urban futures specialist Paul Chatterton: 'this is not a dress rehearsal; we don't get another chance with this' (Mayo, 2019). Hampton further notes that conditions for union engagement with climate issues have deteriorated in the UK with the Conservative government's anti-union legislation purported to promote economic growth policies, both of which objectives can conflict seriously with environmental protection priorities. Alongside employer resistance to union intervention in management's decision-making prerogative, government obstacles appear to be a major factor constraining union aspirations for closer involvement in shifting industry toward emission reduction policies.

Nevertheless, UK unions are taking steps to confront the compound challenges presented by the climate emergency. The TUC in 2019 demonstrated concern at the 'lack of a comprehensive just transition policy or coherent industrial strategy' by the government, which would ensure that new jobs needed in clean energy sectors and low-carbon production are at least equal in terms of pay, security, skill acquisition and retirement benefits to those they replace. The TUC also argues that affected workers must be involved in enterprise discussions over these issues. This, of course, would require a complete reversal of government policy from the past 30 years, whether in terms of support for processual collective bargaining or participative approaches in the European social partnership tradition. Earlier attempts by workplace unions in the UK to encroach on managerial terrain have met with little success. At Lucas

Aerospace, unions prepared a well-thought-out Corporate Plan aimed at utilising workers' technological knowledge, skills and jobs to diversify production away from military hardware to produce a range of socially useful and environmentally desirable products. Despite obvious benefits in protecting jobs and society, the plan gained only a mixed response from other unions (Cooley, 1980), and crucially a weak one from government and a highly negative one from management (Wainwright and Bowman, 2009). Reflecting on the undermining of the plan, Wainwright and Bowman consider that 'a renewed Green New Deal that involved such painstaking attention to grass-roots participation would be a worthy successor', if support were forthcoming. Revisiting the Lucas plan during a pandemic that has caused thousands of people to lose jobs whilst the economy demands products and services that prioritise the public good, Holman (2020) argues that it is 'an idea whose time has come'. Nevertheless, encouraging institutional investors to divest from socially and personally harmful – but profitable – enterprises faces major obstacles (Wander and Malone, 2007).

The tragic lesson of Lucas Aerospace is that mechanisms by which GNDs and JT policies can translate into reality are discernible but likely to face resistance from the same sources, even when the consequences of such resistance are unthinkable. Other UK labour movement initiatives intent on moderating management prerogative have either been traduced by political opponents, resisted by management interests or treated with suspicion by unions not wishing to collaborate with management, with whom they traditionally engage in oppositional negotiation, with the ill-fated attempt to introduce European-model worker directors in the late 1970s offering a pertinent example. Even when participative legislation was introduced, as with reserved places for union trustees on pension boards administrating investment funds on behalf of their members, their role became subsidiary to financial specialists and were opposed (successfully) in the courts when union trustees refused to invest in companies whose financial interests competed with those of their members, with a key case being the National Union of Mineworkers resisting attempts for its pension fund to invest in competing industries (Sackers, 1984). Even in Sweden, with its history of high union density and close union ties with the Social Democratic party, the Meidner plan to raise social ownership and control over industry through incremental transfer of equity to worker funds ultimately failed owing to sustained political and employer opposition (Minns, 1996; Cumbers, 2020).

These examples are not presented to denigrate current union efforts to protect both employment and the environment. Equally, it is important not to underestimate the significance of and obstacles to securing inter-union solidarity, establishing secure union links with other parties, including employers and the importance of gaining unequivocal and decisive political action, which, as we have seen, is not yet forthcoming, notwithstanding all the warnings. The level of uncertainty suggests that unions need to take the lead in proposing positive and concrete plans for collaboration whether at workplace, sector, national or international levels in order to obtain firm commitments for adopting sustainable work and environment. In order to accomplish this, as the 2019 TUC report states, workers must be involved in the key decisions 'at every level decisions are made'.

North American initiatives

Since the end of the Second World War, North America has experienced mixed and complex relations between unions and environmentalists, often influenced by popular – or populist – political direction. As the economy became increasingly liberalised in the 1970s, environmental activists were seen by many workers as elitists (Muir, 2007: xviii) with an agenda of impeding economic growth and hence worker earnings and security (Dewey, 1998). By the early 1980s, there were bitter disputes over territorial issues, such as over oil pipelines. According to some accounts, nearly half of AFL-CIO (see below) members abandoned the Democrats to vote in presidential elections in support of Ronald Regan, arguably attracted by his anti-environment and strongly pro-business stance. But this account fails to reveal the more intricate nature of relations between unions and environmental defenders between the end of the war and the 1970s, where instances of local pollution encouraged environment-union cooperation, often confronted by corporations and pro-business political interests. As early as 1948, when a 'killer smog' engulfed the steel town of Donora in Pennsylvania, killing 20 and poisoning thousands of others, industry-friendly Pennsylvania officials failed to launch a full enquiry, leading the United Steelworkers Union to call for a federal investigation against the United States Steel Corporation. As with the Bhopal tragedy many years later, no direct responsibility for the pollution was attributed to the company, despite considerable attention from the media and from other unions (Dewey, 1998: 47). From the 1960s onward, many unions drew attention to local pollution issues, such as water contamination, pesticide infiltration and campaigns for the preservation of wild spaces. With words that prefigure the anxieties of teenage climate activist Greta Thunberg 50 years later, there was no more eloquent statement of union fears than those expressed by a regional director of the United Auto Workers:

> Better we tear the factories to the ground, abandon the mines, plug the petroleum holes and fill the tanks of our cars with sugar than continue this doomsday madness … We demand that uncompromising and irreversible standards and controls be established to preserve our environment, no matter what the cost, no matter how great the violation of property rights, no matter what the effect on dividends and no matter what the effect on our own bold plans for collective bargaining.
>
> (Reported in Dewey, 1998: 56)

Although North American politicians have historically adopted rejectionist policies of varying vigour toward binding global regulation (Stevis, 2013: 179) this approach was moderated by Obama's Clean Power Plan in 2014. Stringent mandatory restrictions on GHG emissions were imposed on sources that together contribute 40 percent of US emissions, namely power plants and factories, accompanied by a range of other positive ecological initiatives (*Scientific American*, 2013), policies that were subsequently abandoned under Obama's climate-change denying successor (Stevis, 2019: 30). For American unions, as for those in other countries with abstentionist governmental policies, this has meant 'contested and often contradictory approaches to the environment and climate change' (Stevis, 2013: 179). The author identifies a number of general tendencies affecting the US labour movement: (i) policy pragmatism in the face of an absence of social dialogue with employers, (ii) manufacturing unions often have membership across different sectors, with differing affinities toward environmentalism and (iii) a spectrum of orientations toward binding regulation, alliances with environmental agencies and routes towards greening the economy. American unions have often been divided according to the anticipated impact of environmentally sound policies on employment prospects for their members, with a classic example provided by the bitterly contested Keystone XL Pipeline project actively supported by some unions who are also members of the BlueGreen Alliance between unions and environmentalists. But job preservation in the face of competitive and global pressures and underpinned by 'a notoriously hostile business sector and a largely neo-liberal state' (Stevis, 2013: 192) is an understandably major consideration for unions who know that state and federal support for environmental transition is traditionally dominated by growth considerations, upon which political prospects largely depend.

In his updated review, Stevis revisits the issue of whether US unions, 'declining in numbers and divided in climate policy', have addressed policy initiatives to address climate mitigation (2019: 3). US unions now operate in an economy where concession bargaining, a process in which unions surrender previously earned gains in return for increased security, has appeared 'with great intensity and frequency' (Chaison, 2012: 35) at the start of the twenty-first century, with airlines being at the forefront of this regressive movement. In these circumstances, unions tend to be submissive and not surprisingly, findings for ecological proactivity are equivocal. Stevis' research examines some 50 recent initiatives from which it is possible to distinguish the main tensions faced between union and environmental activism in the world's largest economy and one of its biggest polluters. The roles of both federal and regional governments are critical. Federal economic policy under the Trump administration operated in the context of what Stevis terms a hyperliberal state, in which deregulatory policies predominate, including downgrading of the Environmental Protection Agency and the message this conveyed to both sides of industry. Nevertheless, individual state support for environmental protection appears to vary according to political complexion and industry presence. Indeed, the intricacy of America's political economy is demonstrated in the tensions between central government and individual states, several of which are following climate mitigating policies, notwithstanding the federal government's withdrawal from the Paris Agreement and opposition towards the 2014 Clean Power Plan, policies that are expected to be reversed following the election of Joe Biden. At the other extreme, Stevis identifies fully hostile states to climate policy, which have also failed to take steps towards adopting renewable energy. In between, he identifies states that aggressively pursue 'renewable energy sources as part of an economic development policy – rather than a climate policy – to which they may be actually opposed' (Stevis, 2019: 31). Union policy follows similar complex and fragmented lines, ranging from those convinced that protective actions and renewable energy sources are necessary, to those like the powerful North American Building Trades Union (NABTU) who recognise the problematic implications of climate change, but are 'opposed to meaningful climate policy' (*ibid.*: 7). Stevis even considers that 'NABTU opposition has rendered the AFL-CIO silent or led it to support fossil fuel policies' (*ibid.*: 7). The American Federation of Labor and Congress of Industrial Organizations (AFL-CIO) is the largest union federation in North America, composed of some 55 major unions and representing upwards of 12 million members. Union divisions over responses to climate change are never far from the surface. Though its political orientation and policies tend to favour the Democrats, the Federation is clearly divided over the recently released Democrat-sponsored GND as being 'not achievable nor realistic' by members of the federation's Energy Committee. In an open letter to the GND's sponsors, the Presidents of the United Mineworkers of America and International Brotherhood of Electrical Workers, denounced the proposals: 'we will not stand by and allow threats to our members' jobs and their families' standard of living go unanswered' (Itkowitz, Grandoni, and Stein, 2019).

Sectoral representation is another important factor in helping to determine a union's stance toward sustainable production. Stevis notes that three unions, including the Mineworkers, sued the EPA over the Clean Power Plan. These unions are among many that have organic linkages to other unions, and their stated prime objective is to protect their industry, jobs and communities. On the other hand, many unions are internally fragmented with different divisions with members in potentially competing sectors and services, leading to cleavages in union policy orientation toward climate change. For example, the Teamsters Union have 23 fairly autonomous divisions, representing workers in different sectors, leading to different and potentially competing priorities. A union can represent workers in different occupations, leading to internal tensions. Within the Teamsters, electrical workers are leaders in renewables, but the union also represents workers in coal-fired utility plants, and is consequently a strong supporter of coal (Stevis, 2019: 24).

Differences in ideology and practice stem from unions' pragmatic need to protect the jobs of their members, and clearly some unions and their members have more to win or lose in any substantive shift to environmentally friendly production. A 'just transition' for one group of workers may disadvantage other groups. These differences and internal tensions create difficulties for large and diversified unions and their federations to adopt a unified stance regarding the environment and unity among workers will be vital if effective policies are to be implemented. In this they were not helped by the Trump administration, which denied the reality of the climate emergency, nor by corporations who can manifestly benefit from lax environmental standards, lack of regulation and union divisions. On an international scale, the problems are magnified through the USA's role as the world's biggest and most influential economy. Though many were horrified by Trump's unscientific and hostile stance towards the environment, there are political leaders who appear happy to follow similar ecologically dysfunctional and destructive directions.

A major question facing trade unions is whether to build alliances with environmental groups, and if so, with whom and with what objectives, bearing in mind that environmental and union aspirations may well diverge. Stevis (2019) notes a wide range of environmental support organisations in America, some with a broad agenda, others more focused; some conservative, others more radical; some concentrating on issues of local interest, others international in outlook. The most recent, established in 2017 in response to Trump's scheduled withdrawal from the Paris Agreement, is the US Climate Alliance, now supported by the governors of some 25 states covering over half of the US population. Its website (www.usclimatealliance.org) affirms commitment to reducing GHG emissions consistent with the goals of the Paris Agreement. The Alliance has three core guiding principles: recognition of the serious threats posed by climate change; a conviction that state-level action for clean energy and clean air can benefit economies and communities; and third, commitment to the Paris Agreement. Though unions are not specifically mentioned on its website, many of the projects established by the Alliance are undertaken by union members and major Alliance States, such as California and New York, are home to the strongest labour movements in the country.

The BlueGreen Alliance (BGA) is probably the best known and most active in bringing together concerns over work, equality and the environment. The BGA consists of a coalition of six environmental organisations and eight major unions, including the United Steelworkers. Stevis considers that the BGA fulfils a critical role because no unions have internalised environmentalism within their own policy documents. According to the BGA's website, its objectives are to reconcile environmental enhancement with stable economic growth 'in ways that create and maintain quality jobs and build a stronger, fairer economy'. The BGA's charter, *Solidarity for Climate Action*, sponsored by the United Steelworkers Union, sets out the visions, principles and policies by which it intends to achieve these objectives. Though principally an inspirational and aspirational document it nevertheless highlights concrete achievements, such as constructing offshore wind energy projects. Significant weaknesses of the new economy are challenged in the document: 'not enough of the new jobs that have been created or promised in the clean energy economy are high-quality, family-sustaining jobs, nor are these jobs in the same communities that have seen the loss of good-paying union jobs'. In other words, secure jobs have been displaced by unsustainable precarious ones, a significant feature of the 'hyperliberal' economy's operations (Kalleberg, 2018). The central aims of the charter are to reduce inequality and promote good jobs through offering strong support for union rights to organise throughout the economy, including clean energy sectors; commitment to raise union density; expanding public sector employment and providing training for clean jobs. The document also stresses the environmental havoc already visited on the USA through wildfires, droughts, and coastal erosion.

Obstacles to union intervention

Confronting the climate emergency clearly presents complex challenges for the union movement, raising questions of whether unions exist primarily to serve members' 'vested interests' or to act as 'swords of justice' in the interests of a wider community (Flanders, 1970). Many unions in developing countries face oppression in the face of government growth policies that also aim to encourage incoming transnational investment from corporations, whose outsourcing practices provide local jobs but rarely high-quality sustainable employment or independent trade unions who could contribute toward this.

Unions in mature economies have been in long-term decline in terms of membership and influence. At workplace level, unions are attempting to combat insecure and precarious work, though with only marginal success in the face of deregulated economic policies and accompanying repressive legislation, McDonald-ised work regimes and inflexible management. At national and sectoral levels, addressing environmental issues faced by workers and their unions can be detrimental to solidarity and clearly some workers will be affected more directly than others by environmentally positive policies: consequently, affected unions may be expected to act in defence of their membership even if this hinders the provision or success of sustainable development policies. The problem is, of course, that the emerging emergency is planetary-wide, but determining, let alone acting on, global policies to match the scale of the emergency has yet to be accomplished. Nevertheless, GND and JT policies increasingly recognise that collaborative efforts are required and that labour unions need to be included as part of that effort. Whilst unions can make rational stakeholder arguments for their inclusion, they must rely on support from political and other sympathetic agencies for this to happen.

At the workplace, notwithstanding the practical problems noted above, workers and unions can exercise leverage and have gathered experience in dealing with employers and, possibly for this reason, a number of environmentally progressive initiatives have been presented by and for unions. In his review of UK practice Hampton (2018) notes a number of union-sponsored schemes, often designed to familiarise workplace representatives with environmental issues and to indicate ways representatives may contribute through health and safety committees, cost-saving energy efficiency moves and other adaptation measures. His survey also found examples of negotiated agreements with some evidence of small-scale positive impact of union interventions, though the study by Vitols *et al.* (2011) regarded effects of social partner interventions as 'barely measurable'. A UK study by Campaign against Climate Change (2017) was based on interviews and document search of 17 prominent UK trade unions. Not surprisingly, all the unions expressed commitment to limiting a maximum 2°C rise in global temperatures. The report also points out that some UK unions, such as the Fire Brigades Union, are already being directly affected by climate change through flooding and moorland fires. Massive rises in summer temperatures experienced across Europe in 2018 placed huge pressures on emergency and medical services, often depleted through austerity measures. The 2017 study also confirms policy differences between unions in the UK; for example, one large union, UNITE, expressed concern with the government's 'dash for gas' approach, while others with members employed in the gas industry welcome such an approach. Divisions over fracking have been raised internally between unions' National Executive Committees and the lay membership.

Differences among and within unions were further exposed over airport expansion offering jobs to many union members but at potential environmental risk. A clear example of the environment-jobs tensions facing many unions is offered by the Community Union, created in 2004 through amalgamation between the steelworkers' union, ISTC, and the Knitwear, Footwear and Apparel union, both representing workers in sectors facing intense global competition. Though the union has now become a general union to attract members, its interests in steel-making continue and to retain a facility in Britain, has called upon the government to provide green investment for its remaining plant and to scrap the government-imposed carbon tax, which reduces UK competitiveness, especially against newer steel-making countries. The union supports Heathrow Airport expansion (along with a number of other major unions, again in support of jobs). It is estimated that an enlarged Heathrow would require some 370,000 tonnes of steel and would create 700 additional jobs (UKOOG, 2015). Especially controversial among unions has been hydraulic fracturing ('fracking'), to which the UK government has inclined. Some unions have shown concern over potential water pollution and methane leakage, coupled with potential health risks to workers exposed to dangerous chemicals, while other unions, such as the GMB, with members working in gas, have pronounced strongly in favour of fracking. The Campaign report also found that whilst environmental issues were frequently on the agendas of trade unions in Britain, they were not always treated as priority, constrained by more immediate threats to jobs posed by public sector services cuts, continuing austerity, and increased workplace automation and robotisation (*ibid.*, 2017).

In circumstances of government non-intervention, it would be difficult for trade unions to achieve a more proactive political contribution towards emissions strategy. With large employers also not making major interventions (Carbon Credentials, 2018; Dauvergne, 2018) nor seeking strategic union participation (Dobbins and Dundon, 2017), it is claimed that climate campaigning by some UK trade unions has become less committed and environmental activity by the TUC has reduced appreciably (Hampton, 2018). Though reports on the criticality of environmental degradation proliferate, it is worrying that the three dominant partners (government, employers and unions) to commercial activity appear to be adopting a 'business as usual' approach to the environment.

Concern about the lack of concerted action at the international level has also been articulated by the International Trade Union Confederation (ITUC), who voiced their calls at the 2017 UN Climate Conference for a more focused implementation of the Paris Agreement, phasing out fossil fuels and provision of funds for a JT to a low-carbon economy. The ITUC confirms the difficulties of securing unified action for national and sectoral unions in countries with different government and economic policies and systems of industrial relations. Certainly, the EU has made firm commitments to reduce GHG emissions; EU countries, whilst largely expressing commitment to EU targets, adopt different measures, according to differences in culture, traditions, priorities, and indus- trial relations systems, which present problems for coordinated action. So, Germany, recognised for its environmental awareness, is one of Europe's largest producers of GHG emissions, and has failed to meet UN targets for emissions reductions (see below).

Environmental participation at the workplace

If employees are to be more closely involved in strategic management decisions regarding the environment, the key question is how? There are obvious con- straints in that unions in many countries are no longer, or may never have been, regarded as social partners, whether by employers or by state. Second, the means, scope and depth of participation are limited and often determined by management priorities. Individual employee involvement schemes largely oper- ate at the behest of management and are aimed to serve management needs. A similar argument can be made for the various domestic joint consultation pro- grammes on offer. Collective bargaining at a time of labour weakness is usually restricted by managers to a limited range of topics. Legislation underpins board-level employee representation in 18 EU countries and though valued for its concern about corporate accountability, its impact on strategic decision- making is questionable (Gold and Waddington, 2019). Even co-determination has lost some of its original labour influence. When dealing with transnational enterprises, the range of options open to labour are further restricted as domestic unions have little scope to engage with international companies, which may establish competing sites to encourage union rivalry rather than collaboration. Moreover, TNCs may have complex and difficult to identify patterns of ownership and control, which in Europe has presented representa- tive problems for European Works Councils. When we consider the means by which employees and unions can contribute to strategic environmental matters, we are largely dealing with potential, or 'theoretical' (Gunderson, 2019: 39) interventions, which would require determined action by state or (highly unli- kely) voluntary steps by employers to be activated.

One proposal that has attracted attention is that of economic democracy. An immediate problem is one of definition. Indeed, commentators often use the terms industrial and economic democracy interchangeably (Foley, 2014). As Pateman argued in her seminal study 50 years ago, the concept of democracy is applied very loosely when referring to the workplace. She points out that democracy may be used as an expression of organisational culture or climate, with no reference to underlying authority and power relations (1970: 71), a claim reinforced by Archer (2010: 605), who considers that to enable genuine economic democracy, 'direct voice control' should be transferred to workers, for whom ownership is not a necessary requisite. Nevertheless, some commentators regard possession of shares by employees in their companies as economic democracy, especially if employees are granted high proportions of equity, such as with the John Lewis Partnership. But ownership of shares does not necessarily confer control: senior managers at the Partnership make strategic decisions (Cathcart, 2013; Foley, 2014: 74). At a more fundamental level, economic democracy is defined by one authority as offering direct governance by workers over the affairs of the enterprise (Archer, 1995), but more broadly by others to cover regulatory authority over the economy (Gunderson, 2019). In either case, democracy is notable by its absence, whether at work, where employees have little voice in the direction of enterprise affairs, or in the broader political sphere, where even in mature economies full political democracy is under threat with subversive attacks on civil liberties, reduced commitment to pluralism and a growing abrasive political culture (Cumbers, 2020). A further largely unseen but ubiquitous threat to democratic governance is posed by the ability of owners of mobile capital to undermine national government attempts to impose regulatory regimes by threats of capital flight (Pettifor, 2019: 80–81).

But how might economic or industrial democracy be implemented and practiced if it is to exert the necessary impact on environmental issues? Critics point out that most initiatives in workplace democracy have failed to disturb workplace authority relations (see e.g. Hyman, 2015; Streeck, 2016) leading Hyman to question whether democracy at work and 'financialized monopoly capitalism' can indeed coexist. The role of unions as representatives of workers and society writ large is vital and whatever form is implemented, governmental support to union participation at workplace and policy levels will be crucial in confronting the growing environmental emergency. In particular, effective economic democracy would have 'the potential to subordinate the economy to environmental goals' as well as confront questions of inequality and poverty (Gunderson, 2019: 40). In his important intervention for constructing an economy based on democratic principles, Cumbers (2020) fully recognises the obstacles presented by 'rapacious neoliberal capitalism'. He builds a comprehensive case for radical reform founded on three pillars, comprising individual economic rights over one's own labour; diverse forms of collective ownership, both of which are underpinned by the creation of 'deliberative and knowledgeable publics' founded on legislated rights to collective action, a genuine pluralist political community and state funding for political parties and their supportive institutions. Writing before the coronavirus pandemic, he notes that the foremost crises facing society today, namely, the climate emergency, economic inequality and democracy, in all of which the state is implicated, require action *from* the state for resolution, including radical strengthening of support for trade unions.

In an earlier study, Cumbers (2018) examines the positive effects of interaction of these pillars in the case of Denmark. The case illustrates the particularity of each country, for whilst being a small Nordic country of 5.8 million inhabitants, it had no oil deposits or hydro-electric energy resources and, until the 1970s oil shocks, was highly dependent on imported oil for is energy needs. From the 1980s, it was also economically decentralised, which helped in popular rejection of politically favoured nuclear power. The country has subsequently embarked on a programme of renewable energy, based strongly on wind power, a process which has created some 20,000 jobs (*ibid.*: 179). It is not just about what has been done but also *how* that is relevant. As Cumbers points out, environmental policy was driven by values of social democracy aimed at satisfying 'social need, the common good and solidaristic values' (*ibid.*: 180). The process, which as always with energy provision and potential for private profit has not been without tensions, has largely succeeded in phasing out reliance on fossil fuels and increased renewable energy consumption. Further, the process has relied strongly on the public engagement and knowledge formation pillar that informs the pluralistic, localised and cooperative forms of ownership and welfare orientation evident in Denmark. Cumbers contrasts the inclusive Danish approach with the UK government's decision to build new nuclear facilities, which 'involved almost no democratic accountability or critical public engagement' (*ibid.*: 192).

If, however, workers are to have opportunities to exercise direct governance over their organisations, this will not occur through voluntary sharing, let alone demission, by corporate executives, fearful of the impact on profitability, share price as well as impact on their own privileges. So active state intervention is required – and this can be effective. German worker directors were originally legislated for as a means to oversee the activities of Germany's post-war strategic industries. Despite their shortcomings in the present less supportive climate, there is no reason that an expanded and reinforced worker director system in those mature economies that have been most responsible for GHG emissions would not be effective in ensuring that larger enterprises take appropriate actions to mitigate their emissions. Though unlikely to find favour in the USA, such a system would also help safeguard the rights of employees to be provided with the training, security and decent conditions for transition to greener production and services. For these transitions to take place, governments must recognise the legitimacy of unions to be involved in these economic democracy exercises and to put in place the appropriate legislative apparatus to ensure that this occurs. Moreover, bearing in mind the urgency and potential implications of the climate crisis, at governmental level, tripartite bodies involving employers, unions and government agencies should establish appropriate guidelines and monitor performance. Under the original 1930s New Deal, union membership soared with governmental support, and similar accommodative approaches should be adopted to deal with the even bigger and wider emergency faced today.

A similar case could be made for enhancing the responsibilities of European Works Councils, which were originally designed to provide a cross-national platform for formal consultative and informational voice for employees in TNCs with operations in the European Union. EWCs are established through agreement between the parties and formally involve annual joint meetings between representatives and management, provision for consultation on management proposals with serious consequences for employees and representation from each national establishment covered by the enterprise. EWC representatives are also required to convey information to their establishment counterparts. Studies confirm that despite revisions to the system, EWCs have failed to meet the expectations of European unions and their sympathisers in influencing TNC behaviour (Conchon and Triangle, 2017; Waddington *et al.*, 2016) and it is widely acknowledged that these bodies have not realised their voice potential, operating more as management mouthpieces (Hann *et al.*, 2017). Problems include variations in practice linked to dominance of national systems of industrial relations taking precedence over European-wide procedures; influence of company headquarters and its country of origin; and agenda, objectives and proceedings controlled by unified management facing disparate representatives from different countries and employment relations traditions.

The 2008 global financial meltdown put these shortcomings into sharp relief as companies restructured and downsized operations with little obvious contribution or resistance from their EWCs. In 2009 the EWC Directive was revised following reservations from unions about the quality of information provided by management and weaknesses in consultation and training for members (Picard, 2010). There were regular complaints of problems for representatives in articulating between national and European levels of participation and in the timeliness of information provided for consultation purposes. Reactions of union commentators to the revision have not been favourable, with claims that changes have been minimal and that problems of navigating between national and European procedures persist. The quality, timing and extent of consultations continue to be a matter of concern (Waddington *et al.*, 2016). Managers, on the other hand, have faced few threats to their decision-making capacities, and have contrived to maintain control of the participation process and direct it into innocuous or informative channels (De Spiegelaere and Waddington, 2017). Moreover, less than half of eligible TNCs had established an EWC by 2015 and the numbers of EWCs established among the ten Eastern and middle Europe states admitted to the EU in 2004 and 2007 has been far lower, with just eight having their head offices in new member states. Nevertheless, EWCs have the infrastructural design to offer employee input into the challenges presented by the climate emergency. Many major international corporations in the metals, chemicals and airlines sectors already operate EWCs and with heightened supervisory powers these could ensure that their companies not only comply with emissions targets but also respond positively to employees' transitional needs. There is also scope for the EU to legislate for the establishment of compulsory EWCs and to provide them with mandatory powers to monitor company efforts to comply with designated emissions and pollution requirements to meet European emissions targets and to ensure that employees are not materially disaffected by changes to organisational production methods or to products and services.

Whilst not usually considered a form of economic democracy, collective bargaining between recognised trade unions and employers has been analysed as a forum for workplace democracy, notably by Hugh Clegg, who amidst considerable debate about work democratisation in the context of post-war nationalisation programmes, argued that 'joint regulation [i.e. collective bargaining] ... was the key to organizational participation' (Ackers, 2010: 62). Primacy of collective bargaining underpinned objections by many large UK unions to worker directors on company boards as proposed by the Bullock Committee on industrial democracy, whose ill-fated report was published in 1977. These debates took place at a time of considerable union involvement and influence at political, economic and organisational levels across much of Europe. Many social-democratic parties such as in Sweden were backed financially and constitutionally by strong and influential union federations. In Britain, the Labour Party was strongly supported by the TUC, which for many years was also involved in tripartite economic policy with employers and government. While the creation of nationalised industries did little to enhance industrial democracy, union membership was welcomed (and indeed informally enforced) so that in many of these economically and socially crucial sectors unions derived bargaining power from very high union density in an era of historically low unemployment. Indeed, many protective employment reforms (Redundancy Payments Act; Equal Pay Act; Sex Discrimination Act among others) that took place in Britain in the 1970s were attributable to union bargaining power.

Of course, the days of powerful unions and bargaining potency have long gone (Kalleberg, 2018: 45–48). Union density across 36 OECD countries slumped from 33 percent in 1975 to 16 percent in 2018 (OECD, 2019: 15) and as the OECD Report emphasises: 'collective bargaining can only contribute to labour market inclusiveness and have a significant macroeconomic effect if it covers a large share of workers and companies' (*ibid.*: 107). The proportion of workers covered by collective agreements declined from 45 percent in 1985 to 32 percent in 2017. In two-thirds of OECD countries, collective bargaining, where it does occur, is at the level of the firm, though sectoral bargaining is still prominent in mainly Northern European countries (*ibid.*: 16) an important factor as sectoral bargaining may help by 'facilitating job transitions and providing workers with the skills needed in a changing world of work' (*ibid.*: 107). Moreover, in bargaining systems with high co-ordination between different bargaining units, such as between sector and firm, there tends to be higher employment levels, more protection for vulnerable workers and lower wage inequality. The OECD Report makes a strong case for collective bargaining to protect against vulnerabilities, combat inequality and enhance the quality of the working environment, all of which as earlier chapters have shown are in a direct organic link between consumption, competitive commercial pressures and wider ecological concerns.

Trends of falling union membership, collective bargaining decline and precarious work are strongly shown in the UK where bargaining is now restricted to narrow channels of usually pay-related issues and collective discussions often instigated by management to advance cost-cutting and efficiency agendas. In the UK private sector, union recognition and collective bargaining have become minority practices and when negotiations do occur, the range of topics has been much diminished (van Wanrooy et al., 2013: 80). Nevertheless, there is a strong case for governments to support the revitalisation of trade unions and collective bargaining as an element of a JT. Many negotiable aspects of transition, notably ensuring secure, well-paid employment, equality of opportunity and offering opportunities for training for new responsibilities link well, at least in countries with a tradition of treating these topics through joint regulation. And, of course, collective agreements can provide the framework through which transitionary arrangements can be made. It will be recalled also that the ILO's Core Conventions include the right for all workers to be organised in unions and to have access to collective bargaining, an especially relevant factor, for example, for agricultural workers exposed to poor working and unhealthy conditions (Rossman, 2013).

The situation for unions in the public sector is generally more favourable, though privatisation policies have reduced the size of the sector and restrictive legislation, such as the 2016 Trade Union Act in the UK, has specifically targeted public sector unions. Nevertheless, union density has tended to remain relatively stable, even in free-market economies, in key sectors of the economy under government control (such as, in many countries, public transport) and collective bargaining remains an established practice. In the UK, a national study found that nearly two-thirds of public sector workplaces with union members negotiate over a range of topics (van Wanrooy et al., 2013: 85). Also, of course, the government is both *legislator* and *employer* and therefore in a position both to introduce progressive green policies whilst supporting good employment conditions. Offering support to unions also fits well with state or regional governmental remunicipalisation projects, in which unions are often collaborating partners.

The prerequisites for any union involvement

Interconnected patterns become clear: at the workplace, union representatives are unable to improve working conditions if their role is not legitimised or recognised by management; similarly, representatives and union officials cannot participate in discussions at enterprise or sectoral levels without recognition. Governments pursuing laissez-faire economic policies are unlikely to engage with unions at national or international policy levels. Representative participation, whether through negotiation or consultation, with the partial exception of works councils in Europe, is under long-term threat. Contemporary employee relations practice is dominated by direct (non-representative) employee involvement (EI) practices, usually instigated voluntarily by the employer, and consisting of communicating with employees or in inviting individual employees or teams to take responsibility for task-based activities under managerial supervision. There is also a parallel history of indirect representative participation (EP) such as joint consultation, often conducted through trade union channels. Some governments, possibly under pressure from labour movements or through sympathy for them, have passed legislation encouraging union membership and recognition, or by establishing specific structures in which unions are directly involved. This was the case in the UK with the Health and Safety at Work Act, which was passed in 1974 following a major chemical plant explosion that resulted in 28 deaths. The Act later established the right of recognised unions to appoint safety representatives, empowered 'to make representations to the employer on general matters affecting the health, safety and welfare of employees'. The success of the Act can be seen in subsequent dramatic and lasting reductions in both workplace fatalities and injuries. From this model, which has been replicated in many other countries, it can be seen that legislative intervention can be effective in supporting unions and managers to identify and resolve common problems affecting workers and the workplace.

Markey and his colleagues note in their review of environmental participation (2019) that organisations' GHG emission mitigation programmes can potentially implicate or affect employees in several ways: they may be responsible for implementing management-initiated programmes, for which worker cooperation is required. Work arrangements, skills and training may well be affected by changes that are being introduced, especially in the case of more fundamental JTs, where employee contributions to scheme design could be highly mutually beneficial, though ambitious high-level inputs may not be welcomed, as evidenced by the fate of the Lucas Aerospace plan. There has been little research into EI and EP with regard to sustainability, primarily because few organisations have embarked upon far-reaching decarbonification initiatives. Early studies tended to focus on links between environmental activities and companies' economic performance, with little attempt to define or locate the role of what has later become known as employee 'voice'. These more modest schemes were often assessed, by management respondents and interviewees, in terms of their contribution to cost reduction as much as to environment enhancement. One early study was conducted at Unilever, where interviews indicated that promoting individual environmental awareness based on effective communication was regarded as key, but with no mention of more strategic representative inputs (Hunton-Clarke *et al.*, 2002). A 2007 study of 110 Spanish factories surveyed the views of management respondents on the impact of environmental action and organisational competitiveness did find a positive link with individual or group involvement though neither the nature nor depth of the involvement processes was clearly specified (del Brío *et al.*, 2007).

Markey *et al.*'s (2019) telephone survey of 682 Australian organisations again relied solely upon input from managers but offers valuable insights. It indicates the potential for EI and EP to contribute to organisations' carbon remissions policies. The researchers found that representative EP forms are especially important, though their role is enhanced in organisations with dedicated environmental structures such as joint consultation committees and green representatives, though these were only found in a small proportion of companies investigated. They also found that increasing the range of EI/EP mechanisms broadened emissions reduction behaviours. The authors conclude by arguing that public policy aimed at reducing emissions would benefit from deeper participative initiatives by unions and employees.

The political and economic context

Views among national union federations and sectoral trade unions on policies to reduce CO_2 emissions vary according to perception of threats to industries and jobs as well as environment. In Western democracies, there are distinctions between liberal (LME) and coordinated (CME) market economies. To these we can add a distinctive third model, the so-called Nordic or Scandinavian approach, which is of especial interest through its system of social partnership, which aims to achieve high levels of economic performance with strong environmental protection by means of articulation between state, employers and active trade unions. In LMEs, state intervention in employment relations has served to encourage employers and employees to define and negotiate over their own interests with political intercession limited to remedying interpretations of imbalance, a policy dominated by concerns for maintaining the flow of free and flexible movement of labour in the service of economic growth. More extreme forms of liberalism, which formally or informally restrict union intervention, can be seen in a number of developing and post-command economies (Soulsby et al., 2017). Further distinctions to the LME model can be identified: for example, unions in the USA have followed a business unionism approach that has tended to focus on improving workers' terms and conditions within an accepted market relations framework, though when local communities and domestic lives are at risk, American unions effectively organise in their support. To mobilise wider union support for environmental controls, the need therefore is to demonstrate that global environmental damage emerges from specific spatial and sectoral sources and that global impact cannot be separated from the risk presented to those local communities that unions serve to defend. In the UK, wider aspirations to societal change have long formed part of the union narrative, though constant governmental and enterprise pressure in recent years have seen a narrowing in union ambitions as they grapple with austerity and growing precarity of employment (van Wanrooy et al., 2013: 100; Hyman 2018: 136). Despite determined calls for revitalisation of the union project, attempts to date have not met with spectacular success (Ibsen and Tapia, 2017). Though unions in LMEs have been instrumental in formulating policies intended to meet the challenges of climate change, responses from employers and deregulatory-facing governments have ranged from dismissive to obstructive. As many of these countries are also the highest polluters and carbon emitters, the need for centralised tripartite dialogue is critical, but dismayingly absent.

In contrast, CMEs aim for an 'institutional balance of power between labour and capital' (Wailes and Lansbury, 2010: 575) with the balance maintained through governmental superordination. Hence, in Germany, shared influence between employers and unions is provided through a government endorsed system of co-determination involving, in larger enterprises, worker directors, formalised systems of collective bargaining and enterprise works councils. Problems facing CME countries in reducing emissions through consensus are demonstrated by Germany. Following the 2011 Japanese Fukushima nuclear power catastrophe, Germany committed to phase out nuclear energy and revert quickly to renewable energy. Goldstein and Qvist (2019: 29–30) point out that Germany has doubled production of energy from renewables in recent years, such that by 2016, over a quarter of electricity production and nearly 15 percent of total energy production was attributable largely to wind and solar energy. Unfortunately, emissions reductions have remained stubbornly constant (*ibid.*: 30) and failed to reach targets amidst pressure from mining unions to adopt a slow phase-out of coal production and, rather than close down the sector, to introduce cleaner technologies to maintain its operation. Meanwhile, high carbon-emitting lignite mines continue to operate, contributing nearly a quarter of Germany's energy supplies (*ibid.*: 30) The largest lignite-based power plant is at Jänschwalde, burning on average 50,000 tonnes and emitting some 60,000 tonnes of CO_2 daily. Air-borne pollution from this single plant is estimated to be responsible for 650 deaths annually and a far greater number of serious ailments (*ibid.*: 32–33).

Even in a country as environmentally aware as Germany, difficulties in reaching consensus over energy policy are striking. Surveys suggest that a large majority of Germans want the government to stop coal production as soon as possible. Four-fifths of GHG emissions in the country are estimated to arise from burning fossil fuels, among which lignite is the dominant culprit. To reach Paris Agreement targets, the country must limit future emissions from lignite-based power plants, from which a quarter of electricity production originated in 2017. The government appointed a multi-interest commission to take responsibility for determining a policy to discontinue dependence on fossil fuel production. Though only about 20,000 people are employed in the lignite industry (compared with 340,000 in renewables), the sector is strongly unionised with active links to local political parties, an important consideration when regions with lignite mines face regional elections (Wecker, 2018). According to Kraemer (2018), 'preparing workers for the transition to a low-carbon economy with reduced CO_2 emissions is not on the agenda of German trade unions. Nor do they criticise German companies for high CO_2 emissions'. Kraemer comments that only one of the eight affiliates of the German Confederation of Trade Unions is a member of the German Climate Alliance, though a number of major unions have called upon the government to take proactive steps in managing transitions between reducing emissions and maintaining growth. There is evidence that this is happening: in their review of regeneration efforts in the once predominantly mining Ruhr region, Stroud *et al.* (2018) point to the roles of private companies and local government acting in partnership arrangements with unions and other stakeholders in shifting from traditional coal mining to renewable energy activities.

Nevertheless, doubts persist. A recent report examining Germany's response to meeting the UN's 2030 Agenda for Sustainable Development Goals suggests that environmental policy is not only operating too slowly but moving in the wrong direction. Hence, road and air traffic are given funding priority over socialised transport and German agriculture is accused of catering for global markets and paying less attention to agro-environmental protection. Energy transition has been slowed by the 'dismantling' of the Renewable Energy Sources Act and above all hangs the spectre of maintaining economic growth as a priority, accompanied by increasing inequality, heightened worker precarity and growing numbers of low-paid workers (Social Watch, 2018). Conversely, there have been positive developments involving the social partners. In 2016, the federal government adopted Climate Action Plan 2050, which establishes an emissions reduction target of 80 to 95 percent in 2050 from 1990 levels following broad consultations with social partners and communities embracing energy, buildings, transport, trade and industry, agriculture and forestry (ETUC, 2018: 31).

France was one of the main inspirations behind the 2016 Paris Agreement to limit global heating. In May 2019, an independent panel of 11 experts, the *Haut Conseil pour le Climat* (HCC) was appointed to oversee progress toward reducing emissions after failing to meet eight out of nine targets for 2017. According to the HCC, emissions fell by 1.1 percent against a targeted reduction of 1.9 percent. The main problems were identified as transport, where emissions have remained virtually unaltered for ten years, and from buildings. A proposed fuel tax was dropped following prolonged demonstrations from the so-called *gilets jaunes* in 2018. An election commitment to close coal-fired power stations by 2022 has also been dropped. The government has since committed itself to a net zero emissions target for 2050, following the UK's example. However, the Chair of the HCC harbours considerable doubts, indicating that while France's commitments are ambitious, they are unlikely to be met while actions to reduce emissions remain marginal to public policies. At the same time that the HCC with its consultative authority was established, President Macron also created the *Conseil de Défense Écologique* (Ecological Defence Counsel, EDC), whose members include the President, Prime Minister and other relevant ministers with executive powers to define guidelines and set priorities for environmental transition. The HCC is expected to liaise with the EDC and present annual progress reports on meeting emissions targets.

Setting environmental policy in France is highly centralised and politicised though any role for social partners is not immediately clear. Nevertheless, the French constitution does appear to allow for unions and employers to engage in national social policy and, despite declining density levels, French unions still enjoy considerable latent influence. They are recognised as joint managers with business delegates of the country's health and social security system and though union density is the lowest among mature European economies, standing at below eight percent, the labour movement retains tenuous policy influence through state support for industry-wide agreements. Nevertheless, recent labour law reforms gave support to decentralisation of collective bargaining to enterprise level, giving greater emphasis to company-level social dialogue. At workplaces with 50 or more employees, elected union delegates represent the interests of all employees, irrespective of union membership, on works councils and health and safety committees. By law, employers are required to consult regularly with elected union representatives on a wide range of issues and these would certainly embrace environmental policies. The indicative experiences of France and Germany demonstrate that CME underpinnings do offer scope for union involvement in environmental policy-making, certainly at sectoral and enterprise levels, but government policy in both countries is compromised by short-term expediency, which, notwithstanding formal compliance with the Paris Agreement, is affecting attempts to mitigate carbon emissions.

Corporatist social partnership is traditionally a defining feature of the Danish (Kalleberg, 2018: 44) and Swedish models, which are based on a number of fundamental premises: a flexible labour market with support available to assist adjustments to change; promotion of decent work; universal welfare policy 'from cradle to the grave'; and economic policy that promotes equality, openness and stability as well as progressive development (Bergman, 2015: 1030). Under the Swedish system, labour market conditions are largely regulated through collective agreements between the social partners and according to the Government of Sweden Office, unions have maintained high, though diminishing density levels, and 'often play a central role in the implementation of major reforms', including 'structural transformation', a feature regarded as being 'unique to the Nordic countries'. The country has experienced high levels of employment for men and women, even following the 2008 recession. JT in Sweden is an active policy, supported by effective retraining programmes, subsidised employment and unemployment insurance. In 2015, an investment company report placed Sweden as the most sustainable country in the world, ahead of 59 other countries, based on a range of environmental, social and governance factors. Other Scandinavian countries also featured highly (RobecoSam, 2018). Nevertheless, according to an authoritative commentator on the Scandinavian economy, Åke Sandberg, there have been important changes to the Swedish (and by extension, to the Nordic model generally) system; the most relevant from a sustainability perspective is that the period of austerity and conservative retrenchment has encouraged 'a backdrop to a political and ideological crisis within social democracy in Scandinavia' in which

> formal corporatism with open tripartite cooperation has been abandoned with the exit of the employer side and has been replaced by more informal, in part, secret ways of influencing political decision-making ... Compared to the former tripartite arrangements, the labour movement in this new

informal system is much weaker, with fewer financial resources and less media access than the capital side.

(2015: 1037)

Moreover, criticism toward the Nordic countries' emissions controls has also emerged from teenage environmental activist Greta Thunberg when refusing an award from the Nordic Council, explaining that 'the climate movement does not need any more awards. What we need is for our politicians and the people in power to listen to the current best available science' (reported in Graham, 2019).

While there may be some loosening of the partnership model, the Swedish government has commissioned an independent delegation to oversee progress and prepare an action plan to meet the SDG goals laid down in the UN Agenda. Trade unions have been involved in consultations with the delegation and provided input toward the action plan. Klein (2014: 179) argues that the ascendancy of sustainable green power in Scandinavian countries is attributable to its political embrace of social democracy, a system that 'extends the principles of equality and inclusion beyond the political realm ... into the social and economic realms so that ordinary people (non-elites) have more choice over how they live their lives' (Joshi and Navlakha 2010: 73). The authors suggest that social democracy is typified by a number of characteristics, all of which are fulfilled by Nordic countries. These include: a proportional representative electoral system: high union membership; strong social-democratic parties; and high tax revenues to fund public projects. Taking the example of Sweden, there is a 'strong belief that public social services should be established on the basis of democracy and solidarity to promote economic and social security as well as equal living conditions and active participation in community life' (*ibid.*: 79). Clearly, the relative success of the Nordic model in dealing with environmental issues gives strength to assertions that incorporating relatively minor socio-economic reforms will do little to unsettle corporate hegemony and dominance over governmental institutions.

Conclusions: what can unions do and what's stopping them?

A review of large UK unions showed that a number have appointed environment representatives, but these have no statutory backing and time spent on environmental issues is at the employer's discretion, unless the role is incorporated into formal health and safety representative duties. Nevertheless, there is little information on the impact of green representatives (Campaign against Climate Change, 2017). The ETUC (2018) report confirms that unions across Europe are taking steps to address the climate emergency through mobilisation campaigns, alternative action plans and coalitions with other partners. There is less evidence of involvement in actual policy, whether politically or organisationally, and a number of barriers have been identified concerning union involvement in long-term decarbonisation strategies. A significant barrier is presented by unions themselves: for example, while the IG Metall union in Germany is pursuing effective member mobilisation campaigns, it is less assertive about confronting GHG emissions from its core industries. Similarly, the Industrial Mining, Chemistry and Energy Union is resistant to calls to end lignite mining. The dilemma facing these is well stated: 'Job defence lies at the foundation of union politics. Where job defence comes in contact with industries that are seen as damaging and threatening to the environment is where unions' commitment to the environment is truly tested' (Snell and Fairbrother, 2010: 97).

A second major problem is presented by employers who are either unwilling or unable to confront the reality of the climate emergency (see Chapter 5). Dauvergne (2018: 55) makes the very credible claim that deregulatory economic policy *already* virtually hands environmental standards to major commercial extractors and producers who publicise their sustainability credentials whilst using their resources and influence to benefit their commercial interests, though doing little to address the crisis. He also argues that this approach is scarcely compatible with the interests of wider society whilst big business seeks to continue to 'extract, exploit, expand'.

In these circumstances, the findings from the 2018 ETUC report are no surprise: when representatives were asked about the main obstacles they faced, the most significant was the lack of priority given by organisations to transition issues while 60 percent questioned the ability of their organisations to participate in relevant issues. There are, of course, more fundamental problems for unions and workers. At an existential level, there is risk to life and limb for union activists confronting deforestation and related environmentally catastrophic commercial activities, with threats not just from proprietors and their representatives but from state agencies themselves, whose expected role is to uphold the law and protect citizens faced by law transgressors. In mature economies, we have noted the decline in union membership and recognition that affects workplace, organisational and political power. Unions, moreover, face everyday struggle with the cumulative impacts of years of austerity on actual and potential members. With the decline of unionisation, work relations have become increasingly individualised with human resource techniques that purport to offer individual control but whose effect is to consolidate that of managers (see e.g. Laaser, 2016). Lacking the status of employees, increasing numbers of insecure, freelance and low-paid workers, are often located in workplaces that directly or indirectly contribute to environmental problems, but are distanced from union membership owing to their non-employed status. However, there is some evidence of positive links between work-based employee participation and carbon emission reductions by companies. Of strategic relevance is that major generic programmes proposed to combat GHG emissions reductions, namely JTs and GNDs, specify the centrality of union involvement in programme design, application and monitoring.

Therefore, for unions, and for the planet, to have a collective voice heard at political levels is crucial. Many of them, individually or in coalition with other unions and environmental protective groups (such as Trade Unions for Energy Democracy) make representations to politicians, often supported by independent research, on the urgency of delivering the means to make transitions to low-carbon economies and the value to society of developing secure, satisfying and well-paid employment, but all too often these calls are drowned by advocates of continuous high growth and from those business interests who most benefit from it.

The reversal of policies and rhetoric that define and restrict unions as a problem to one where they may be offered positive steps to support government actions and targets through agreement or collaboration with employers would be an important step. Clearly, unions no longer represent the majority of employees in many countries, but history and experience confirm the contributions that unions can make to policy-making at enterprise, sectoral and governmental levels. We have reviewed their positive impact towards the health, safety and welfare of employees, contributing to the success of the original New Deal and involvement in contemporary GNDs. There is plentiful evidence from several countries that social dialogue between employers, governments and unions has served to deliver effective policy responses to a range of labour market challenges accompanying the Covid-19 pandemic (OECD-ILO, 2020). Additionally, unions help to educate members over links between consumption, production and global transport, giving support to JTs and GND manifestos, though as accumulated scientific evidence has shown, very little time remains. Immediate issues, of course, are to consider how workers may be effectively represented on decision-making forums, and a more profound one is to ensure that all governments position themselves to regulate the activities of major commercial corporations. We next consider how new technology is impacting on employment, the experiences of workers, and the environment.

Bibliography

Ackers, P. (2010), 'An industrial relations perspective on employee participation', in A. Wilkinson *et al.* (eds.), *The Oxford Handbook of Participation in Organizations*, Oxford, Oxford University Press; 52–75.

Araújo, H. (2019), 'Save the Amazon from Bolsanaro', *The New York Times*, 13 May.

Archer, R. (1995), *Economic Democracy: The Politics of Feasible Socialism*, Oxford, Oxford University Press.

Archer, R. (2010), 'Freedom, democracy, and capitalism', in A. Wilkinson *et al.* (eds.), *The Oxford Handbook of Participation in Organizations*, Oxford, Oxford University Press; 590–608.

Arrowsmith, J. (2003), 'Post-privatisation and industrial relations in the UK rail and electricity companies', *Industrial Relations Journal*, 34 (2); 150–163.

Bartlett, C. and Ghoshal, S. (2002), *Managing Across Borders*, 2nd edn, Boston, Harvard Business School Press.

Bastin, J-F, Finegold, Y., Garcia, C. *et al.* (2019), 'The global tree restoration potential', *Science*, https://science-sciencemag-org.ezproxy.st-andrews.ac.uk/content/365/6448/76. full.

Bergman, A. (2015), 'Book review symposium: Nordic Lights (2013)', *Work, Employment & Society*, 29 (6); 1029–1037.

Bloodworth, J. (2018), *Hired: Six Months Undercover in Low-Wage Britain*, London, Atlantic Books.

Boxall, P., Haynes, P. and Freeman, R. (2007), 'What workers say in the Anglo-American World' in R. Freeman, P. Boxall and P. Haynes (eds), *What Workers Say: Employee Voice in the Anglo-American Workplace*, Ithaca, ILR Press, Cornell University; 206–220.

BlueGreen Alliance (2019), 'Solidarity for climate action', www.bluegreenalliance.org/work-issue/solidarity-for-climate-action, June.

Boucher, D., Roquemore S and Fitzhugh E (2013), 'Brazil's success in reducing deforestation', *Tropical Conservation Science*, 6 (3); 426–445.

Bureau of Labor Statistics (2019), *Union Members Summary*, Washington DC, Department of Labor.

Campaign against Climate Change (2017), 'Trade unions in the UK: Engagement with Climate Change', London.

Carbon Credentials (2018), 'Carbon commitment report', https://info.carboncredentials.com/carbon-commitment-report2018.

Carley, M. (2005), 'Board-level representatives in nine countries: a snapshot', *Transfer: European Review of Labour and Research*, 11 (2); 245–248.

Casado, L. and Londoño, E. (2019), 'Under Brazil's far-right leader, Amazon protections slashed and forests fall', *The New York Times*, July 28, https://www.nytimes.com/2019/07/28/world/americas/brazil-deforestation-amazon-bolsonaro.html.

Cathcart, A. (2013), 'Directing democracy, competing interest and contested terrain in the John Lewis Partnership', *Journal of Industrial Relations*, 55 (4); 601–620.

Chaison, G. (2012), *The New Collective Bargaining*, New York, Springer.

Committee on Climate Change (2019), *Reducing UK emissions: 2019 Report to Parliament*, London, July.

Conchon, A. (2015), 'Workers' voice in corporate governance: a European perspective', *Economic Report Series*, London, TUC.

Conchon, A. and Triangle, L. (2017), 'IndustriAll European trade union: over 20 years of working with European Works Councils', *European Journal of Industrial Relations*, March.

Cooley, M. (1980), *Architect or Bee? The Human Price of Technology*, London, The Hogarth Press.

Cumbers, A. (2018), 'The Danish low carbon transition and the prospects for the democratic economy', in P. North and M. Scott Cato (eds), *Towards Just and Sustainable Economies*, Policy Press, Bristol; 179–194.

Cumbers, A. (2020), *The Case for Economic Democracy*, Cambridge, Polity.

Dauvergne, P. (2018), *Will Big Business Destroy Our Planet?*, Cambridge, Polity.

D'Art, D. and Turner, T. (2008), 'Workers and the demand for trade unions in Europe: still a relevant force?', *Economic and Industrial Democracy*, 29 (20); 165–191.

DBEIS (2017), 'Trade Union Act measures come into force to protect people from undemocratic industrial action', Department for Business, Energy and Industrial Strategy, London.

DBEIS (2020), 'Trade union membership 2019', Department for Business, Energy and Industrial Strategy, London.

De Spiegelaere, A. and Waddington, J. (2017), 'Has the recast made a difference? An examination of the content of European Works Council agreements', *European Journal of Industrial Relations*, 23 (3); 293–308.

del Brío, J., Fernandez, E., and Junquera, B. (2007), 'Management and employee involvement in achieving an environmental action-based competitive advantage: an empirical study', *International Journal of Human Resource Management*, 18 (4); 491–522.

Dewey, S. (1998), 'Working for the Environment: Organized Labor and the Origins of Environmentalism in the United States, 1948–1970', *Environmental History*, 3 (1); 45–63.

Dobbins, T. and Dundon, T. (2017), 'The chimera of sustainable labour-management partnership', *British Journal of Management*, 28; 519–533.

ETUC (2018), 'A guide for trade unions: Involving trade unions in climate action to build a just transition', www.etuc.org/sites/default/files/publication/file/2018-09/Final%20FUPA%20Guide_EN.pdf.

Eurofound (1994), 'Environmental protection in Europe – The effects of cooperation between the social partners', European Foundation for the Improvement of Living and Working Conditions, Dublin.

Flanders, A. (1970), *Management and Unions: The Theory and Reform of industrial Relations*, London, Faber.

Foley, J. (2014), 'Industrial democracy in the twenty-first century', in A. Wilkinson *et al.* (eds.), *Handbook of Research on Employee Voice*, Cheltenham, Edward Elgar; 66–81.

Fox, M. (2019), 'Brazil's labor unions prepare for war with far-right President Jair Bolsonaro', *In These Times*, www.inthesetimes.com/article/21770/jair-bolsonaro-war-on-brazils-unions.

Furåker, B. and Bengtsson, M. (2013), 'On the road to transnational cooperation? Results from a survey of European trade unions', *European Journal of Industrial Relations*, 19 (2); 161–177.

Global Witness (2018), *At what cost? Irresponsible business and the murder of land and environmental defenders in 2017*, London.

Gold, M. and Waddington, J. (2019), 'Board-level representation in Europe: state of play', *European Journal of Industrial Relations*, 25 (3); 205–218.

Goldstein, J. and Qvist, S. (2019), *A Bright Future*, New York, PublicAffairs.

Graham, F. (2019), 'Daily briefing: Rising seas threaten three times as many people as previously thought' *Nature Briefing*, 30 October, nature.com/articles/d41586-019-03336-3.

Gumbrell-McCormick, R. and Hyman, R. (2010), 'Trade unions, politics and parties: is a new configuration possible?', *Transfer: European Review of Labour and Research*, 16 (3); 315–331.

Gumbrell-McCormick, R. and Hyman, R. (2013), *Trade Unions in Western Europe*, Oxford, Oxford University Press.

Gunderson, R. (2019), 'Work time reduction and economic democracy as climate change mitigation strategies: or why the climate needs a renewed labor movement', *Journal of Environmental Studies and Sciences*, 9 (1); 35–44.

Gustin, G. (2016), 'Beef companies failing in effort to slow Amazon deforestation, study says', *Inside Climate News*, https://insideclimatenews.org/news/17102016/beef-companies-failing-effort-slow-amazon-rainforest-deforestation-climate-change-mcdonalds-burger-king-walmart.

Hall, M., Purcell, J. and Adam, D. (2015), 'Reforming the ICE regulations: What chance now?', *Warwick Papers in Industrial Relations*, 102, Industrial Relations Research Unit, University of Warwick.

Hampton, P. (2018), 'Trade unions and climate politics: prisoners of neoliberalism or swords of climate justice?', *Globalizations*, 15 (4); 470–486.

Hann, D., Hauptmeier, M. and Waddington, J. (2017), 'European Works Councils after two decades', *European Journal of Industrial Relations*, 23 (3); 209–224.

Heathrow Expansion (2018), 'Union leaders: We urge Parliament to take positive action on Heathrow expansion', www.heathrowexpansion.com/press/union-leaders-we-urge-parliament-to-back-heathrow-expansion.

Hilgert, J. (2019), 'Article 23(4) trade union rights and the United Nations policy of devolution on labor relations', *Labor History*, 60 (5); 503–519.

Hodson, R. (2001), *Dignity at Work*, Cambridge, Cambridge University Press.

Holman, K. (2020), 'The right to socially useful work', *Social Europe*, 6 November.

Hunton-Clarke, L., Wehrmeyer, W., Clift, R., McKeown, P. and King, H. (2002), 'Employee participation in environmental initiatives', *Greener Management International*, 40, Winter; 45–56.

Huws, U. and Podro, A. (2012), 'Outsourcing and the fragmentation of employment relations: the challenges ahead', ACAS, Future of Workplace Relations Discussion Paper, London.

Hyman, J. (2018), *Employee Voice and Participation*, London, Routledge.

Hyman, J., Lockyer, C., Marks, A. and Scholarios, D. (2004), 'Needing a new program: why is union membership so low among software workers?', in G. Healy *et al.* (eds), *The Future of Worker Representation*, Basingstoke, Palgrave Macmillan; 37–61.

Hyman, R. (2015), 'The very idea of democracy at work', *Transfer: European Review of Labour and Research*, 22 (1); 11–24.

Ibsen, C. and Tapia, M. (2017), 'Trade union revitalisation; where are we now? Where to next?', *Journal of Industrial Relations*, 59 (2); 170–191.

International Commission for Labor Rights (2013), 'Merchants of menace: repressing workers in India's new industrial belt', New York, ICLR.

Itkowitz, C., Grandoni, D. and Stein, J. (2019), 'AFL-CIO criticized Green New Deal, calling it "not achievable or realistic"', *The Washington Post*, 12 March.

ITUC (2019), *ITUC Global Rights Index*, Brussels, International Trade Union Confederation.

ITUC (2020), *ITUC Global Rights Index*, Brussels, International Trade Union Confederation.

Joshi, D. and Navlakha, N. (2010), 'Social democracy in Sweden', *Economic and Political Weekly*, 47; 73–80.

Kalleberg, A. (2018), *Precarious Lives: Job Insecurity and Well-Being in Rich Democracies*, Polity, Cambridge.

Kalleberg, A. and Hewison, K. (2013), 'Precarious work and the challenge of Asia', *American Behavioral Scientist*, 57 (3); 271–288.

Klein, N. (2014), *This Changes Everything*, London, Penguin.

Kraemer, B. (2018), 'Germany: Trade unions' approach to climate change policies', *Eurofound*, www.eurofound.europa.eu/publications/article/2018/germany-trade-unions-approach-to-climate-change-policies.

Laaser, K. (2016), '"If you are having a go at me, I am going to have a go at you": the changing nature of social relationships of bank work under performance management', *Work, Employment & Society*, 30 (6); 1000–1016.

Lewis, S., Wheeler, C., Mitchard, E., Koch, A. (2019), 'Restoring natural forests is the best way to remove atmospheric carbon', *Nature*, 568; 25–28.

Markey, R., McIvor, J., O'Brien, M. and Wright, C. (2019), 'Reducing carbon emissions through employee participation: evidence from Australia', *Industrial Relations Journal*, 50 (1), doi:10.1111/irj.12238.

Maslin, M. (2014), *Climate Change: A Very Short Introduction*, 3rd edn, Oxford, Oxford University Press.

Mather, K. and Seifert, R. (2014), 'The close supervision of further education lecturers: "you have been weighed, measured and found wanting"', *Work, Employment & Society*, 28 (1); 95–111.

Mayo, N. (2019), '"Sleepwalking" to climate disaster', *Times Higher Education*, 11–17 July, p. 9.

Minns, R. (1996), 'The social ownership of capital', *New Left Review*, 1/219; 42–61.

Muir, J. (2007), *My First Summer in the Sierra* (First published 1911), Introduction by R. Macfarlane, London, Canongate.

Noronha, E. and D'Cruz P. (2009), 'Engaging the professional: organizing call centre agents in India', *Industrial Relations Journal*, 40 (3); 215–234.

OECD (2019), 'Negotiating our way up', Paris.

OECD-ILO (2020), 'Social dialogue, skills and Covid-19', https://www.theglobaldeal.com/social-dialogue-skills-and-covid-19.pdf.

Oreskes, N. and Conway, E. (2012), *Merchants of Doubt*, London, Bloomsbury.

Pateman, C. (1970), *Participation and Democratic Theory*, Cambridge, Cambridge University Press.

Peetz, D. (2010), 'Are individualistic attitudes killing collectivism?', *Transfer: European Review of Labour and Research*, 16 (3); 383–398.

Pettifor, A. (2019), *The Case for the Green New Deal*, London, Verso.

Picard, S. (2010), 'European Works Councils: a trade union guide to Directive 2009/38/EC', Report 114, Brussels, ETUI.

Purdy, S. (2017), 'Brazil's June Days of 2013: mass protest, class and the left', *Latin American Perspectives*, 46 (4); 15–36.

Pyper, D. (2017), 'Trade union legislation (1979–2010), Briefing Paper No. CBP7882', House of Commons, London.

Rankin, J. (2019), 'EU climate goals "just a collection of buzzwords" say critics', *The Guardian*, 10 June, https://www.theguardian.com/world/2019/jun/10/eu-priorities-climate-buzzwords-critics

RobecoSam (2018), *Country Sustainability Ranking*, November.

Rossman, P. (2013), 'Food workers' rights as a path to a low carbon agriculture', in N. Räthzel and D. Uzzell (eds), *Trade Unions in the Green Economy*, London, Routledge; 58–63.

Sackers (1984), Cowan v Scargill 13 April 1984, www.sackers.com/pension/cowan-v-scargill-high-court-4-april-1984/.

Sandberg, Å. (2015), 'Book review symposium: Nordic Lights (2013)', *Work, Employment & Society*, 29 (6); 1029–1037.

Scholz, R. and Vitols, S. (2019), 'Board-level determination: a driving force for corporate social responsibility in German companies?', *European Journal of Industrial Relations*, 23 (3); 233–246.

Scientific American (2013), 'Barack Obama's climate change initiative', www.scientificamerican.com/article/barack-obamas-climate-change-initiative/.

Singh, H. and Saini, D. (2016), '*Private sector manufacturing organizations and workplace industrial relations: Towards a new employment relationship*', XVII Annual International Seminar Proceedings, January; 850–873.

Snell, D. and Fairbrother, P. (2010), 'Toward a theory of union environmental politics: unions and climate action in Australia', *Labor Studies Journal*, 36 (1); 83–103.

Social Watch (2018), 'Germany and the global sustainability agenda', www.2030reportde.

Soulsby, A., Hollinshead, G. and Steger, T. (2017), 'Crisis and change in industrial relations in Central and Eastern Europe', *European Journal of Industrial Relations*, 23 (1); 5–15.

Stevis, D. (2013), 'Green jobs? Good jobs? Just jobs? US labour unions confront climate change', in N. Räthzel and D. Uzzell (eds), *Trade Unions in the Green Economy*, London, Routledge; 179–195.

Stevis, D. (2019), 'Labour unions and green transitions in the USA: contestations and explanations', *Working Paper* 108, Toronto, ACW (Adapting Work and Workplaces to Respond to Climate Change).

Streeck, W. (2016), *How Will Capitalism End?*, London, Verso.

Stroud, D., Fairbrother, P., Evans, C. and Blake, J. (2018), 'Governments matter for capitalist economies: regeneration and transition to green and decent jobs', *Economic and Industrial Democracy*, 39 (1); 87–108.

Tait, C. (2017), *Future Unions*, London, Fabian Society.

Towers, B. (1997), *The Representation Gap: Change and Reform in the British and American Workplace*, Oxford, Oxford University Press.

TUC (2010), 'Arab trades unionists on trade unionism in the Arab world', tuc.org.uk/research-analysis/reports/arab-trade-unionists-trade-unionism-arab-world.

TUC (2019), *A Just Transition to a Greener, Fairer Economy*, London, Trades Union Congress.

UKOOG (2015), *Joint Charter on Shale Gas*, Dublin, United Kingdom Onshore Oil and Gas, June.

Union of Concerned Scientists (2016), 'Cattle, cleared forests and climate change', www.ucsusa.org/beefscorecard.

Van Wanrooy, B., Bewley, H., Bryson, A. *et al.* (2013), *Employment Relations in the Shadow of Recession*, Basingstoke, Palgrave Macmillan.

Veiga, J. and Martin, S. (2013), 'Climate change, trade unions and rural workers in labour-environmental alliances in the Amazon Rainforest', in N. Räthzel and D. Uzzell (eds), *Trade Unions in the Green Economy*, London, Routledge; 117–130.

Vitols, K., Schütze, K. *et al.* (2011), *Industrial Relations and Sustainability: The Role of Social Partners in the Transition to a Green Economy*, Dublin, Eurofound.

Waddington, J.Pulignano, V., Turk, J. and Swerts, T. (2016), 'Managers, BusinessEurope and the development of European Works Councils', *ETUI Working Paper*, 2016, Brussels.

Wailes, N. and Lansbury, R. (2010), 'International and comparative perspectives on employee participation', in A. Wilkinson *et al.* (eds), *The Oxford Handbook of Participation in Organizations*, Oxford, Oxford University Press; 570–589.

Wainwright, H. and Bowman, A. (2009), 'A real green deal', *Red Pepper*, 7 October.

Wander, N. and Malone, R. (2007), 'Keeping public institutions invested in tobacco', *Journal of Business Ethics*, 73 (2); 161–176.

Wecker, K. (2018), 'Germany's coal exit; jobs first, then the climate', *Deutsche Welle*, https://p.dw.com/p/2yob2.

Wood, G. (2010), 'Employee participation in developing and emerging countries', in A. Wilkinson *et al.* (eds), *The Oxford Handbook of Participation in Organizations*, Oxford, Oxford University Press; 552–569.

World Bank (2016), 'The cost of fire: an economic analysis of Indonesia's 2015 Fire Crisis', Washington DC, February.

9   The role of technology

Technology: threat or saviour?

Running like a thread through the arguments and debates concerning both the future of work and the future of our planet and its climate is the perceived role of technology – its design, the consequences of its use and the social values it incorporates.

For example, in our earlier examples of contemporary low quality, low paid and precarious jobs we noted the role of technology as an enabler of these new kinds of employment: internet-based retail has created the demand for both warehouse and distribution and delivery jobs, while web-based platforms enable on-demand taxi services and food delivery and the same technology has also made possible enhanced managerial surveillance of worker performance against ever more stringent performance targets. The most recent debates around technological change have focused on the potential for employment of current developments in robotics and artificial intelligence (AI), frequently giving rise to apocalyptic media headlines and dire predictions of a drastic shrinkage in those jobs still requiring human beings to do them.

In our examination of the increasing threat to climate stability we saw that the overwhelming evidence from climate science is that it has been the historical link between industrialisation and the use of fossil fuels as motive power for the technologies of extraction, production and transport that is the dominant cause of planetary heating. Following from this, environmental campaigners are agreed that solutions must be sought through development of alternative and sustainable technologies.

The crystal ball as technology

Debates about the socioeconomic benefits and costs of technological change have been a recurrent theme since the first water-powered mills displaced domestic spinning in the mid-eighteenth century. Ever since this first example of 'techno-shock', technology-enabled labour displacement has always generated both negative and positive scenarios. In the short-term, large scale job shedding in declining or outmoded sectors has historically created emiseration for whole occupations and communities and it is understandable that the prospect of technological change is seen with some foreboding by many sectors of the current workforce. Optimists argue, however, that to focus on this immediate effect is to assume that there is only a fixed amount of work to be done and that replacement of labour by technology will yield a zero-sum result: the machines will get a bigger slice of the economic activity cake but the cake will remain the same size. Realistically, they argue, we should adopt a positive-sum model as in the long-term increases in productivity stimulate the economy and create jobs, often in totally unforeseen sectors: the 'cake' gets bigger. However, even if true we should note that this is not quite comparing like with like: when occupations die the negative effects (loss of income, career, status and hope) are often concentrated in specific sectors (such as printing) or geographical localities, whereas the consequences of growth are usually more widely dispersed. To evaluate the usefulness of warnings of 'the end of work' or 'robotisation' we first need to understand the relationship between a given technology and the society that chooses to develop and use it.

When attempting to predict the development trajectory of any work technology and its likely interaction with the path of economic activity, we often have little choice but to start by observing and analysing current and emergent trends in technological, social and economic development and projecting them into the future. Yet the record of such attempts is rather chequered (Geels and Smit, 2000): change has gone in a different direction from that anticipated or been faster or slower than expected. To illustrate this, we simply need to review popular predictions about the coming 'Information Society' being made in the early 1980s. We find both gloomy forecasts about the 'collapse of work' (Jenkins and Sherman, 1983) and, conversely, talk of a coming bright 'Computopia' (Masuda, 1985); bold projections into the future were common – according to 'futurologist' Alvin Toffler (1981) between a third and half of the working population would be teleworking by the 1990s and we would only need a four-hour working day. Forty years later we know the outcome was neither all black nor all blue sky but rather a mixture of both. Admittedly these were, in the main, populist accounts rather than measured academic assessments but, like current alarms over AI and robotics, these are the stories that grip the public imagination.

Using the strategic advantage of looking with hindsight at these broad utopian and dystopian predictions, we see that many were based on the analytical fallacy of 'technological determinism'. Technological determinism is essentially the view that, firstly, technological development has marked historical 'progress' of our societies and should therefore be welcomed and encouraged and, secondly, this development will then have an inevitable 'impact' on the rest of our social and economic lives: technological development thus becomes both an imperative and a determinant. In this view, our agency is necessarily reduced to those behaviours that ensure speedy uptake of the new technological possibilities and any objections tagged as 'standing in the way of progress', usually with some (erroneous) reference to the Luddites of the early nineteenth century. In reality, the course taken by take-up and diffusion of a technological development *is* steered by human agency, and in most cases that agency is restricted to those in society with the social and economic power to decide what is 'useful' and what is a problem worthy of a technological solution. Not all problems get technology thrown at them: in our society we can conveniently get money from a hole in the wall while many of our fellow Earth citizens in the global South have trouble getting a clean water supply.

While avoiding such technological determinism we must also be wary of the view that sees technology *solely* as a social product: to say that technologies, once developed, have no effects beyond those that have been planned is equally delusory as the 'there is no alternative' view. While technologies are undoubtedly shaped by the processes of social and economic decision-making, they then, in turn, affect the policies of the users and the infrastructure associated with them (MacKenzie and Wajcman, 1987). While the initial path or trajectory a technology takes may be influenced by non-technological factors, once a step is taken for whatever reason at the time, further developments will often follow on from it in a pattern of irreversibility and path dependency, thus limiting future possibilities for choice (Beirne and Ramsay, 1992). Past mistakes in forecasting often stem from this failure to acknowledge this mutual interaction between the technical and the social.

We can say therefore that technological trajectories are constructed by societies in an interactive process between scientific innovation, social demand, the intervention of relevant actors, such as companies, governments or workers, and the responses of economic and legal institutions (Peláez and Kyriacou, 2008). If we see the ways in which new technologies are developed and adopted as hoped-for answers to current problems, then the *context* of their adoption and diffusion becomes important in understanding how they will be used. If the nature of our current economy stays the same (that is, there is no new economic paradigm), then we are still looking at a context where the private ownership of productive wealth requires the generation of surplus or profit. Pressures on profit margins and share prices make any systems that offer management enhanced control over the productive process increasingly attractive; similarly, as we shall see, the exchange and gathering of personal data from internet users is more likely to be monetised as a source of revenue than create the online democracy envisaged by the internet's early proponents.

Robots, AI, the internet and the degradation of work

What can we therefore say of working in a digital economy? When we talk of 'work' we can mean a number of things – the *amount* of work (employment levels), what *type of jobs* we will be doing (the occupational structure); how will that work be *organised* (patterns of working time, mobile working, teleworking, platform-based work) and, lastly, what will be the *experience* of work – its satisfactions, dissatisfactions, rewards and stresses?

Automation is quite simply the substitution of capital for labour, making a machine that will do work previously done by humans, and is hardly a new phenomenon. Historically it has taken place where task activities, whether manual or cognitive, are routine, repetitive or predictable. Current technological developments, however, are making increasingly possible the automation of non-routine activities. Robotics, for example, have the advantage over single function automation of task flexibility as they can be programmed, and reprogrammed, to do a range of tasks in a changeable sequence (Upchurch, 2018). Another focus of current debates and future speculation is on the rapid development of machine learning and AI, seen by many as taking over those activities, particularly mental and cognitive activities, which have been regarded as uniquely human – the ability to learn and to take innovative decisions.

These two developments together have given rise to a fresh wave of 'the end of work' predictions, reminiscent of those of the 1980s mentioned earlier. Frey and Osborne's (2013) conclusion that nearly half of occupations in the United States are at risk of automation over the next two decades has received much media attention, as has Susskind's predictions of a 'world without work' (Susskind, 2020). The economic journalist Paul Mason has taken this even further and predicted 'the end of capitalism' and market-based work, based on the diminishing marginal cost (and therefore diminishing profitability) of information (Mason, 2016).

Conversely, we note that a survey of 21 countries by the OECD (Arntz *et al.*, 2016) predicted that only nine per cent of jobs are at risk. The OECD analysis was based on the more realistic observation that a job comprises multiple tasks, only some of which may be replaceable and therefore, while jobs may change their task composition, they will not necessarily disappear. Further, Lloyd and Payne (2019) point out that generalisations about employment effects fail to take into account the societal context of innovation; in their comparison of the UK and Norway they found that Norway had a much more supportive environment for long-term investment, in which the country's high wage economy and strong tradition of employee rights in the labour market helped to give employer organisations and unions a common interest in automation in the context of a high-wage, high-welfare model. For the UK, in contrast, business short-termism and the availability of relatively cheap and flexible labour have contributed to a significantly slower rate of diffusion of robotics/AI.

One example of where there may well be significant job loss is in the back-office part of retailing and logistics. Here the huge warehouses and 'fulfilment centres' that supply both the high street and online customers *can* successfully use robotics for picking and packing. Given that current human holders of these jobs are in the low wage and low-skilled end of the employment spectrum the potential consequences for increased social and economic inequality are significant. However, while this is possible, its likelihood is limited in the UK by the societal context mentioned above and by a more fundamental critique of the 'march of the robots' scenario.

Using figures from the International Federation of Robotics, Martin Upchurch (2018) points out that three-quarters of robots are sold in just five countries: China (with 30 percent share), South Korea, Japan, the USA and

Germany. These patterns reflect the dominance or absence of specific industrial sectors, in particular automotive (35 percent of all purchases), and electrical/ electronics (with 31 percent). However, there are currently only 68 installed robots in China for every 10,000 employees in manufacturing, and the number of new robots installed in 2016 was 87,000 (the Chinese labour force is estimated to be 800 million). In terms of robot density (outside auto manufacture), South Korea led in 2014 with 475 robots per 10,000 employees, followed by Japan with 214, Germany with 181 and Sweden with 164 units. As Upchurch concluded, 'the reality is that robot density remains tiny'.

In a similar manner we can throw the cold water of reality over speculation concerning AI. Firstly, there has been a common failure in the media treatment of AI to distinguish between the two main types of algorithms: a rule-based algorithm is simply a detailed recipe for a computer to undertake prescribed actions, whereas a machine learning algorithm refines the machine's strategies for achieving a set goal by incorporating feedback on its own decisions – it learns from doing the job. This is what is most commonly referred to by the terms 'machine learning' or 'artificial intelligence'. The mathematician Hannah Fry has put the panic over AI 'taking over' from, and surpassing, human intelligence into the context of the current state of the science:

> For the time being, worrying about evil AI is a bit like worrying about overcrowding on Mars ... Frankly we're still quite a long way from creating hedgehog-level intelligence. So far, no-one's even managed to get past worm.
>
> (Fry, 2018: 13)

As Fry has noted, humans are really good at understanding subtlety, analysing context, applying experience and distinguishing patterns. We are really bad at paying attention, precision, constancy and being fully aware of our surroundings: 'We have, in short, precisely the opposite set of skills to algorithms' (Fry, 2018: 139).

In talking of emergent high tech or knowledge intensive sectors of the economy we should remember that this by no means implies that *all* jobs in those sectors will be empowered and knowledge-based nor, conversely, that all jobs will be automated. A work technology is rarely directed at all work and all workers in a given organisation in the same way – there may be core workers in the organisation whose skills are in scarce supply and whose long-term commitment is valued, while for other sections of the workforce the demands for flexibility and competitiveness may result in the technology being used for the intensification of work through the setting of ever more stringent performance and measurement targets and various forms of labour market flexibility and insecurity.

What digital technology has already enabled is the blurring of the old industrial distinctions between work and home and indeed the very idea of a fixed workplace. While it is true that millions of workers still leave home to go to 'the factory' or 'the office', it is also true that the last two decades have seen an expansion in types of work performed in non-standard locations. The end of the twentieth century saw an expansion in home working (Bryant, 2000; Mann and Holdsworth, 2003), while for many semi-professionals internet-based mobile devices mean that, even while at home, they are 'always on' and accessible to colleagues and customers (Douglas, 2020). Mobile devices mean that, for many, the workplace may be the car or the train, perhaps 'hot desking' every so often back at company premises (Hislop and Axtell, 2007). For our food delivery workers, couriers and Uber drivers, there is literally no workplace, with work tasks and performance standards being set, directed and monitored by their 'employer's' digital platform. These trends have significantly accelerated in the Covid-19 crisis during which it has been government policy in many countries to actively encourage employees to work at home, to the extent that, even with easing lockdown restrictions, many workers feel safer continuing to use the 'home-office' rather than risk the inevitable social contacts and interaction of the actual office. However, the ensuing flood of articles and journalistic comment predicting 'the end of the office' severely underestimates the difficulties in extending these practices. While journalists and well-paid media commentators are likely to have room in their home for work-space, this does not apply to the majority of white-collar workers who have to juggle the twin demands of time and space between work and domestic responsibilities.

One notable development, which incorporates many aspects of both home and remote working, has been the rapid rise of crowdsourced labour. Here, what could for example be the development of a complex software program, is broken down into minute sub-tasks, each to be performed by a separate individual; this has been described as a 'digitised assembly line' featuring thousands of discrete 'Human Intelligence Tasks' designed to be completed in seconds and rewarded in cents (Pein, 2017). As we shall see, this largely invisible aspect of the gig economy, while offering the usual claims of flexibility and worker control over the job, also has consequences for the experience of work, which could be interpreted as coming close to the complete alienation of the worker from both co-workers and the end product. This complete atomisation of labour is achieved by crowdsourcing the 'human cloud' – a huge transnational pool of freelancers who are available to undertake digital work on demand from remote locations. Platforms such as Amazon's Mechanical Turk or Upwork aim to match employers (or 'requesters') to the freelancers (or 'taskers') who bid against each other for a range of requested tasks, from writing code to accounting services. Employers thus get instant access to a huge pool of casual but dispersed labour, each member of which could be undercutting the others to get the job.

The platform that has received the most analytical attention is Amazon's 'Mechanical Turk', named after a chess-playing automaton of the same name that was the wonder of eighteenth-century Europe (but that actually concealed within it a chess master of restricted stature). Organisations who sign up to the MTurk platform can make an open call on the internet for bids for tasks ranging from data entry, statistical analysis, to image modification and tagging. The taskers (or, in Amazon-speak, 'Turkers'), each of whom is given an anonymised personal ID number, then bid for the job in terms of price or completion time. For Amazon and similar platforms such as MobileWorks and CloudFactory, despite the fact that Turkers can be working for more than 30 hours a week for an average wage of under \$2 per hour (Pein, *op.cit.*: 78) they are treated as independent contractors, in other words, self-employed. Such 'cloud workers' largely exist outside the scope of labour laws and labour market regulation; instead of off-shoring digital jobs to low-cost locations, now the employer has access to a global on-demand workforce that can be expanded or contracted at will with no overheads, sick pay, or national insurance costs (Bergvall-Kåreborn and Howcroft, 2014). Crowdsourced work can also be seen as the reverse side of the low-pay coin: two thirds of crowd employment workers in this study had a primary job and were undertaking crowdsourced work to supplement their income (*ibid.*: 220).

The work is promoted by Amazon and similar platforms with familiar promises of 'work from home', 'choose your own hours' and 'get paid for doing good work', all reminiscent of the early literature on 'telecottaging' in the 1980s (Bergvall-Kåreborn and Howcroft, *ibid.*; Baldry, 1988). Lehdonvirta (2018) found some evidence of taskers gaining some time flexibility but also that much time was spent hovering over their devices in case a new invitation to bid was posted. This kind of work is different from online freelancing where the workers are likely to be professionals and paid by the hour or on completion of a job. Crowdsourced labour is piecework on largely standardised tasks and more akin to a return to industrial (or pre-industrial) homeworking (Lehdonvirta, 2018). Unlike factory-based piecework there is no 'shop floor' or 'workgroup', and so no opportunities for either workplace socialisation or to develop those informal collective practices to restrict work intensity; there is similarly no chance of collective representation or collective bargaining, although there is some evidence that some taskers, particularly in the USA, have created their own online networks for exchange of information (Lehdonvirta, *ibid.*). The taskers are literally atomised – the majority have no contact with any others, have no idea what the product is that their contribution will be part of and, taken together, form another strand in the growing army of precarious labour.

As indicated above, the path taken by, and consequences of, technological change cannot be understood without reference to its socioeconomic context. As the seasoned analyst of technology and society, Judy Wajcman, points out, as long as the current economic system is more interested in profit than abstract notions of progress, we can expect these technologies to facilitate 'not less work but more worse jobs', an aspect that is largely ignored in most of the predictions on employment:

> They seem blind to the huge, casual, insecure, low paid workforce that powers the wheels of the likes of Google, Amazon and Twitter. Information systems rely on armies of coders, data cleaners, page raters, porn

filterers and checkers, subcontractors who are recruited through global sites such as Mechanical Turk and who do not appear on the company payroll.

(Wajcman, 2017)

Surveillance at work and at home

The capacity of digital technology for surveillance and capture of data about our behaviour and actions has always been there since the first word processors counted typists' keystrokes but, with the development of the internet, has increased exponentially in the past decade to the extent that Shoshana Zuboff (2019) has argued it can be seen as a further development of capitalism – 'surveillance capitalism'. As we use their services, Google, Facebook and the other tech titans regularly collect more data from us than they need to improve the services they offer; this 'behavioural surplus' is then used by Big Tech to produce *predictions* of behaviour that are the real product that they sell and the source of their wealth, power and influence. In addition to offering Big Tech our personal data as we access their services through our mobile devices, it is the spread of so-called 'smart' technology into our homes and even what we wear that has, according to Zuboff, created an almost limitless web of the monitoring of our movements, thoughts and deeds.

As she points out, given our occasional resistance when the incursions on our privacy become too blatant, the digital giants can always turn to the one place where there is a captive population – the workplace. Since the beginning of industrialism, employers have monitored the performance of their employees: even early socialist Robert Owen used different coloured blocks of wood to indicate performance levels of his workers at New Lanark. Originally performance management was by visual surveillance by a supervisor, but the arrival of digital technology towards the end of the twentieth century meant that such surveillance could be achieved remotely, away from the actual location of work. For example, studies of call and customer service centres have chronicled capturing data on agents' call times and the facility with which team leaders could covertly listen in to phone conversations to assess whether agents were sticking to the script and maintaining the correct 'mood' when talking to customers (Baldry *et al.*, 2007).

The more recent development of portable tracking and monitoring devices also offers huge opportunities to employers for gathering data on employees' performance and even mental and emotional state, such that wearing such devices is often a condition of employment. In Chapter 4 we encountered examples of technology as work surveillance. The van drivers studied by Moore and Newsome were subject to close monitoring and surveillance through hand-held Personal Digital Assistants (PDAs) that electronically tracked and traced the movement of not just the goods but of the drivers and vans (Moore and Newsome, 2018). James Bloodworth and his fellow pickers in the Amazon warehouse carried hand-held devices that tracked their movements and pick-rates and also received electronic warnings from line managers if their rate was deemed to be dropping too low. Amazon has been granted a patent for a wristband that tracks workers' location in their delivery centres and can also 'read' their hand movements, buzzing or pulsing if they are detected reaching for the wrong item. In 2017 the US company Three Square Market micro-chipped dozens of its workers (most of them voluntarily); the chips could be used to open security doors, log onto computers and to make payments at the company's vending machines.

In the US a proponent and developer of such systems has been Alex Pentland, Director of the Human Dynamics Lab at MIT. Zuboff sees Pentland as a major prophet and high priest of the vison of techno-utopia whose idea of a discipline of 'social physics' will solve most of society's problems, which he sees as stemming from 'incorrect' behaviour (Zuboff, 2019: 418). Among his commercial develop-ments are a number of platforms offered to corporations that are based on employees wearing 'unobtrusive' sociometers that measure things like commu-nications with peers and managers, voice tone and body language. The smart badges then convey behavioural data in order that action patterns can be judged for their 'correctness' and interventions made where necessary to change 'bad' action to 'correct' action (Zuboff, *ibid.*: 424–6). This apparently will enable employers to 'form a team of employees with harmonious social behaviour and skills'. For Pentland and his followers, the HRM mantra of 'people are our great-est asset' is simply wrong: people, unless their behaviour can be monitored and corrected, may be the cause of organisational incompetence. The major effect in terms of the direct experience of work would seem to be the deliberate creation of what industrial sociologist Alan Fox called a 'low trust' workplace.

Several European unions have expressed concern that, in addition to control over work allocation and performance exercised by digital platforms such as those described above, many key personnel and human resources functions are being incorporated into algorithms, including recruitment, the organisation of working time and disciplinary action. The Spanish union UGT is therefore campaigning for 'algorithmic justice at work' and requesting legislation that uses the EU's General Data Protection Regulation to ensure that decisions based on algorithms are explicable, accessible and reliable and that workers and their representatives should have rights to information and consultation over these issues (Varela, 2020).

Big Tech's dislike of democratic regulation

One thing that becomes very clear from the analysis of Zuboff and others is that Big Tech does not relish, and often goes out of its way to frustrate, any attempts at governmental regulation of its activities. The argument most com-monly advanced by the leaders of Google and Facebook is that technology is developing so fast and government, and indeed democracy in general, is so slow that there should not even be any attempt by political processes to regulate the tech companies: any problems that emerge will themselves be solved by tech-nology. Through their significant ability to wield political influence, particularly in the US, questions of privacy, information and anti-social content, have largely been left to the indistinct processes of 'self-regulation'. The companies themselves seem to have developed what Zuboff calls a 'cyberlibertarian' ideology, which, in its opposition to government regulation, has led Google for example to contribute to the funding of several far-right groups known for their anti-state agenda, including support for climate change denial (such reluctance to play by the same rules as everyone else is also reflected in the inventive financial structures Big Tech uses to avoid taxation). This anti-democratic undertone in Big Tech's social and economic pronouncements is reflected in the claim that governments should not bother to work out policies for current problems as the tech companies are smarter and faster at finding solutions (Zuboff, *op.cit.*: 105). (This assumes of course that the tech companies will identify the same problems as democratically elected governments.) In the arrogant 'techno knows best' standpoint we can detect the familiar 'inevitabalist' theme that follows from a technological determinist world-view and, in direct line from the information futurologists of the 1980s, the use of quasi-historical terminology in welcoming a new 'age', 'era', 'wave' or 'phase' of human development, which it would be foolish to obstruct. As a consequence, recent attempts by various governments to regulate either information privacy, or the employment conditions of platform-based work have had little impact.

Technology and the environment

We have seen how recent developments in internet-based technology have, rather than usher in an era of progress, enabled the worsening of employment conditions and security of work for many workers in many different sectors. An important aspect of current technological development, though one usually overlooked, is the parallel theme of this book: the energy requirements of particular technologies and their consequences for the climate. While the early water-powered textile mills used a renewable energy source, once the promise of steam power with its more regular supply and greater productivity had been realised, each new phase of technological change has made different demands on energy resources and, of course, when that power is steam power or coal and gas-generated electricity, we are talking of the energy locked up in fossil fuels.

Perhaps coming as a surprise is the fact that this is no less the case with digital internet-based technologies. The clean lines and silent running of digital technologies seem a far cry from the stereotypical smoking chimneys of fossil-powered industry, yet the invisible behind-the-scenes story of our online activities is that the servers and data centres that service the net worldwide consume significant amounts of conventionally produced energy and are consequently responsible for significant and growing amounts of carbon emissions. In order to provide reliability, data centres need to store the same information on multiple machines, which need to be constantly on and accessible and served in their turn by cooling systems. Worldwide, data centres currently consume about two percent of total electricity generated and this is estimated to rise to eight percent by 2030 (Carroll, 2020). A good example of the conflict of priorities this induces is given by Ireland's current economic strategy of becoming the data centre capital of Europe.

By early 2020 Ireland was hosting over 50 data centres with a combined power capacity of 642MW; others planned or in the pipeline will add another 600+MW (*ibid.*). This presents problems for Ireland to meet reduced emissions targets, particularly as they are, at the time of writing, one of the EUs worst emissions offenders. While most Big Tech companies operating in Ireland such as Google, Amazon and Microsoft, claim that their data centres will soon be supplied by renewable energy, the Irish Academy of Engineering (IAE) predicts that 30 percent of the projected increase in energy demand created by the expansion in data centres will still come from thermal generation and this will add at least 1.5 million tonnes to Ireland's carbon emissions by 2030 – about a 13 percent increase on present electricity sector emissions (Irish Academy of Engineering, 2019).

Representatives of the Big Tech companies have argued that the alternatives to the internet, such as transportation, are far greater sources of emissions; as a Google senior executive was quoted: 'It's much more efficient to move electrons than to move atoms' (Johnson, 2009). While this has a superficial, if glib, plausibility it is nonetheless a spurious argument: most internet traffic, via online searches, Facebook, Snapchat, or Instagram for example, is not an alternative to transport. If anything, it's a replacement for face-to-face social interaction. But don't worry – for the technological determinists all problems, including the climate crisis, are capable of a 'techno-fix'. Those who wish to deny the immediacy of the climate crisis share a great belief in the future development of 'planetary-scale and highly speculative Negative Emissions Technologies' (CAT, 2019). Suggestions for geoengineering the planet have ranged from dimming the sun's rays by spraying sulphur droplets to reflect light back into space, to designing carbon extracting machines to take billions of tonnes of CO_2 from the atmosphere (Klein, 2014: 258). The former has the problem that greenhouse gases will still accumulate under the sulphur blanket, while the latter has yet to be successfully developed anywhere, despite Bill Gates' financial support for the idea. Alas, swift as recent technological developments have been, environmentalists argue that the climate crisis is so urgent that we cannot place all our hopes of taking the heat out of the planet on the future development of some carbon extracting device that has yet to be designed, let alone constructed.

The promise of sustainable technology

Most climate-aware commentators however are not looking to Big Tech to save the planet but to the substitution of fossil-based energy with that derived from renewable sources – particularly wind, tide and solar power.

Over a decade ago David MacKay (2009) calculated the total energy demand of a typical European against the potential energy supply per head from renewables in order to answer the question 'could we live on renewables?' i.e. could we maintain our current levels of socioeconomic activity with sustainable energy alone? He did this from a purely engineering perspective, deliberately ignoring questions both of aesthetics (do we want beauty spots housing wind-farms or solar arrays?) and economics (rather than asking 'is wind cheaper than nuclear?' we should be asking 'how much wind is available?' and 'how much uranium is left?') (*ibid.*: 22). He calculated that the average British or European power consumption (in 2009) was around 125 Kilowatt Hours per day (125 KWh/d) per person (in the USA this would be about 250KWh/d), whereas the total contribution made by all renewables could amount to just over 30KWh/d per person. He came to the regretful conclusion that, even when adding up the most optimistic estimates of the contributions from all sustainable sources, it is unlikely that Britain or Europe could live on its own renewables – 'at least not in the way we currently live'.

These calculations, however, were based on figures available in 2008, since when the relative cost of renewables has dramatically improved faster than predicted: costs of solar panels have dropped, offshore wind turbines are larger and some can now float, making deep water arrays more possible and there has been a massive change in energy storage technologies. As a result, the Centre for Alternative Technology, in their comprehensive 2019 report, could conclude:

> Our work clearly demonstrates that we already have the tools and tech-nology needed to efficiently power the UK with 100% renewable energy, to feed ourselves sustainably and so to play our part in leaving a safe and habitable climate for our children and future generations.
>
> (CAT, 2019: *x*)

This is based on a target of net zero carbon emissions by 2030, which the authors claim is now the necessary date to aim for, rather than the UK gov-ernment's legislated goal of zero carbon by 2050. They reach their conclusion by looking at a range of measures that by 2030 would enable the UK both to 'power down' energy demand and 'power up' sustainable supply.

The two biggest targets for powering down our current energy demand are buildings and transport, each of which account for around 40 percent of energy use, and, in the case of transport, for 37 percent of UK GHG emissions (*ibid.*: 47). The British building stock is old and poorly insulated so that 25–30 percent of the energy used in the heating, cooling, cooking, lighting, providing hot water and powering electrical appliances in our houses, offices, shops and public buildings is wasted (MacKay, 2009). CAT calculate that demand for heating alone could be reduced by around 50 percent through measures such as having high 'Passivhaus' standards for new buildings and retrofitting existing buildings. While this is a low technology solution, it would be cost-effective and also create localised employment.

Reducing the amount of travel and changing methods of travel, with more use of public transport, walking, cycling, switching to efficient electric vehicles and two-thirds less flying – would reduce energy demand for transport by 78 percent. By 2030 the CAT model sees around 30 percent of road freight switching to rail, with HGVs, other heavy commercial vehicles and shipping powered by a mixture of electricity, carbon-neutral synthetic liquid fuel and hydrogen.

Much debate has been generated by the focus on the development of electric vehicles. A German economist, Sinn, has controversially argued that electric vehicles should not be seen as emissions-free as CO_2 is still created but at a remove – in the power station. As long as coal- or gas-fired power plants are needed to ensure energy supply during the 'dark doldrums' when the wind is not blowing and the sun is not shining, EVs, like conventional vehicles, run partly on hydrocarbons. And even when they are charged with solar- or wind-generated energy, enormous amounts of fossil fuels are used to produce EV batteries in China and elsewhere, offsetting the supposed emissions reduction (Sinn, 2019). This has, however, been refuted by several detailed transnational studies, which find that current and future life-cycle emissions from EVs are on average lower than those of new petrol cars – not just on the global aggregate but also in most individual countries. Even if future end-use electrification is not matched by rapid power-sector decarbonisation, the use of EVs almost certainly reduces emissions in most world regions, compared with fossil fuel-based alternatives (Knobloch *et al.*, 2020).

However, perhaps Sinn's critique of the supply chain for the production of sustainable technology should not be totally dismissed. As we saw in Chapter 4 with our example of the mobile phone, much modern technology depends on increased mining of metals such as cobalt and coltan. Sustainable technology is no different, as solar panels, batteries and wind turbines incorporate mined materials such as lithium, cobalt, copper and aluminium. Recent research on the ecological consequences of metals extraction found that mining areas around the world often threaten biodiversity conservation sites, not just by the mines themselves but also by the associated infrastructure including roads and power lines; increased demand for these metals could see expansion of mining operations into increasingly fragile environments (Sonter *et al.*, 2020).

Whereas a decade ago MacKay suggested that a sufficient carbon-neutral supply would have to include either nuclear generation or using other people's deserts for solar generation, the CAT model concludes that, having reduced demand, it will be possible to supply 100 percent of the UK's 'powered down' energy level with renewable and carbon-neutral energy sources, without fossil fuels and without nuclear. The UK, as a collection of islands in the path of the westerly wind-belt, is potentially well-endowed with different renewable energy sources. Offshore wind generation is currently the most promising (although on-shore turbines are easier to install, land wind speeds are generally lower) but the full sustainable mix would include solar, geothermal, hydro, tidal and biomass.

However, the key variable in a sustainable energy system is not the amount of energy produced but the stability of the supply. Energy demand is always variable, both during the day and through the seasons of the year and the current energy network has been designed to respond to this, with (fossil-fuelled) power generation capable of being increased or decreased to meet the peaks and troughs of demand. In a system reliant on renewables, however, energy supply is also variable but in largely unpredictable ways, dependent on the amount of sunshine or wind strength. Finding ways to balance renewable energy supply and even a 'powered down' demand, through a combination of energy storage and demand management, will be key to powering the UK on 100 percent renewable energy. The heatwave of August 2020 provided a reminder of this when output from the UK's windfarms, which had generated 30 percent of the country's electricity in the first quarter of the year, fell to 4 percent due to low wind speeds, causing the National Grid to start up a coal-fired power plant for the first time in 55 days; up to that point Britain had operated for over 60 percent of the year without coal.

For critics doubting the efficacy of renewables the proposed answer to reducing emissions is carbon capture and storage (CCS) at fossil fuel power stations or industrial plants. This itself has severe limitations, however, as not all GHG emissions can be captured at plant level, particularly as fossil fuels are becoming dirtier as reserves become harder to find. The storage usually proposed for carbon captured through CCS and other systems is usually old oil and gas fields (on land or under the sea), which must be monitored indefinitely to minimise leakage. This, like nuclear waste, implies unknown costs and effective risk management long into the future, effectively pushing the problem onto future generations. Such schemes also, of course, do nothing to reduce the CO_2 already in the atmosphere.

Waste is only one issue facing nuclear power, presented by some advocates as the 'safest energy ever' (Goldstein and Qvist, 2019: 87) but by others as inherently risky. Chernobyl chronicler Serhii Plokhy (2018) notes that whilst the West is recognised for the 'relative safety of its reactors', many are now being developed in less stable regimes with questionable regulatory and safety procedures. In the West, too, there are regular reports of safety breaches and leakages: one UK plant received a large fine for incorrect disposal of waste (Safety and Health Practitioner, 2013) whilst the same plant was fined for exposing a worker to radiation (Safety and Health Practitioner, 2019) and of course, as the 2011 Fukushima incident demonstrated, whilst nuclear power offers significant decarbonisation benefits, the potential for widespread radioactive contamination can never be discounted.

Although sustainable energy models such as that developed by CAT would seem encouraging, there remain significant obstacles to their full realisation. Some of these are economic: we have already noted the powerful economic lobbies that contest the whole notion of planetary heating and actively campaign against policies designed to ameliorate it. Although some major companies have responded to shareholders' requests for a 'green audit' of their activities, the world's financial major institutions, against all the scientific evidence, continue to invest in fossil fuels. Perhaps because of this, there has been a notable lack of urgency about the practical climate policies adopted by governments, including in the UK (see Chapter 6).

There are also signs of confusion in the public's mind, which may slow down a shift to a sustainable energy balance. It is perhaps understandable that proposals for new windfarms or solar arrays often meet strong objections if these look like being in areas of natural beauty or those seen as an asset by the local tourist economy. At a deeper level, however, it could be argued that a significant shift in psychological perspective and values will be necessary for the sort of radical changes required. Fossil fuel-based industrialism has supported vast changes in lifestyle and standards of living and a consequent expectation of continued consumerism. Changing norms and expectations concerning energy use would essentially involve changing our organisational and social life. At present, for example, while there may be growing acceptance of the case for replacing a petrol-powered car with an electric one, this only goes halfway towards rethinking our whole attitude to transport along the lines suggested by the sustainable energy modellers; the majority of people still expect to continue to have personal and individualised control over their means of transport. Attitudinal norms and values seem to be seldom considered in energy planning: a study of UK government energy plans and modelling found that very few predictive models made any reference to social norms as a demand-side variable (Hardt *et al.*, 2019).

Clearly, there is no single technology, policy or action that can prevent climate breakdown. It will require a concerted bringing together of all the above, plus some significant changes in how we organise ourselves for both work and leisure. We will return to this in later chapters.

Bibliography

Arntz, M., Gregory, T. and Zierahn, U. (2016), 'The risk of automation for jobs in OECD countries: a comparative analysis', OECD *Social, Employment and Migration Working Papers No. 189*, Paris, OECD.

Baldry, C. (1988), *Computers Jobs and Skills; the Industrial Relations of Technological Change*, New York, Plenum.

Baldry, C., Bain, P, Taylor, P. *et al.* (2007), *The Meaning of Work in the New Economy*, Basingstoke, Palgrave Macmillan.

Beirne, M. and Ramsay, H. (1992) 'Manna or monstrous regiment? Technology, control and democracy in the workplace', in Beirne, M. and Ramsay, H. (eds.) (1992), *Information Technology and Workplace Democracy*, London, Routledge; 1–55.

Bergvall-Kåreborn, B. and Howcroft, D., (2014), 'Amazon Mechanical Turk and commodification of labour', *New Technology, Work and Employment*, 29 (3); 213–223.

Bryant, S. (2000), 'At home on the electronic frontier: work, gender and the information highway', *New Technology, Work and Employment*, 15 (1); 19–33.

CAT (2019), *Zero Carbon Britain: Rising to the Climate Emergency*, Machynlleth, Wales, Centre for Alternative Technology.

Carroll, R. (2020), 'Why Irish data centre boom is complicating climate efforts', *The Guardian*, 6 January, theguardian.com/environment/2020/jan/06/why-irish-data-cen tre-boom-complicating- climate-efforts.

Douglas, I. (2020), *Is Technology Making us Sick?*, London, Thames and Hudson.

Frey, C. and Osborne, M. (2013), 'The future of employment: how susceptible are jobs to computerisation?', *Oxford Martin Programme on Technology and Employment Working Paper*, September, Oxford, Oxford University.

Fry H. (2018), *Hello World: How to be Human in the Age of the Machine*, London, Transworld.

Geels, F. and Smit, W. (2000), 'Lessons from failed technology futures: potholes on the road to the future', in Brown, N., Rappert, B. and Webster, A. (eds.) *Contested Futures: A Sociology of Prospective Techno-Science*, Aldershot, Ashgate; 129–155.

Goldstein, J. and Qvist, S. (2019), *A Bright Future*, New York, Public Affairs.

Hardt, L., Brockway, P., Taylor, P. et al. (2019), *Modelling Demand-side Energy Policies for Climate Change Mitigation in the UK: A Rapid Evidence Assessment*, UK Energy Research Centre, https://ukerc.ac.uk/publications/modelling-demand-side-policies/.

Hislop, D. and Axtell, C. (2007), 'The neglect of spatial mobility in contemporary studies of work: the case of telework', *New Technology Work and Employment*, 22 (1); 34–51.

Irish Academy of Engineering (2019), *Electricity Sector Investment for Data Centres in Ireland*, July 2019.

Jenkins, C. and Sherman, B. (1983), *The Collapse of Work*, London, Eyre Methuen.

Johnson, B. (2009), 'How much energy does the internet really use?', *The Guardian*, 14 May, www.theguardian.com/technology/2009/may/14/internet-energy-savings.

Klein, N. (2014), *This Changes Everything*, London, Penguin Books.

Knobloch, F., Hanssen, S., Lam, A. et al. (2020), 'Net emission reductions from electric cars and heat pumps in 59 world regions over time', *Nature Sustainability*. https://doi.org/10.1038/s41893-020-0488-7.

Lehdonvirta, V. (2018), 'Flexibility in the gig economy – managing time on three online piecework platforms', *New Technology, Work and Employment* 33 (1); 13–29.

Lloyd, C. and Payne, J. (2019), 'Rethinking country effects: robotics, AI and work futures in Norway and the UK', *New Technology, Work and Employment*, 34 (3); 208–225.

MacKay, D. (2009), *Sustainable Energy: Without the Hot Air*, Cambridge, UIT Cambridge.

MacKenzie, D. and Wacjman, J. (eds), (1987), *The Social Shaping of Technology*, Milton Keynes, Open University.

Mann, S. and Holdsworth, L. (2003), 'The psychological impact of teleworking: stress, emotions and health', *New Technology, Work and Employment*, 18 (3); 196–111.

Mason, P. (2016), *Postcapitalism*, London, Penguin Books.

Masuda, Y. (1985), 'Computopia', in T. Forester (ed.) (1985), *The Information Technology Revolution*, Oxford, Blackwell; 620–634.

Moore, S. and Newsome, K. (2018), 'Paying for free delivery: dependent self-employment as a measure of precarity in parcel delivery', *Work, Employment & Society*, 32 (3); 475–492.

Pein, C. (2017), *Live Work Work Work Die: A Journey to the Savage Heart of Silicon Valley*, New York, Metropolitan Books.

Peláez, A. and Kyriacou, D. (2008), 'Robots, genes and bytes: technology development and social changes towards the year 2000', *Technological Forecasting and Social Change*, 75; 1176–1201.

Plokhy, S. (2018), *Chernobyl: History of a Tragedy*, London, Penguin.

Safety and Health Practitioner (2013), 'Nuclear firm receives huge fine for incorrectly disposing of radioactive waste', www.shponline.co.uk/chemical-hazards/nuclear-firm-receives-huge-fine-for-incorrectly-disposing-of-radioactive-waste.

Safety and Health Practitioner (2019), 'Sellafield investigation: £380k fine after safety breach', www.shponline.co.uk/news/sellafield-investigation-update.

Sinn, H-W. (2019), 'Are electric vehicles really so climate friendly?' *The Guardian*, 25 November, www.theguardian.com/environment/2019/nov/25/are-electric-vehicles-really-so-climate-friendly.

Sonter, L., Dade, M., Watson, J. *et al.* (2020), 'Renewable energy production will exacerbate mining threats to biodiversity', *Nature Communications*, 11, 41–74, https://doi.org/10.1038/s41467-020-17928-5.

Susskind, D. (2020), *A World Without Work: Automation and How We Should Respond*, London, Allen Lane.

Toffler, A. (1981), *The Third Wave*, London, Bantam.

Upchurch, M. (2018), 'Robots and AI at work: the prospects for singularity', *New Technology, Work and Employment* 33 (3); 205–218.

Varela, J. (2020), 'For a law of algorithmic social justice at work', *Social Europe*, 9 July.

Wajcman, J. (2017), 'Automation: is it really different this time?', *British Journal of Sociology* 68 (1); 119–127.

Zuboff, S. (2019), *The Age of Surveillance Capitalism*, London, Profile Books.

10 Towards a just transition

Introduction

Historically, periods of technological change and economic restructuring have invariably been to the benefit of some but the detriment of many others, as whole classes of jobs have no longer been required, or the geographical focus of economic activity has shifted from one region or country to another. The current combination of the climate crisis with increasing structural inequalities in national and particularly global economies make this possibility even more likely. This threat is why many of the major campaigns for environmental policies have adopted the concept of just transition (JT).

The concept has a surprisingly long history, with early formulations traceable to the 1970s and given impetus through the work of American labour activist Tony Mazzocchi in the early 1990s in a campaign for redesigned jobs for his members in the fossil fuel industries, and thereafter associated largely with union concerns that GHG emission mitigation policies should not be enacted at the expense of those who work in emitting sectors. Mazzocchi later collaborated with environmentalists, an unusual move for a union official in the USA at the time, as unions tended to be deeply suspicious of environmental objectives. These collaborations developed the argument that combatting environmental degradation called for a 'just transition' enabling workers to shift to non-polluting employment. The concept has been taken up and broadened by climate campaigns as 'the twin crises of inequality and the biosphere feed one another' (Laurent, 2020) and puts social justice at the heart of policies necessary to combat climate change, a goal best summed up in the phrase 'nobody is left behind'.

The concept has rapidly moved beyond advocacy by the green fringe of campaigners, adopted as policy by both the European Trade Union Confederation (ETUC) and the International Trade Union Confederation (ITUC) who are adamant that, in combatting the climate crisis, 'workers must not pay the price with their jobs and livelihoods' (Voet, 2020). In 2016, the two bodies launched the Just Transition Centre and the President of the European Commission, Ursula von der Leyen, has since set up an EU 'Just Transition Fund' with (albeit modest) funding of €4.8bn. As the concept of JT usefully brings together the two concerns of this book, and for the first time expresses them in proposals for socio-economic policy, we will adopt it as the framework for this and subsequent chapters.

However, JT 'is a concept defined in multiple and contested ways ... for often different and competing political purposes' (Snell and Fairbrother, 2013: 158). It can be quite narrowly interpreted as a means to ensure that workers in threatened industries, over whose activities they have no control, are adequately protected and compensated in terms of alternative comparable employment. Broader interpretations go beyond work reformulation and challenge absence of worker agency by incorporating the *justice element*, providing worker, community and unions with closer involvement in transition decisions. More progressive interpretations propose that unions be centrally involved in planning arrangements alongside government bodies and employers, or further, that alternative economic models to conventional neoliberalism should provide the foundations for the transition process (*ibid.*: 148–149). The more ambitious models are associated with Green New Deals and emphasise and address the interrelated crises of societal and international inequality and environmental degradation.

A further aspect is whether the JT approach should specifically target a particular sector associated with high emissions, which might also be related to a defined region, for example, coal-mining in the Ruhr valley (Gärtner, 2019). Alternatively, broader and more societal interpretations can be applied at national or even supra-national levels, as evidenced by the EC's recent decision to commit JT funds over the next seven years to support European regions still heavily dependent on fossil fuels to integrate affected industries and jobs into the European green agenda. COP 21 in 2015 offered a global perspective to JT with its declared and agreed imperative to apply 'a just transition of the workforce and the creation of decent work and quality jobs in accordance with nationally defined development priorities'. This accord was followed by the ILO's operating guidelines toward JT, which consist of ten points specifying transitional actions needed to move toward a greener economy (van der Ree, 2015). As the limitations of 2019 COP demonstrated, translating these admirable intentions into actual policy is proving a difficult challenge. The first ILO point is to 'send a strong political signal' of transitional shift and to involve workers and employers in the process. In the years following the Paris accords, few governments have embarked on this process. Little action has been taken towards promoting inclusive social dialogue between workers, employers and government or developing environment-friendly economic growth policies. The guidelines emphasise that those most exposed and vulnerable to both climate emergencies and sustainable transitions should receive priority protection by their governments. Sustainable businesses, especially SMEs, should be made more resilient through financial incentives and technical advice and workers should receive necessary training to support their transition to sustainable employment. Further points specify the integration of social policy measures to protect populations affected by economic and environmental disruption, while policies also need to be in place to anticipate and prepare for changing labour market conditions. Finally, the ILO argues for commitment to decent work, which overlays the bulk of the above actions to progress a JT.

Suggestions for a green economy

There are inherent contradictions between the political and economic goal of continued economic growth and ability of the planet's ecosystems to accommodate this. Regrettably, few authors have attempted a connection between the degradation of work and the degradation of the planet. One author who does make the connection is economist Kate Raworth (2017) whose 'Doughnut' economic model we encountered in Chapter 1. The model brings together the twin themes of this book in recognising there are upper and lower limits to sustainable socio-economic activity. The upper limit is the planet's ecological ceiling, above which the Earth's systems suffer increasing damage, while below the lower limit it is human wellbeing that becomes damaged through increasing inequalities in life chances. Environmental safety and social justice can only occur between these 'inner and outer rings' of the doughnut.

The question of where the doughnut's boundaries are – whether some sort of economic growth is still compatible with maintaining the planet's ecosystems, at a level that is sustaining of human activity, is one that is currently dividing both economists and environmentalists and is of course of particular relevance to those countries in the global South that are attempting to improve the living standards of their citizens. The 1987 Brundtland Report *Our Common Future* defined sustainable development as that which 'meets the needs of the present without compromising the ability of future generations to meet their own needs' (World Commission on Environment and Development, 1987). While, as Daly and Cobb (1994) suggest, this rather depends on what we mean by 'needs' (widescreen colour TVs or clean drinking water?), the solution most commonly put forward by those economists who now accept the reality of climate change is the concept of 'decoupled' growth, that is, growth that has zero or negative ecological impacts.

If, through the use of new technologies and energy efficiency, a country's growth occurs at a faster rate than its use of finite resources, this is 'relative decoupling' and is the green growth model of most interest to the world's poorer countries and for which there is some, if uneven, evidence (Jackson, 2011). However, for the planet's richest countries, who already consume unsustainable amounts of resources, what would be required is 'absolute decoupling', where there is an absolute fall in resource use at the same time as economic expansion; at present the debate between economists and some climate scientists is over what rate of reduction in resource use would constitute a 'sufficient absolute decoupling' to bring the global economy back within planetary boundaries (Raworth, 2017: 260). In reality, there is little evidence so far for absolute decoupling in any of the major economies; one reason for this, and for the fact that the figures are hard to decipher, is due to the relocation of significant sectors of manufacturing from the first industrial nations to the newcomers. Thus, while a major industrial nation may claim that there has been a reduction in its resource use, if account is taken of the carbon and resource content of its imported goods – its global material footprint – then, as we saw in our examples of phone and clothing production, any claimed progress in reduction starts to look very flimsy (Office of National Statistics, 2019). This reminder that we cannot measure global environmental progress by looking at a single nation's figures similarly undermines the suggestion that, as employment in the rich nations moves from manufacturing to services, this will have a positive effect on environmental pollution; as we have seen, all this does is export the environmental 'externalities' overseas, letting someone else, somewhere else, do the dirty work.

Obstacles to environmental protection progress

Previous chapters demonstrate the efforts made by the international community of scientists and advisory bodies to convince policymakers, companies and public of the impending catastrophes presented by carbon emissions and pollution. Despite warnings and pledges of support from many countries, the pace of carbon emission mitigation has scarcely altered. Moreover, many of the commitments are presented as long-term targets, which in practice has meant that year-by-year action has been desultory and piecemeal. Individual countries are able to act according to their own perceived interests in meeting internationally established targets and national sovereignty often results in inaction or even opposition if short-term economic interests are perceived to be compromised. An associated problem is that political party ambitions are wedded to policies of economic growth, despite these being inimical to environmental protection. The USA is the world's largest and most influential economy but, as shown earlier, has a long history of opposition to environmental arguments and pressures (Oreskes and Conway 2012), an approach that has continued and been exacerbated in recent years. Adler (2019) points out that the hegemonic influence of large and transnational organisations, underpinned by financial interests, over individual governments cannot be under-estimated. We have already noted the warnings by former governor of the Bank of England Mark Carney of the serious consequences for planetary heating of a failure by the capital markets to switch from continued investment in carbon-intensive sectors of the economy (Carney, 2019). Clearly, meeting scientifically determined international targets for emissions control and pollutant reductions requires coordinated actions that are effective at all levels in mitigating their production.

Though in a number of enterprises and establishments, unions and workers have collaborated effectively with managers in initiating sustainable projects, the central problem is that ownership and control over the financing and means of production are concentrated in the hands of executives answerable to investors rather than to society (including employees) as a whole. Yet it is society that pays for and is clearly threatened by the environmental crisis, which in turn is largely the 'neighbourhood' outcome of the activities of cost-minimising companies and state enterprises. These also aim to optimise profits through stimulating damaging (and often gratuitous and excessive) consumption. Together, these activities draw upon irreplaceable resources, poison the atmosphere, pollute cities and damage the health of citizens through encouraging unhealthy life cycles and supported by mind-numbing and insecure jobs. From the obstacles noted above, the challenges facing the global community are complex, contested by powerful vested interests, are multidimensional and vary across countries. Some of these are chief contributors to the crisis, while others, especially in the global South, face the consequences.

These observations suggest that despite the negative ecological forecasts of most research and policy teams, and their consistent and increasingly urgent recommendations for treating the root causes of environmental degradation, much work remains to be undertaken to remove the obstacles to progress. One overriding message is that in the absence of national and international binding regulation and agreements, the prime sources and drivers of environmental and work degradation will remain relatively untouched. In subsequent passages we explore ways in which formal regulation could be applied, though of course, polluting processes and governmental controls vary significantly across countries. Deforestation may represent an especially acute originating factor in some regions, whilst direct fossil fuel extraction may feature heavily in others. Whatever the specificity of the source, CO_2 emissions and other pollutants have no respect for national boundaries, infecting the atmosphere, international oceans, waterways and soil, though their effects especially ravage the poorest and most vulnerable countries and their inhabitants. The common factor is that these outcomes emerge largely from the world of work and it is our contention that the workplace, whether in a factory or a forest, is one locus where workers, their unions and communities can make positive contributions to protecting the local environment whilst enhancing the working lives of workers, providing the political will to engage with necessary reforms is present. The authors of this book are neither climate nor environmental scientists and, for this reason, we have not attempted to intervene in the more technical aspects of mitigation, for example, through geo-engineering measures (e.g. Romm, 2018). Our speciality as industrial sociologists is to consider how organisational behaviour, which so far has been largely oblivious to environmental demands, can be shaped to benefit communities and workers in ways that help support the wider environmental policies laid down by the international scientific community.

Two conclusions emerging from the environmental crisis are already clear: first, to prevent much of the planet becoming uninhabitable, radical mitigation policies are needed; 'business-as-usual' can only lead to devastating climatic consequences (FitzRoy and Papyrakis, 2016: 159). Second, the lack of sustained action since the first IPCC warnings were announced 30 years ago means that speed of response is now of the essence (Klein, 2019: 39). Some 15 years ago, the Stern Review warned that early action was imperative and that failure to act by governments was cited as evidence of extreme market failure (Stern,

2006). Three main approaches to GHG mitigation can be identified. First, the market approach favoured by conservatives and business interests; second, regulatory controls such as required insulation standards or vehicle speed restrictions; and third, the more comprehensive and radical Green New Deals.

Market-based approaches

The principal market options to reduce carbon emissions are through carbon taxes, tradeable permits, and subsidies for energy saving and developing renewables. A carbon tax is relatively straightforward to design, if less so to apply, and it could gather popular support if the resulting tax revenues are redistributed to citizens or disadvantaged communities, or if they are used to help boost employment, training and infrastructural improvement. There are problems with carbon taxes though. First, they will be strongly opposed by business, political and media interests wedded to a low-tax economy. Second, setting the tax rate would not be straightforward as it needs to be high enough to deter consumption but not so high as to provoke social disorder and flexible enough to be adjusted as circumstances alter. Third, applying carbon taxes will tend to exert a lagged effect on emissions: applied on their own, they would be unlikely to exert the 'shock' impact needed to attain emissions targets. Complementary regulatory and market measures would almost certainly be required. FitzRoy and Papyrakis (2016: 153) point to successful carbon tax schemes operating in British Columbia and Ireland. Per capita emissions in the Canadian province declined by nine percent since the introduction of the redistributive tax. In Ireland, emissions have also declined and the policy broadly accepted as less disruptive than traditional tax impositions. One major problem is that reforms introduced in one country will not necessarily inspire coordinated action elsewhere. Carbon taxes are also a highly sensitive political issue as President Macron discovered after imposing fuel taxes in 2018 shortly after offering generous tax reductions for high earners in France. This sparked the infamous and spontaneous *gilets jaunes* reaction, which led to the abandonment of the scheme, a huge aid package for the poorest but also a lingering feeling of popular resentment, which undermined Macron's authority and popularity. There have been other setbacks too. An Australian carbon tax was dropped after just two years by an incoming conservative government and an early British scheme was abandoned after protests led to fuel shortages and panic buying. Thiele (2016: 162) estimates that if all externalities were to be included, consumer prices for petrol and diesel would be 'three to four times higher' in the USA than they actually are.

Carbon trading, or cap-and-trade, is a market-based policy that allows trading of emissions credits. The system establishes national limits on emissions for producers and industry. The cap is reduced over time so that overall emissions will fall. Within the cap, companies receive or buy emission allowances, which they can trade with one another. Participants in the system are allocated permits that define their allocation, which, if exceeded, requires them to purchase additional allowances from producers that have not used their full allocations. The system is designed to encourage producers to keep emissions below their allowance to avoid the cost of purchasing additional allowances (Leichenko and O'Brien 2019: 120). The best known – and much debated – carbon trading scheme has been operated through the EU since 2005 as the Emissions Trading System (ETS), which is compulsory for some 11,000 carbon-intensive installations and for airlines operating between member states. The ETS system has been revised frequently since its inception and will be strengthened again for the ten-year period from 2021 in line with the 2030 climate and energy policy framework. A review of the first ten years of the EU initiative by Muûls and her colleagues (2016) was generally favourable, indicating that emissions in industrial carbon had declined without noticeable adverse effects on economic performance. They also suggest that the system has helped raise the profile of clean technology such as wind farms and solar heating. Some environmentalist bodies and commentators are not so sanguine, pointing out that there is nothing to prevent corporate interests manipulating the trading system to their advantage while passing on costs to consumers (CANE, 2018). Klein (2014: 220) points out the ease with which questionable but lucrative offset projects have attracted 'carbon cowboys' to range the world looking for easy pickings, especially in developing countries. There are other potential disadvantages: individual consumption is not included in the scheme and with transport comprising 29 percent of Europe's CO_2 emissions in 2016, a proportion that is growing, limited policy emphasis is put on investing in social and electric transport. Finally, political parties are hostages to their specific constituencies – conservatives to big business and socialists to the demands of trade unions but also to the business interests on whose support left of centre parties are dependent. One consequence of this dependency relationship is the ambiguity and incoherence surrounding government climate policies. In the UK, from 2015 Conservative governments have cut environmental agency budgets and influence whilst missing several interim targets and advocating expansion of London's Heathrow airport, whilst professing commitment to carbon net zero by 2050.

Interventionist approaches

Regulation can take many diverse forms, most, if not all of which are rejected by fundamentalist conservatives as an offence against personal freedom. Nevertheless, regulation and revised standards are becoming more common in the combat against further environmental damage. Steps can be small but with a major public impact, such as charging for plastic carrier bags, which has significantly reduced their usage. Or they can be potentially more significant, such as placing limits on car emissions, increasing taxes for bigger cars and flying or regulating the phased withdrawal of internal combustion vehicles. There can be strong regulations on land use and building construction and insulation. Fishing quotas may be passed to protect the marine environment. Regulation can also take the form of protective legislation for employees at the workplace, for example, to restrict number of hours or days worked, actions that can also impact positively on emissions and pollution through reducing transport pressures. Most, if not all of these regulatory instruments will be met by opposition from industry and from conservative supporters, resentful at interference in commercial activities and people's ability to decide how they spend their money. Further, as we saw with the VW emissions testing scandal, profit-seeking companies may well seek illegitimate ways to avoid regulatory requirements. Unfortunately, we have also witnessed the disastrous consequences of loosening or removing regulations – from inadequate flood defences to mad cow disease.

Green New Deals (GNDs)

The potentially strongest interventionist approach, the GND, has emerged not only as a response to the growing climate emergency but also to address the win-lose extremes of the market economy. Taking its name from Roosevelt's programme of government investment in infrastructure projects, most versions of the GND propose public sector investment in the areas of energy security, low-carbon infrastructure and ecological protection (Jackson, 2011: 107). Such proposals focus on two interrelated responses to the climate emergency – those that attempt to halt or nullify human impacts on climate and those that increase resilience or adaptation to the effects of climate change (Laybourn-Langton *et al.*, 2019). Most GND programmes for infrastructural investment emphasise renewable energy generation, public transport, waste recycling, energy efficiency, sustainable agriculture and retrofitting buildings (Green New Deal Group, 2018; Real World Coalition, 2001; Romm, 2018; Raworth, 2017).

The original New Deal represented a response to the economic calamity of the Great Depression, which following the stock market collapse of 1929 had left a quarter of the USA workforce unemployed, massive increases in homelessness and growing social disquiet. The New Deal essentially consisted of bundles of programmes and reforms designed to stimulate the economy through large-scale public investment and infrastructure programmes to build roads, bridges, dams and public buildings. Progressive tax and social reforms were introduced, and the banking system reformed and stabilised. An essential feature was that large-scale programmes were successfully coordinated and regulated at appropriate governmental levels, from federal down to city. Trade unions were accorded more formal powers to negotiate with employers. By 1933, union membership had plummeted to about three million. Following the New Deal-informed National Industrial Recovery Act (1933), which supported collective bargaining, and especially the Wagner Act (1935), which required employers to negotiate with unions if given majority support by workers, union membership increased rapidly to over seven million by 1940, and with government encouragement, continued to grow for the next seven years, opening the door to unions to make valuable contributions to wartime governance. Lessons of the New Deal are that when faced with a national crisis, radical remedies directed from the centre, but operated at grass roots levels are not just necessary but can be successful. At the same time, objections to the New Deal (Moley, 1939) demonstrate that resistance from ideologues and well-resourced conservative associations, fundamentally opposed to economic intervention, can be anticipated (Klein, 2019: 275). The second point arising from the original New Deal is that government support for trade unions can lead to union revival and positive contributions from unions helped to address the crisis. Similarly, solid political and union support is crucial for any form of Green New Deal to succeed. We therefore examine the GNDs on offer so far and attempt to assess the level of support they may have or can reasonably anticipate and the level of opposition they may face.

Klein (2019) makes the point that what is different about GND programmes for action, compared to previous suggested initiatives (change your lightbulbs, recycle your paper), is that they represent a programme for systemic change. The market-oriented thinking that has dominated political and economic policies for the past half century has conditioned us (as 'consumers') to think in terms of *individual* responses, the logical conclusion of which is that if the planet burns up it is because we personally were not trying hard enough. Even at national economy level, policies such as carbon taxes and cap-and-trade are still limited to a continuation of the 'business as usual' model. The reference back to historical initiatives such as the New Deal or the Marshall Plan reminds us that it is possible to think of systemic change.

GNDs have been proposed by the European Union, the British Labour Party and most prominently by US Democrat Congresswoman Alexandria Ocasio-Cortez and Senator Ed Markey who emphasise the need for a just transition policy to 'be developed through consultation, collaboration, partnership with frontline and vulnerable communities, labor unions, worker co-operatives, civil society groups, academia and businesses' (House of Representatives, 2019). A central goal is to create 'high-quality union jobs'. As America is responsible for a high proportion of global GHGs, it is worth considering these proposals in depth, as successful adoption and implementation is certain to impact on other countries' motivations to adopt a GND. The proposals are aspirational and wide ranging, to be operated as a federal programme, devolved as necessary to states and civic communities as appropriate for ratification. Like the original New Deal, the GND aims to extend beyond environmental restructuring to embrace social and human rights as well. The main aims of the USA programme can be summarised as: to build resilience against climate change-related disasters; eliminate GHGs/pollution through natural ecosystems e.g. (re) forestation; provide universal access to clean water; meet all USA power demand through clean, renewable and zero-emission energy sources; construct and maintain energy-efficient power grids; upgrade buildings for maximum energy efficiency; develop growth in clean manufacturing; work with farmers to remove pollution (e.g. fertiliser drainage) and GHGs; invest in public transport and electric cars; provide funding for community-based projects to reduce pollution; enhance bio-diversity; clean up hazardous waste; and promote international exchange of technology.

This is clearly a highly ambitious project and already the usual suspects are lined up to undermine it: from its relentless negative commentary on the Deal, the Cato Institute describes it as a front to nationalise the US economy with unclear benefits, which would crush liberal values and moreover, cannot be afforded. There was no holding back from the Heartland Institute spokesperson: 'the socialist green new deal threatens freedom, prosperity, and the environment' (Heartland Institute, 2019). Mainline business-supporting Republicans have also demonstrated opposition, arguing that it would discourage economic growth, undermine funding for other social causes such as Medicare for all, would involve tax increases not just for the rich but also potentially for middle-earners, and make US exports less competitive, especially if other countries failed to follow a similar route. As might be expected, major corporations, including oil companies, are deploying millions of dollars to actively campaign against a GND, publicly advocating a market-based carbon tax in which polluters would be charged for the amount of carbon gases they emit. The reason? 'They want a carbon tax because it doesn't threaten the industry's very existence and allows them to keep polluting – so long as they pay for it' (Atkin, 2019).

While opposition would be expected from pro-free enterprise and ultra-libertarian sources, more worrying is the division among Democrats and unions over the proposed deal. There are genuine concerns, especially among affected unions, that the GND represents a broad statement or wish list offering no budgetary details or few practical remedies to confront the crisis. There are also differences among unions and individual states: while some states, such as Maine have signed in law a modified version of the GND, with the full support of the unions (Irfan, 2019), the President of the largest union federation, the AFL-CIO, representing upwards of 12 million members, has opposed the Ocasio-Cortez/Markey deal, complaining that the union was distanced from the drafting process and would seek significant changes to ensure that worker interests be fully protected in a revised version. Some unions, including the AFL-CIO, are concerned that an abrupt break with fossil fuels will disadvantage workers in high-union sectors and that insufficient attention has been given to carbon capture and sequestration, which would allow a more protracted transition from fossil fuel to cleaner energy. These concerns were fully spelt out in a letter to the GND's sponsors by members of the powerful AFL-CIO's Energy Committee: 'We will not accept proposals that could cause immediate harm to millions of our members and their families. We will not stand by and allow threats to our members' jobs and their families standard of living go unanswered' (in Irfan, 2019). Irfan, an official with a national media union, expressed concern that opposition could not only threaten the GND, but divisions among unions could also undermine Democrat efforts to regain the Presidency, a fear that was not realised in the 2020 election. A further potential problem is that the fossil fuel sector is more strongly organised than clean energy and that many clean energy companies like Tesla are allegedly strongly anti-union in their management policies (White, 2018).

Nevertheless, the Maine deal was accepted by the state-level AFL-CIO, representing some 40,000 members, mainly because a more gradual transition to renewables was adopted and also because 'labor groups were closely involved in drafting the Maine New Deal from the outset' and were able to ensure that union interests and bargaining rights were protected (Irfan, 2019). Other states are also actively pursuing local green deals. The lesson appears to be that unions need to be involved with the drafting process at the earliest stages: as the Maine AFL-CIO executive director pointed out: 'We know that the energy transition is coming. It either happens to us, or with us, and we think "with us" is a much better process' (*ibid.*, 2019). Nevertheless, though GNDs are being negotiated at individual state levels, resistance from Republicans and big employers will make federal progress more complex and contested.

Underlying business objections to GNDs lurks the ever-present Friedmanite doctrine of free markets as the expression of a free society, reflected in govern-ment antipathy to any form of preventative economic regulation, even when faced by devastating evidence about the environmental and health impacts of DDT, smoking, chlorofluorocarbons and acid rain. The current crisis, though, is unprecedented 'in failing the majority of people on multiple fronts' (Klein, 2019: 26): through its wide range of causative factors, the urgency of the required response, the global implications of environmental damage to civil society, the accumulated and constantly growing body of scientific evidence and finally, the lived experience of climate change in terms of floods, wildfires, drought, famine and risk of mass population displacement, melting sea ice and the impacts these have on the lives of the world's citizens. Moreover, it is becoming clear that the struggle for the environment is increasingly becoming a struggle over economic and political ideology, as witnessed on one hand by growing calls for democratic socialism and on the other, preference for 'market solutions' from Western conservatives and their supporters and highly negative responses to GND models from conservatives on both sides of the Atlantic. This tension was demonstrated by Senate rejection of the Democrat-initiated GND resolution, identified by some Republican Senators as 'radical left-wing ideology', arguing that environmental problems should be treated through pri-vate sector innovation and application of technology. The new President, though prepared to invest in the green economy, is not wedded to more ambi-tious GND interpretations: moreover, any prospective GND would be harried by powerful lobbyists and would have faced obstruction if the Republican Party had retained control of the Senate in 2021. Nevertheless, legal challenges could still be addressed by the conservative-rich Supreme Court.

Such a backlash greeted the UK Labour Party's own GND proposals (2019), put together by a coalition of Labour and union organisers together with environmental campaigners. The Deal offers a broad programme based both on rapid decarbonisation and radical social and economic change. The programme consists of nine essential elements, including commitment to zero carbon emis-sions by 2030, based on phasing out fossil fuels and long-term investment in renewables. This programme would be complemented by a shift to well-paid unionised green jobs, extending public ownership and control and developing green public transport, underpinned by a shift to an internationalist socialist economy. In his review, the Assistant Editor of the right-of-centre *Daily Telegraph* (2019) newspaper pulled no punches, describing Labour's version of the GND as 'puerile, delusional nonsense dressed up as industrial strategy'. His remedy? 'Decarbonisation of the UK economy is perfectly feasible, but it has to be done in a market-driven way'.

Like the Republicans, the Conservative Party has not committed itself to any formal comprehensive green programme, though the Conservative Home website suggests that 'the conservative movement has never been as enthused by environmental issues as it is now'. The website confirms that this commitment is more aspiration than actual policy: 'We must harness this energy and translate it into an enduring conservative legacy through market-led solutions to these pressing environmental challenges' (2019). The market-driven approach clearly has a narrower focus than the proposed GNDs and will not directly address more transformative social justice elements. Moreover, recent analysis casts considerable doubt on green Conservative credentials, reporting that their MPs, including Cabinet ministers, are five times more likely to vote against climate action than any other Party parliamentary representatives. Also, ministers were more likely to receive gifts and donations from oil companies, airports and other commercial sources with interests in fossil fuels than representatives from other parties (Watts and Duncan, 2019). Moreover, the Conservative government (as well as significant numbers of the Opposition), persuaded by the business case, is committed to airport expansion at London's Heathrow, already one of the world's busiest airports. Campaigners are concerned that toxic NO_2 emissions around the airport, already high from current vehicular traffic levels, will increase significantly; that numbers of residents exposed to constant noise levels will grow to 1.6 million; and take-offs and landings would grow from under half a million to three-quarters of a million annually, increasing CO_2 emissions proportionately. According to campaign group Friends of the Earth, Heathrow is already the biggest single source of GHGs in the UK, and these will rise appreciably with a third runway and new terminal, casting doubt on the country's ability to meet its already ambitious and slipping emissions targets.

In contrast, the suggestion for funding in the UK Green New Deal Group's Report (2018) applies the concept of 'quantitative easing', or printing money, last used to rescue banks from the 2008 international financial crisis they had created. This time, the proposal says, the way out of the environmental crisis could be for the central banks to create money by lending to invest in renewables and job-creating energy efficiency schemes. Because such green quantitative easing would create jobs in places that need them most – old and neglected industrial regions – rather than swelling banks' coffers, it is argued that it would have positive social and political advantages in addition to environmental ones.

The EU's GND has been afforded priority by the European Commission, and the new President, Ursula von der Leyen, has framed ambitions for the project in broad terms, with the aim to become the world's first climate-neutral continent by mid-century. Objectives include reducing emissions from a previously targeted 40 percent to at least 50 percent by 2030, though by late 2019 only a minority of member states had committed themselves to this target. The intention to be carbon-neutral would be endorsed in a European Climate Law, requiring unanimous support from all member states, which to date has not been forthcoming from a number of accession states like Hungary and Poland, which are strongly wedded to fossil fuel production. A proposed JT fund is aimed to support coal-mining regions to phase out dependence on fossil fuels and provide training and resources to lubricate shifts to greener energy and more energy-efficient homes and workplaces. Like other GNDs, the European version does have a wider ecological and social agenda to reduce pollution, promote the circular economy and advance social justice. But German Social Democrat MEP Delara Burkhardt (2019) warns that whilst the EU establishes the framework for action, national governments are responsible for implementation and for this, 'strong welfare states with robust social systems are essential' as these 'enable equal opportunities, access to the labour market, fair working conditions, social security and inclusion', which, following the austerity years and broad free-market policies pursued by the EU, includes just a handful of countries. Moreover, as critic Albena Azmanova points out about the EU, 'the combination of pledges for bold action and conspicuous failure to act has become a trademark of environmental policy' (2019). She argues that action on the environment is being restricted by a capital-labour alliance, which blocks the environment-justice agenda. She claims that this alliance, founded on Keynesian policy of stimulating demand and then redistributing some of the gains, 'relied on intensified production and consumption' (*ibid.*: 2019), with accumulating negative all round effects on the environment, on health and on working conditions. This creates double pressure, on the environment and on social justice, which the neo-liberal EU consensus is ill-adapted to address.

Some of the above GND approaches have been designed through collaboration between parties with a direct interest in reducing poisonous emissions and with equivalent aims to consolidate social partnerships at work, reduce inequality and improve life chances for the most vulnerable citizens and communities. While some unions feel distanced from GND policies, it should not be too problematic to engage in dialogue with those who feel that they have been insufficiently engaged in the drafting process. The bigger problem will be to ensure that employers buy into the greening programme. We have seen how the US fossil fuel lobby has helped to obstruct environmental progress even in the light of the most persuasive evidence for the impact of fossil fuel originated GHGs on the environment and climate. The same influence can be observed in the EU as well. Research commissioned by a number of environmental NGOs found that since 2010, five of the largest oil and gas companies and associated lobby groups spent at least a quarter of a billion Euros lobbying the EU to support industry's demands to dilute climate policy (Corporate Europe Observatory, 2019). Clearly, though, employer commitment to combatting climate change through a GND would be essential, and this could be gained through offers of transition funds and guarantees that their interests can be protected in the transitional process to environmentally positive production.

GNDs and trade unions

As unions are centrally involved in protection of jobs, communities and society generally, their support for GNDs is unquestionably vital. Without union support, it would be difficult, if not impossible, to construct or implement GNDs. It is useful therefore first to comment on the roles of unions, identify the obstacles that are inhibiting enactment of these roles and the potential effects of unions on the viability of introducing and maintaining GNDs. Throughout industrial history, trade unions have been the pursuers and sometimes generators of social change, but previous chapters have noted the challenges and tensions facing unions today. First, we are in an era where unions, heralded not so long ago as on a progressive forward march toward constructing a more socially responsible capitalism, face considerable constraints to their protective, let alone progressive, influence, though the common underlying source of these constraints is the hegemony of neo-liberal economic doctrine. The scale of constraint, as we have noted, differs. Unlike their counterparts in some developing and post-command economies, though occupational 'blacklisting' occurs (Chamberlain and Evans, 2019), union activists in Western countries are rarely in danger of their lives or liberty. Nevertheless, the chief obstacle facing Western unions to participation in both political and enterprise governance is lack of recognition to act on behalf of members. Yet recognition and representation are central to the sustainability role that unions envisage for their members and representatives. Here, the role of government is crucial: their policies help or hinder union organisation, recruitment of members and meaningful engagement with employers and politicians (Bain and Price, 1983).

Notwithstanding their current weaknesses, there are strong justifications for promoting trade unions as necessary and beneficial agents for change. First, those countries where unions have an established and legitimated role are often those at the forefront of both positive environmental and social justice programmes. Second, unions are socially and economically connected to employees, members and their communities, from whom union officials and lay representatives are often drawn. To maintain employment security and quality, they also have a direct concern for supporting the viability of their sector and of individual companies. Third, unions also represent a progressive social movement with a mission to protect and advance the interests of vulnerable members of the community or to promote egalitarian social and economic policies, whether through their own actions or in collaboration with sympathetic non-governmental and political parties (Cock and Lambert, 2013). One of the first steps that authoritarian and reactionary governments take in acceding to office is to discourage or curtail union activity and, in extreme cases, outlaw unions completely, as was the case in Germany in the 1930s and Chile under General Pinochet in the 1970s and 80s, where radical neo-liberal reforms were implemented under military rule by free-market economic ideologues under the tutelage of Milton Friedman. This fiercely repressive tradition is continued in Latin America, Saudi Arabia and other Gulf States today (ITUC, 2020). Fourth, and especially relevant to environmental protection, trade unionism has recognised internationalist traditions and networks, and while these have been undermined in recent years through competitive globalisation, internationally collaborative unionism is still well established through organisations like the European Trade Union Confederation (ETUC), which in turn is affiliated to the International Trade Union Confederation (ITUC).

In their thinking on transformational as opposed to surface sustainability, many unions are ahead of enterprise leaders who, as we saw in previous chapters, either fail to recognise or to act on the warning signals provided by the scientific community. These union manifestos are significant as well for their recognition that the climate emergency affects everyone, requiring collaborative action at different levels, but also acknowledgement that worker needs in terms of redeployment, skills acquisition and security will vary across occupations and sectors. For example, manual workers in fossil fuel extraction, transport and some areas of manufacturing may understandably feel more immediately exposed to transitional change than their colleagues in education or health services. The important message is that unions do have a successful record in protecting and advancing their members' working lives, usually in agreement with or in partnership with management, which as we have seen, is an essential element of meeting the challenge addressed by the climate emergency. Also, as was seen with the Lucas Aerospace plan, unions have the capability to offer and deliver plausible alternative strategic solutions to societal problems when conditions require them. The capacity to switch from one form of manufacture to another has been well-demonstrated in periods of wartime emergency. With union support, automobile workers swiftly adjusted to assembling tanks and aeroplanes in place of motor vehicles in the Second World War (Wilson, 2016) and indeed, American unions made notable contributions to strategic federal war planning (Adler, 2019: 119) as did British unions through the tripartite Joint Consultative Council, which became the wartime 'primary instrument of government industrial policy' (Middlemas, 1980: 278).

If unions are to have the central strategic role that they are demanding as essential to confronting the environmental crisis, a role to which they successfully contributed in earlier national emergencies, central governments and employers must recognise that the transformational changes, the need for which is becoming increasingly evident, must be prefigured by providing unions with tripartite status alongside employers and policymakers. This has been the case in a number of countries with regard to safeguarding the health and safety of workers. When legislation accords unions a more strategic role in securing and maintaining organisational and industrial safety, evidence shows a positive union response and that employee (and community) welfare increases and accident rates are lowered. Between 1974, when the UK Health and Safety at Work Act came in to force and 2018, fatal accidents to employees dropped by

84 percent and non-fatal accidents by over 70 percent (HSE, 2019). Under the Act, unions are strongly represented on health and safety committees. Though in Bangladesh harassment and sexual violence toward female garment workers is still reported to be rife and the country is one of the worst for working people (ITUC, 2020: 6), overall safety and fatality rates after the Rana Plaza disaster have improved considerably following agreements for factory inspections and remedial actions made between major commissioning companies such as Primark and local unions to improve fire, building and worker safety in participating local companies (Barrett *et al.*, 2018). But despite these welcome advances we have to recognise that there are broad variations of union experience, which transcend the basic varieties of capitalism approach. Developing and emerging countries have long histories of colonialism, oppression and external intervention in their economic and industrial affairs, which, combined with national growth imperatives, cannot be dissociated from lack of state enthusiasm to engage in transitional reform in collaboration with unions.

In all cases, if unions are to exercise a more proactive role in transforming the parameters of work for the benefit of both workers and community, it is clear that a different mind-set from employers and governments is needed. Scandinavian style corporatism can support the environment through union involvement with political and commercial bodies. Also, in coordinated economies like Germany, though there are signs of strain, unions can and do work successfully alongside employers in both tripartite policy bodies and with employer partners through worker directors and social partnership. This does lead us to ask, therefore, what potentially can be the roles of trade unions in helping to secure the changes towards just transition to sustainable work in sustainable workplaces.

Our argument is that when offered legitimacy by governments as valuable partners, labour movements would establish themselves as valuable agents in combatting the emissions and pollution emergency. National crises have often been met with interlinked responses: first, through direct intervention by the government, which was critical in helping America to recover from the Great Depression in the 1930s and second, by public ownership. Following the 2007 finance collapse, nationalisation of banks helped to mitigate the subsequent acute economic crisis (Tooze, 2018: 293). An associated crisis response has been through incorporation of labour into strategic decision-making bodies, whether at industrial or governmental levels. During the First World War, union leaders were appointed to the British War Cabinet, and the Webbs, chroniclers of trade unionism, insisted that during the war the movement achieved 'distinct recognition as part of the social machinery of the State'. The UK government also introduced centralised Joint Industrial Councils of union officials and employer representatives to oversee industry-wide collective bargaining (England and Weekes, 1981: 15) to establish 'greater cooperation between employers and workers through discussing all aspects of the business' (MacInnes, 1987: 13). In World War II, we saw above that in North America and Britain, unions were involved with both employer and governmental bodies in planning and coordinating activities (Adler, 2019: 117ff). Following the war, British unions were instrumental in supporting the regeneration of German industry and in establishing the extant system of monitoring and controlling capitalism's excesses through worker directors in key sectors of the German economy (Bullock Report, 1977).

A number of questions arise from these developments. First, as we have seen, governments have adopted varying degrees of adherence to the global neo-liberal economic project, which has uncontestably contributed to the climate crisis. Former US government advisor Gus Speth concludes that 'most environmental deterioration is a result of systemic failures of the capitalism that we have today and that long-term solutions must seek transformative change in the key features of this contemporary capitalism' (in Oreskes and Conway, 2012: 255). Another establishment insider, Paul Adler, argues that competitive private ownership and control of society's productive sources are the root causes of the environmental crisis of capitalism, a crisis that can only be overcome through social ownership and democratic management of industry under a democratic socialist regime (2019). There are of course profound practical and ideological obstacles to such proposals, especially in the USA where even environmental science has been dismissed as a socialist ploy (Oreskes and Conway, 2012: 253) and eminent scientists caricatured as 'watermelons', green on the outside, red on the inside (Lynas, 2012). Whether radical measures could or would be contemplated among established capitalist societies in the short time available is questionable: light-touch sustainability measures may be less threatening to established interests but are unlikely to meet growing environmental challenges.

Conclusions

The obstacles to GHG mitigation confirm that (i) despite the urgency of the crisis, there will be no easy solutions to curb anthropogenic emissions and pollution; (ii) national governments must take a central determining role, advised, aided and supervised by authoritative and universally acceptable scientific bodies such as the IPCC or through international systems of governance; (iii) various and coordinated actions will need to be explored to confront the crisis e.g. development and use of technological alongside socio-economic policies; (iv) governmental intervention in the economy at national level requires cooperation and coordination with employers, employees and their representative bodies; (iv) while some countries are taking serious steps to address the multiple sources of the crisis, major polluters such as America and Brazil are ambivalent toward the environment and doing little to embark on transformations needed to meet internationally agreed targets; and (v) there are several levels in which structural intervention could be envisaged, ranging from radical departures from free-market capitalism to voluntary persuasion (possibly backed by sanctions for infringement) within an otherwise unchanged market structure. Whilst market-driven approaches favoured by conservatives and their commercial backers have been available and adopted by governments and employers since the first IPCC meetings 30 years ago, these appear to have made only a limited impact on global emissions reductions. Certainly, if these approaches do remain the prime focus, which potentially will be the case if neo-liberal doctrine continues to dominate economic policy-making, unions, their members, employees and local communities will continue to be estranged from decisions affecting not just the environment but their work, social and domestic lives. Cumulatively, lack of planned centrally directed action can lead to missed mitigation targets with resultant humanitarian disasters impacting across the whole planet.

Domestically, Brexit could also have a derogatory impact on Britain's GHG emissions. As we saw above, a GND is likely to be afforded priority status within the EU, whilst in pursuit of market freedoms, the exiting UK could be heading for light-touch or even deregulatory environmental policies. Already, in common with America's EPA, the UK Environment Agency has lost one-fifth of its workforce, some 2,500 full-time staff, since 2013. At the same time key responsibilities have been cut back: fewer industrial site inspections, fewer water pollution inspections and fewer legal actions against polluters (Unearthed, 2018).

GNDs do offer positive and comprehensive means to address these issues but are only likely to be implemented in a minority of countries, and confronting the entrenched vested interests of profit-seeking enterprises and growth dependent governments would risk considerable dilution even when supported by sympathetic governments. Whether such governments can get elected on a platform of radical economic change is at present an open question. And if not, what might be the consequences? Ultimately, this means that multilateralism, so despised by populist politicians, must be defended and actively promoted as an appropriate way forward to ensure protection against cross-national vulnerability, cruelly exposed by the Covid pandemic. The climate, like the coronavirus, has no respect for borders and hence international health and environmental cooperation is vital, as the EU is slowly beginning to recognise through its GND agenda. Global bodies able to support this ambition, such as the UN, ILO and WHO acknowledge interconnections and as we have seen from the lead taken by the UN over climate change and global heating, the role need not just be symbolic.

Bibliography

Adler, P. (2019), *The 99 Percent Economy: How Democratic Socialism Can Overcome the Crises of Capitalism*, New York, Oxford University Press.

Atkin, E. (2019), 'Corporate America is terrified of the Green New Deal', *The New Republic*, https://newrepublic.com/article/153953/corporate-america-terrified-green-new-deal.

Azmanova, A. (2019), 'The big Green New Deal and its little red social question', *Social Europe*, www.socialeurope.eu/the-big-green-new-deal-and-its-little-red-social-question.

Bain, G. and Price, R. (1983), 'The determinants of union growth', in W. McCarthy (ed) *Trade Unions*, Harmondsworth, Pelican; 245–271.

Barrett, P., Baumann-Pauly, D. and Gu, A. (2018), 'Five years after Rana Plaza: the Way Forward', New York, Center for Business and Human Rights, New York University.

Bullock Report (1977), *Report of the Committee of Enquiry on Industrial Democracy*, Chairman Lord Bullock, CMND 6706, London, HMSO.

Burkhardt, D. (2019), 'Europe's Green Deal is turning red', *Social Europe*, November, www.socialeurope.eu/europes-green-deal-is-turning-red.

CANE (2018), Climate Action Network Europe, www.can.europe.org/climate/emissions-trading-scheme.

Carney, M. (2019), Commons Treasury Committee, https://parliamentlive.tv/event/index, 15 October.

Chamberlain, P. and Evans, R. (2019), 'Crossrail hired security firm to monitor trade unionists', *The Guardian*, 29 July, https://www.theguardian.com/uk-news/2019/jul/29/revealed-crossrail-hired-security-firm-to-monitor-trade-unionists.

Cock, J. and Lambert, R. (2013), 'The neo-liberal global economy and nature: redefining the trade union role' in N. Räthzel and D. Uzzell (eds), *Trade Unions in the Green Economy*, London, Routledge; 89–100.

Conservative Home (2019), 'Sam Hall: Conservatives shouldn't ridicule or rage against Extinction Rebellion. Instead, we should trump them with our own ideas', www.conservativehome.com/platform/2019/10/sam-hall-conservatives-should-respond-to-extinction-rebellion-with-an-ambitious-market-based-agenda.html, 13 October.

Corporate Europe Observatory (2019), 'Big oil and gas buying influence in Brussels', 24 October.

Daly, H. and Cobb, J. (1994) *For the Common Good*, Boston, Mass., Beacon Press.

England, J. and Weekes, B. (1981), 'Trade unions and the State: a review of the crisis', *Industrial Relations Journal*, 12 (1); 11–26.

FitzRoy, F. and Papyrakis, E. (2016), *An Introduction to Climate Change Economics and Policy*, 2nd edn, London, Routledge.

Gärtner, S. (2019), 'An attempt at preventive action in the transformation of coal-mining regions in Germany' in C. Calgóczi (ed.), *Towards a Just Transition: Coal, Cars and the World of Work*, Brussels, ETUI; 135–154.

Green New Deal Group (2018), 'Jobs in every constituency', www.greennewdealgroup.org/wp-content/uploads/2018/09/GNDJobsReport9-18.pdf.

Heartland Institute (2019), 'Socialist Green New Deal threatens freedom, prosperity, and the environment', *Climate Change Weekly*, 327, June 21.

HSE (2019), 'Historical picture statistics in Great Britain, 2019 – Trends in work-related ill-health and workplace injury', Health and Safety Executive, hse.gov.uk.

House of Representatives (2019), 'Resolution presented to House of Representatives: recognizing the duty of the federal government to create a Green New Deal', 5 February.

Irfan, U. (2019), 'The Green New Deal is fracturing a critical base for Democrats: unions', *Vox*, June, www.vox.com/2019/5/22/18628299/green-new-deal-labor-union-2020-democrats.

ITUC (2020), *Global Rights Index 2020*, Brussels, International Trade Union Confederation.

Jackson, T. (2011), *Prosperity Without Growth: Economics for a Finite Planet*, London, Earthscan.

Klein, N. (2014), *This Changes Everything*, London, Penguin.

Klein, N. (2019), *On Fire: The Burning Case for a Green New Deal*, London, Allen Lane.

Labour Party (2019), 'The Green New Deal explained', https://labourgnd.uk/gnd-explained.

Laurent, E. (2020), 'Reimagining a just transition', in *Social Europe (2020) Just Transition: a social route to sustainability: Social Europe Dossier*, Berlin, Social Europe/Friedrich Ebert Stiftung; 1–6.

Laybourn-Langton, L., Rankin, L. and Baxter, D. (2019), *This is a Crisis: Facing up to the Age of Environmental Breakdown*, Initial Report, London, Institute for Public Policy Research.

Leichenko, R. and O'Brien, K. (2019), *Climate and Society: Transforming the Future*, Cambridge, Polity.

Lynas, M. (2012), 'Review of J. Delingpole, Watermelons: How Environmentalists are Killing the Planet...' in *The New Statesman*, 16 February.

MacInnes, J. (1987), *Thatcherism at Work*, Milton Keynes, Open University Press.

Middlemas, K. (1980), *Politics in Industrial Society*, London, Andre Deutsch.

Moley, R. (1939), *After Seven Years*, New York, Harper.

Muûls, M. *et al.* (2016), 'Evaluating the EU Emissions Trading System: take it or leave it? An assessment of the data after ten years', *Grantham Institute Briefing Paper*, No. 21, London, Imperial College.

Office of National Statistics (2019), 'The decoupling of economic growth from carbon emissions: UK evidence', www.ons.gov.uk/economy/nationalaccounts/uksectoraccounts/compendium/economicreview/october2019/thedecouplingofeconomicgrowthfromcarbonemissionsukevidence#main-points.

Oreskes, N. and Conway, E. (2012), *Merchants of Doubt*, London, Bloomsbury.

Raworth, K. (2017), *Doughnut Economics: Seven Ways to Think Like a 21st Century Economist*, London, Penguin.

Real World Coalition (2001), *From Here to Sustainability: Politics in the Real World*, London, Earthscan.

Romm, J. (2018), *Climate Change: What Everyone Needs to Know*, Oxford, Oxford University Press.

Snell, D. and Fairbrother, P. (2013), 'Just transition and labour environmentalism in Australia', in N. Räthzel and D. Uzzell (eds), *Trade Unions in the Green Economy*, London, Routledge; 146–161.

Stern, N. (2006), *Stern Review on the Economics of Climate Change*, London, Cabinet Office, HM Treasury.

Thiele, L. P. (2016), *Sustainability*, 2nd edn, Cambridge, Polity.

Tooze, A. (2018), *Crashed: How a Decade of Financial Crises Changed the World*, London, Penguin.

Unearthed (2018), 'Inspections and pollution tests drop as Environment Agency sheds thousands of staff', https://unearthed.greenpeace.org/2018/12/08/environment-agency-pollution-inspections-cuts-rivers.

van der Ree, K. (2015), '10 action points towards a greener economy', Geneva, ILO, https://www.ilo.org/global/about-the-ilo/newsroom/comment-analysis/WCMS_429777/lang--en/index.htm.

Voet, L. (2020), 'A Just Transition Fund: one step on a long march', in *Just Transition: A Social Route to Sustainability: Social Europe Dossier*, Berlin, Social Europe/Friedrich Ebert Stiftung; 7–11.

Warner, J. (2019), 'Labour's Green New Deal? No, just puerile, delusional nonsense dressed up as industrial strategy', *Daily Telegraph*, 5 November, https://www.telegraph.co.uk/business/2019/11/05/labours-green-new-deal-no-just-puerile-delusional-nonsense-dressed/.

Watts, J. and Duncan, P. (2019), 'Tory MPs five times as likely to vote against climate action', *The Guardian*, 11 October, https://www.theguardian.com/environment/2019/oct/11/tory-mps-five-times-more-likely-to-vote-against-climate-action.

White, J. B. (2018), 'Elon Musk anti-union tweet spurs labour law violation charge', *The Independent*, 24 May, https://www.independent.co.uk/life-style/gadgets-and-tech/news/elon-musk-tesla-twitter-union-uaw-organizing-labor-nlrb-fremont-plant-a8368111.html.

Wilson, M. (2016), *Destructive Creation: American Business and the Winning of World War II*, Philadelphia, University of Pennsylvania Press.

World Commission on Environment and Development (1987) *Report: Our Common Future*www.un-documents.net/wced-ocf.htm.

11 Can employee-owned cooperative enterprises and public banks help save the planet?

Introduction: the failures of financial participation

We have seen that the current capitalist model of economic organisation all too often results in unsustainable poorly paid and precarious work and a disregard for the environmental consequences of the processes of extraction, use and waste: a disregard that is now threatening the planet's ability to support human social and economic activity as we have come to know it. This therefore raises the question of what sort of organisation is better suited to support a just transition to socially and environmentally sustainable work?

Workers have much at stake and much to lose when conventional capitalist organisations make decisions that affect them, for example through downsizing or relocation. Few employees are in a position to initiate new directions let alone challenge or propose alternative routes to those adopted by senior management. Even when employees are putative part-owners through company-organised equity allocations, shareholdings are minimal, even if employees retain or collectively pool their successive allocations. With the bulk of equity owned and controlled by institutional investors, whose allegiance to their investments is conditional on securing short-term gains, even aggregated employee shareholdings would be insufficient to influence major strategic corporate decisions. Further, as with the John Lewis Partnership, majority ownership does not confer majority control. Government-endorsed equity schemes offering minority allocations to employees are popular with managements as they are designed to fuse identity of interests between company and employees and for their alleged recruit and retain properties. They are also favoured by employees who are provided with a welcome profit-related bonus if the company performs well. Many fossil fuel companies most implicated in high carbon emissions have established employee share schemes, as have, of course, the financial institutions that do so much to support them. Nevertheless, far-reaching projects that aim to secure employees with more strategic control often face severe and ultimately fatal political and commercial opposition when companies' profit-optimising objectives and managerial prerogatives become compromised. Conversely, influential authorities like the UN (2019) and ILO (2019) maintain that companies owned and run by and for their employees are less answerable to external investors, less driven by profit and are more likely to adopt policies that are more environmentally and employment sustainable. The present chapter examines these claims and policies and evidence for their impact.

Worker cooperatives

Cooperatives (co-ops) are defined by the ILO (2011) as an 'autonomous association of persons united voluntarily to meet their common economic, social and cultural needs and aspirations through a jointly owned and democratically controlled enterprise'. The ILO/ICA (2014) estimates that worldwide about a billion people are involved in cooperative activity with at least 100 million employed in cooperatives. There is an assumption that cooperatives are (or should be) a sustainable form of business through satisfying their socio-cultural interests and protecting the environment (*ibid.*, 2014). Advocates, including the UN Secretary-General, argue that co-ops, as values-driven and principles-based enterprises, 'are natural vehicles to deliver the collaborative partnership and the people-centred and integrated approach required to attain the 17 Sustainable Development Goals' (SDGs, see Chapter 3), goals that include the need to take urgent action to combat climate change and its impacts and to offer sustainable economic growth and decent working conditions (UN, 2017).

The contributions of co-ops to these ambitions are considered below. There is, though, a wide variety of cooperative forms, including those based on consumers, producers, community, finance and workers. Most claim to subscribe to the core principles of cooperativism, namely, to provide voluntary and open membership; democratic control and economic participation by members; autonomy and independence; provide education, training and information for members and community; cooperation among cooperatives; and concern for the community. To which in recent years there have been 'significant efforts within the international cooperative movement to add ... commitment to environmental sustainability' (McMahon, 2019). As our focus is primarily on links between work and the environment and workers have significant investment in both, we concentrate primarily on worker cooperatives. Nevertheless, other forms of co-ops that comply with the principles outlined above may also aspire to environmental and employment sustainability.

Cooperatives are prevalent in Europe, with some 25,000 in Italy, several thousand in Spain and about 2,000 in France. In 2009, there were upwards of 200,000 cooperatives in the EU, in which some 4.7 million people worked (North and Scott Cato, 2018: 6). Indeed, in the global North, co-ops actually outnumber conventional firms (Utting, 2015). Co-ops are strongly represented in Latin America, where they have a long and sometimes turbulent history, and the Caribbean, with some 121,000 based in these two regions. Though all sectors are represented, a quarter of co-ops are associated with rural undertakings (agriculture, forestry, fishing, etc.) with prominent representation also in the finance sector through savings and credit unions. More recently, co-ops have become associated with the broader social and solidarity economy (SSE) movement, which is founded on inclusive growth, solidarity, cooperation and community development, in opposition to conventional competitive neo-liberalism, 'centred on self-interest, profit maximisation and consumerism' (Cooperatives Europe, 2015: 6). Despite endorsement from international and national institutions, there has been critical questioning of cooperatives, from supporters of workers' control of industry and from more conventional sources who contend that diverting from profit-seeking jeopardises optimum performance and that cooperatives may also rely on public funding for viability. Conversely, it could be argued that public support is actually of wider social benefit *because* cooperatives operate toward broader objectives than profit alone.

Worker cooperatives, in particular, have always faced difficulties, both conceptual and practical. From the conceptual perspective has been the challenge, going back 100 years to the Webbs, that rather than progressing working-class ambitions, cooperatives inhibit them. It has been argued that workers lack management expertise and skills, that cooperatives would act only in the selfish interests of members rather than to further class interests, and in a rapidly evolving industrial society, would remain as isolated islands of self-interest (Ackers, 2010). In a debate that rumbles on today, opponents argue that trade unions and collective bargaining are the appropriate participative instruments to advance working people's interests (*ibid.*: 58). Further, worker cooperative viability is highly sensitive to economic fluctuations and faces threat of collapse or potential degeneration into standard profit-orientated business through lack of access to external capital, leading Blumberg to comment that they have 'proved both economically and socially an inappropriate vehicle for workers' management ... In the Western world, they are economically inconsequential' (1968: 3). From a sustainability and social perspective, the need to rely on finance from internal sources and competing effectively with conventional enterprises could pose problems for co-ops when their survival is at stake (Singer and Primavera 2018: 196). Commentators have also argued that workers have little interest in participating in higher-level organisational decisions (Pateman, 1970: 85) and, quoting J. S. Mill, Pateman questions whether cooperatives actually benefit from universal suffrage: 'though everyone ought to have a voice – that everyone should have an equal voice is a totally different proposition' (*ibid.*: 32). Some studies confirm the potential democratic and equality vulnerability of cooperatives. Research into five mixed-sex worker co-ops in Buenos Aires showed that in 2006 all the workers were paid the same. By 2011, only two women-dominated co-ops came close to equal pay; in the other three, masculinised hierarchical pay scales paralleled decline in workplace democracy whilst the workplace became increasingly uncongenial for women (Oseen, 2016).

A further concern has been that where unions are recognised, their functions could be compromised between upholding enterprise objectives (sustainable or profitable operations) and protecting employee interests, for example in securing wage increases (Pendleton, 2005). Cooperatives also face the danger of 'self-exploitation' for workers who may deny themselves adequate rewards in the greater interests of the institution and its members. This may even extend to worker safety. A study of plywood mills worker co-ops in the Pacific Northwest of America found a much higher rate of accidents compared with conventional mills. 'The workers in the co-op are more inclined to compromise on safety in their desire to meet production targets' (Pencavel, 2002: 36). Apart from the plywood experience, worker co-ops in America have led a patchy existence, occasionally revived during periods of economic crisis (Logue and Yates, 2001: 7) or squeezed into small-scale low capital or niche areas. The same authors argue that while cooperatives may be suited for limited scale artisan operations, they pose the rhetorical question: 'how do you start up a cooperative steel mill or auto plant' when access to capital is limited (*ibid.*: 7)? A recent census study of American worker co-ops confirmed their relatively small number, approximately 400 in 2017, with an average size of just nine workers and gross annual revenue of about $600k (Palmer 2019), though, of course, there are far greater numbers of active producer and consumer cooperatives.

Other observers are more sanguine about cooperative performance. Pérotin (2012) argues that cooperative survival rates compare favourably with conventional firms of equivalent size; that they perform equally as effectively; and also tend to protect job security, possibly at the expense of member salaries. Some researchers go further in stating that survival rates for co-ops are higher than those of conventional enterprises (Olsen, 2014; Logue and Yates, 2005). Arguing against the marginality thesis, Pérotin notes that cooperatives are found in all sectors, including capital intensive operations. The main questions are whether, or in what ways, do worker cooperatives behave differently to conventional businesses and whether differences are reflected in approaches to the environment. The largest and most influential cooperative enterprise is the Mondragon Group (MG), based in the Basque region of Spain, and in 2018 with some 82,000 jobs occupied in manufacturing and a range of different services, it is worth examining the trajectory of this flagship cooperative in some detail. As recently as 2012, when it had an annual turnover in excess of €14bn, *The Guardian* reported that Mondragon represents 'a stunningly successful alternative to the capitalist organization of production' (Wolff, 2012), though in recent years Mondragon has ventured increasingly into international markets and with it, questioning accounts of its operations in a competitive global marketplace have gathered force. The history of Mondragon starts with the efforts of one anti-fascist priest to establish a cooperative system in the Basque region after World War II. The project was internally funded by a locally established and worker-owned credit union, which was involved within the cooperative system. The market grew strongly, and the original cooperative became established household appliance manufacturer, FAGOR, with some 8,000 employees at its peak. Over time, the cooperative diversified into four strategic areas: manufacturing, finance, retail and knowledge. In 2012, manufacturing alone consisted of over 120 separate businesses. The overall structure is overseen and governed through ten principles, including: admission open to all; democratic and participatory self-managed governance based on one worker, one vote; capital as the instrument of labour; pay solidarity with levels set by worker-owners; strategic cooperation between member co-ops, including flexible deployment of members to preserve job security; vision to protect and transform local and international communities; education as a core value (Davidson, 2012).

One early criticism was that Mondragon workers became too internally focused and tended not to offer support to trade union or political activism outside the cooperative even though research demonstrated little difference in shop-floor conditions and negative rank and file attitudes compared with a nearby unionised conventional factory. 'Worker-ownership did not shield them from factory regimes that were devised for profit maximization and workplace discipline in the capitalist market' (Kasmir, 1996) and as Cheney *et al*. (2014) commented, survival strategies have led to the establishment of mainstream management techniques. Another question concerns the hollowing out of those solidarity principles between internal and external actors that contribute to cooperative fundamental values. Altuna-Gabilondo (2013) suggests that these values have given way to a form of 'bureaucratically administered solidarity', which closes down rather than opens boundaries between Mondragon and its external communities. On the other hand, in 2009, MG and the United Steel-workers union in America signed a partnership memorandum designed to encourage unionised cooperatives in depressed post-industrial regions of the USA.

In another crucial development, MG embarked on a programme of international expansion in the early 1990s, and by 2018, there were 141 subsidiaries abroad, with a workforce of 14,400 (Mondragon, 2018). Internationalisation was accompanied by subsequent significant and potentially fundamental changes to workforce composition, with Kasmir (2016) identifying three tiers of workers: full co-op members in the Basque region with employment security; temporary workers in Spain; and employed wage earners in foreign subsidiaries. A study of Mondragon subsidiaries in China was revealing (Errasti, 2015). To maintain competitiveness, the cooperative's main decision-making body had opted for an internationalising strategy and acknowledged insecurity and poor working conditions for overseas workers for whom low pay, intensive working regimes and long hours were the norm. Most of the Chinese subsidiaries were SMEs. Errasti introduces the term 'co-optalist' to suggest hybridisation between conventional capitalist multinational management control and localised community-based cooperation based on self-management. In an earlier review, Defourny and Develtere (1999) identified globalisation and heightened competition as the driving force behind co-optalism, in which cooperatives adopt the dominant practices of their sector, including financial concentration, diluting membership and weakening member control in their attempts to survive.

Errasti studied 11 subsidiary companies based near Shanghai. These sites were greenfield, and numbers of workers ranged from 15 to 250. Errasti found that there were no membership rights to workers in these companies or in other foreign subsidiaries. Unions were not recognised and there were no other representative forums for workers. Management practice replicated that of conventional companies with no commitment to cooperative principles or to social development. Working hours were long, often supplemented by weekend overtime working and management faced few obstacles in dismissing workers over alleged poor performance. In classical Taylorist terms, one manager in the study asserted: 'You have to stay on top of them all the time, they are like robots' (*ibid.*: 490). Not surprisingly, these 'robots' tended to resist conventional performance-directed HRM policies and practices (Basterretxea *et al.*, 2019). Errasti concludes that under competitive conditions, these subsidiaries did display characteristics of cooperative degeneration toward conventional cost-primacy management practice. So, though Mondragon presents itself as 'Humanity at Work', it appears that this is restricted to membership only.

The 2018 Annual Report also proclaims commitment to 'sustainable development of our businesses, our society and our planet', but offers little evidence to suggest that environmental practice is of a higher standard than conventional enterprises. Nevertheless, a visit to Mondragon by an American advocate for worker-ownership suggested that though environmental philosophy and practice remain undeveloped there is potential for progress:

> I did not see much traction in Mondragon in either practice or articulated values. There is nothing in cooperativism that is inherently greener than any other structure. On this issue, there is good news on three fronts. First, as environmental responsibility becomes an important consideration for both customers and worker-owners, it is likely to find its way into the mission and function of the cooperatives. Second, as the environmental crisis worsens, it will create many new business opportunities for companies with the technical skills to address them. Mondragon, with its strong technical and industrial skills, will be well positioned to take advantage of these new markets. Third, the sustainability orientation of Mondragon is potentially further along than I could observe on the ground.
>
> (Bamburg, 2017)

Another major test for the movement and employment sustainability was the bankruptcy of FAGOR in 2013, with the loss of some 5,600 jobs, though many members were subsequently redeployed. Following rapid early expansion, the core manufacturing company had been struggling since the 2008 financial crash and collapse of the Spanish housing market. By 2013, FAGOR's debt reached €1.1bn and funding support was terminated by the Central Council of the Cooperative. Street demonstrations and factory occupations followed the closure. Apart from the trauma of closure, bankruptcy mainly affected the jobs of workers in the overseas subsidiaries, who did not enjoy cooperative privileges (Kasmir, 2016). Errasti and his colleagues suggest that the company failed owing to a complex of issues but assert that 'it went too far, too fast and took too many risks'. Dow (2003) also suggests that Mondragon cooperatives can be criticised for focusing on becoming internationally competitive rather than on social values of the cooperative movement. However, Errasti *et al.* (2017)

emphasise that the company had survived for many years in an increasingly volatile and international market that might have led a conventional firm to face survival problems much earlier. Nevertheless, there is no doubt that 'economic crises ... have significantly transformed and transfigured the Mondragón experience', which in turn will test the 'organisation and *ethos* of the cooperative movement' (Barandiaran and Lezaun, 2017: 290). Further, the then conservative Spanish government introduced deregulatory economic policies encouraging competitive product and labour market behaviour from the effects of which cooperatives, alongside eco-unionism, were not immune (Gil, 2013). This was not the first time that cooperative principles were undermined by government policy, as Harnecker (2009a: 335) reminds us: 'It was precisely the establishment of market reforms that diluted cooperation among Yugoslav self-managed enterprises and between them and the rest of society and promoted a behavior that resulted in greater inequalities and unemployment'.

Of course, problems of surviving with limited size, resources and operating in a competitive international arena have affected the Mondragon culture and it would be wrong to generalise on the ethos and practice of cooperatives from the mixed experience of one organisation. Nevertheless, there is no doubt that economic pressures can lead cooperatives to shift from their founding values: for example, a process of gradual cooperative dilution has been observed in the kibbutz movement in Israel, where differential salaries have become common and overall, half the labour undertaken in the kibbutz is by paid non-members (Russell *et al.*, 2011). Other cooperatives operate according to different demands and contexts and in consequence different features may be emphasised, though studies suggest at least formal adherence to common principles. A meta-analysis of websites and reports from cooperatives and federations mainly based in North America and Europe showed that though cooperative principles endorse social dimensions of activity, with strong focus on community aspects, they are relatively silent on environmental and economic features, possibly because these are not differentiated from the more general sustainability principle (Dale *et al.*, 2013).

Latin America has a long and varied history of cooperativism, notably in agriculture. Recent years have seen a resurgence in interest and practice. For example, in Venezuela there were an estimated 877 cooperatives in 1998, which swelled to anywhere between 30,000 and 60,000 ten years later, involving 14 percent of the country's labour force (Harnecker, 2009a). This growth was heavily influenced by legislative support and general endorsement by then President Chávez, hoping to diversify the economy and shift from its dependence on oil production (Larrabure *et al.*, 2011) as part of the 2005 'Bolivarian revolution' to counter the economic and cultural hegemony of American neo-liberalism. Promotion and protection of cooperatives along with other major socialising reforms became the basic organising tool of a new 'social economy' driven by community level enterprise (Buxton, 2019).

According to Larrabure *et al.* (2011: 182), developments toward a 'new cooperativism' based on socially responsible and community responsive Socialist Production Units were stimulated by three factors: offering a populist practical and ideological riposte to neo-liberalism; seeking reduced organisational hierarchy in decision-making and pay; and objectives to reinforce links with communities and social movements. The cooperatives were initially viewed by policy makers as an economic model to promote collective well-being rather than capital accumulation (Harnecker, 2009a: 310). This ambition remained elusive in an economy that became increasingly crisis-ridden as the governments headed by Chávez and then Maduro were confronted by escalating American opposition, with external pressures leading to internal disruption, exacerbated by major strikes and the collapse of oil prices in 2014. Once the region's richest economy, in 2016 growth had collapsed by ten percent and inflation soared by 720 percent amidst rising poverty, inequality and vigorous demonstrations against the government. In these volatile circumstances commentators agreed that many cooperatives were under pressure to behave 'like capitalist enterprises' that seek 'to maximize their net revenue without consideration of the ways they could alleviate the problems of their surrounding communities' (Harnecker, 2009a: 316–317). As early as 2008 problems were being exposed, with cooperatives demonstrating serious motivational, administrative and technical deficiencies compounded by difficulties in competing with conventional enterprises and lack of coordination among cooperatives. Their viability was also compromised by reliance on the state for capital inputs and funding (Harnecker, 2009b).

Hence for worker cooperatives to rise above their 'economically inconsequential' status, support from the state is critical for them to pursue and attain community and environmental objectives. Such support can be justified, bearing in mind that profit-seeking companies are not required to follow environmental objectives and indeed, as earlier chapters show, often disregard environmental and pollution implications of their commercial activities or even oppose attempted regulation of these same activities. Commercial organisations are themselves frequent beneficiaries of state largesse through generous subsidies and governmental funding of overseas climate-damaging fossil fuel developments (Hodgson, 2020).

Other cooperatives

Cooperatives are united by adherence to a number of core principles. One of these, concern for the community, could feasibly extend into environmental care, to which a second democratic control principle should also be relevant as this concerns autonomous governance by the membership, usually reflected in policies based on one member, one vote. Under these circumstances, members can determine environmental policies and also help to maintain secure and meaningful employment. Through member advocacy, at least some producer and retail cooperatives are taking steps to reduce their carbon emissions. The UK consumer Co-op group operates 3,000 stores and has around 4.6 million members. Group Head Office publishes an annual report that goes beyond corporate social responsibility-speak in describing the ways in which it meets core cooperative principles, including that of democratic member control. The 2019 report also identifies ways it is meeting, or aiming to meet, sustainability and environmental responsibilities, stating that the Co-op was complying with the UN's SDGs by halving its GHG emissions between 2006 and 2017. It focuses purchasing on Fair Trade products, sales of which increased by 6.3 percent in 2018. All of its electricity has been produced from renewable, UK wind farm sources. Actions were being taken in ensuring that all palm-oil and soy-based products were from certified sustainable sources (Co-Operative Way Report, 2019).

Banking cooperatives can also have a role in financing sustainable activities. It has been calculated that between US $5–7tn annually is needed in financial support for the UN's SDGs but so far relatively little has been forthcoming (Marois, 2017a). However, whilst funders seeking quick returns continue to reward GHG industries, a number of financial institutions have taken more progressive steps. Marois (*ibid.*) has recently researched two institutions, a banking cooperative in Costa Rica and a public and state-owned bank in Germany. In 2016, the *Banco Popular y de Desarrollo Communal* (BPDC) had total assets of US$5.4bn, with (in 2015) over 4,000 employees located in 103 branches. The co-op is legally owned by its 1.2 million savers, representing about a fifth of the Costa Rican population. Shares are offered to anyone who holds a savings account for at least one year.

The governing body is the Workers' Assembly, consisting of 290 representatives drawn from ten social and economic sectors. The Assembly is guided by five main principles, namely: promotion of the social economy; quality provision of services; competitive management of the institution; regional and local development; and as a facilitator for development. Banking decisions are also guided by principles of gender equity, accessibility, environmental responsibility and the collective welfare of society (Marois, 2019). Though the Assembly makes core decisions, operational control is in the hands of the National Board of Directors, consisting of four members representing the Assembly and three from the government. Committed to gender equality, four of the Board were women in 2016 (Marois, 2017a).

Marois points out that from environmental and community perspectives, the bank prioritises support for cooperative ventures, public institutions and for citizens otherwise denied financial support. The bank's financial sustainability is based on meeting clear environmental objectives, for which it has developed dedicated green lending facilities and produces annual internal sustainability reports. The bank also benefits from having governmental support for its activities. Funding has been utilised to back a programme to purchase and install residential solar panels and another to support local community associations to provide clean water systems. Support is demonstrated by the bank's alliance with a regional energy co-op, COOPELESCA, which supplies about a tenth of Costa Rican energy. In 2013, the energy was certified by the government as carbon-neutral and by 2015 it had fully offset its carbon footprint. The bank has also helped finance an initiative by COOPELESCA to purchase land for conservation purposes when it was at risk of being over-farmed with potential contamination of water supplies.

Nevertheless, descriptors such as 'sustainable' and 'ethical' in commercial and financial undertakings do need to be treated with caution, as the fate of the UK Co-operative Bank, proudly self-proclaimed as the 'ethical bank', demonstrates. The bank was long established as the banking facility of the Co-operative Wholesale Society, which supplied the group's stores, many of which had been experiencing difficulties in competing with the growth of cheaper supermarkets in recent years. Growth by acquisition appeared to offer a solution for the bank's viability, though difficulties accumulated with the takeover of another faltering financial institution, discussions over which the Board was not involved. There was financial misreporting (Gosling, 2018) and it is likely that a 'due diligence' search, which would have revealed the takeover target's substantial shortcomings, was not undertaken. Problems for the Co-op Bank accumulated with the attempted takeover of another bank's branches, which led to eventual exposure of the Co-op's multiple deficiencies. Before and during the crisis, the bank's managers continued to emphasise its ethical behaviour, even though this was later discredited when mis-selling of financial products was also revealed. The bank is now largely in the hands of international hedge funds but continues to trade on its ethical credentials (Gosling, 2019). Though many of the problems of the bank were attributable to the incompetence of its managers, difficulties derived, in part at least, from ambitions for a cooperative institution to succeed in an increasingly competitive financial market.

Diverse forms of cooperatives operating in differing contexts do, however, face similar issues. Cooperatives in Latin American countries have often been associated with socialist ideas and political parties. These links have made co-ops highly vulnerable either to reversion to capitalist operations or to collapse when more reactionary governments take over, often instigated or supported by outside commercial and political interests (Vásquez-Léon 2010). Many cooperatives have had to meet the competitive challenges of globalisation, whether in manufacturing or agriculture, where price and output demands from powerful buyers can impose pressures on cooperatives that can detract from their core principles. A case study by Burke (2010) of the relationship between AmazonCoop, a Brazil-based cooperative linking indigenous nut harvesters and a multinational 'socially responsible business', reveals the fragility of cooperative members in such dependent situations. Pressures on the cooperative increased local member vulnerability; fostered discriminatory practices and hampered self-governance. Burke (*ibid.*: 49) concludes that 'the profit motive continues to top the list of priorities, which means that efficiency, flexibility, rapidity, and image must be the top concerns'. Linked to the corrosive character of globalisation are the deregulatory free-market policies adopted in varying degrees throughout the world, which have impacted negatively on indigenous cooperative farming and agricultural communities. Reductions in subsidies and other forms of protective state support have also constrained cooperatives notwithstanding their sponsorship by the UN and ILO and recognition that these organisations can help mitigate GHG emissions through reforestation and sustainable activities extending from agriculture to manufacture. However, without committed governmental and infrastructural support, cooperatives will always be hostage to the vicissitudes of unsympathetic economic and political forces, pushing them to the margins of commercial activity or forcing them into the same environmentally neglectful cost and control frameworks associated with capitalist enterprises (Gasper, 2014). We are then faced with the question of who cooperatives can rely on for structured and embedded support when they are overlooked by, or meet resistance from, the institutions and powers that financially benefit, albeit in the increasingly narrowing short term, from the existing system?

Public banks

We have seen that financing a green transition is proving highly problematic as conventional investment banks primarily aim to secure profit, not serve the public interest. Despite the Paris targets, leading private sector banks continue to fund fossil fuel companies. A major report headed by a respected group of environmental NGOs showed that 35 leading banks together invested $2.7tn in fossil fuels since the Paris agreement was adopted. Further, overall financing from these banks to fossil fuel interests has increased each year since the Accord as has financing for major oil and gas companies operating in the Arctic (Fossil Fuel Finance Report, 2020). There have been reports that funding for the controversial Keystone XL pipeline, a project that had been blocked through lack of funds and by widespread environmental protest, has received financial and political support to proceed as a 'critical' project during the coronavirus emergency, at a time when protest assemblies would be unlawful. The UN's 2019 Report, *Financing for Sustainable Development*, strongly recommends reshaping national and international financial systems in support of the development goals, otherwise 'we will fail to deliver the 2030 Agenda' (2019, xvii). Among its recommendations the Report confirms that 'governments should continue to strengthen the enabling environment including by considering appropriate financing sources' (*ibid.*: 53). One potential source is through public banks. Referring to Germany's *Sparkassen* (savings banks), the Report identifies that

> as public banks, the *Sparkassen* mandate is to serve the economy and people in the local region. This mandate also includes pursuing economic viability rather than profit maximization … this mandate allow *Sparkassen* to align their business operations more closely with sustainable development.
>
> (*ibid.*: 66)

Clearly, these banks could provide the investment and backing to help reduce emissions, support local communities and endorse environmentally positive enterprises. Public banks have been defined as banks owned and controlled by state authorities or by public enterprises. Public-like banks are cooperative institutions owned collectively and operated broadly in the public interest (Marois, 2018). There are nearly 700 public banks globally, with some US \$38tn assets, representing nearly half of global GDP (Marois, 2019). These banks are found in economically developing countries such as Vietnam as well as mature economies like France with its well-established *La Caisse des Dépôts*. Until recently, the only public bank in the USA was the Bank of North Dakota, established 100 years ago to support the state's farming community, but recent Californian legislation paves the way for cities and other local bodies in the state to establish public banks. The Californian success has led to interest by other municipalities and states such as New York, Philadelphia and New Jersey in establishing public banks (Weinberger, 2020). Marois indicates that the major distinction between conventional financial institutions and these banks is that, like *Sparkassen*, they are mandate, rather than profit, driven. Mandates include supporting developmental goals from municipality to international level. Projects vary in scope, including water infrastructure, agriculture and general pursuit of sustainable development.

Germany's KfW (*Kreditanstalt für Wieder-aufba*, Reconstruction Credit Institute) is an example of a mature long-established public bank, with formal, stated aims to provide 'sustainable improvement of the economic, social and ecological conditions of people's lives' (Marois, 2017a: 8). The bank was established in the aftermath of World War II to help administer Marshall Plan reconstruction funds. In 2016, it was lending €81bn, of which €35bn was directed toward environmental and low carbon projects. Green lending is a prominent feature, for example, in support of renewable energy. KfW supports greening of German municipalities and local authorities through loans supporting energy efficient refurbishment projects. The bank does not aim to maximise returns but supports a range of activities including climate and environmental protection. The Federal Republic owns 80 percent of shares and Federal States the remaining 20 percent. The shareholders are represented by a Board of Supervisory Directors, with senior Federal ministers acting as chair and deputy chair. The Board is composed of 37 additional members, representing industrial groups, housing, agriculture and trade unions.

These examples demonstrate that public banks have potential financial capacity to support the Paris agenda, though experience shows that they can be influenced by the same competitive pressures and managerial ambitions as conventional profit-optimising banks. Moreover, not all public banks are exclusively directed to environmental priorities and some others suffered badly during the financial crisis (Hallerberg and Markgraf, 2018). Critics warn that to serve environmental and social needs, public banks, as demonstrated by *Banco Popular*, need to be democratically accountable. Pressures on democratisation, and hence on meeting the UN's environmental objectives, continue from the neo-liberal axis. Marois (2017b) warns of the vulnerability of the financial services sector from the proposed Trade in Service Agreement (TiSA). Negotiations over this trade agreement were initiated by the USA in 2012 and subsequently driven by America and the EU. Negotiations have been conducted between 23 members of the World Trade Organization, representing about 70 percent of world trade in services. These secretive discussions have continued for a number of years but are presently suspended. The proposed agreement aims to liberalise the economy through global market-based competition for services on the basis that unfettered competition leads to optimum efficiency. As we have seen, public interest banks are not required to be profit-maximisers, which could lead to legalised challenges under TiSA that overseas institutions face unfair market obstacles as public banks rely strongly on government to support and finance social objectives. An eventual trade agreement would prioritise profitability, competition from overseas providers and privatisation over regulation. As Marois (2017b: 2) cautions: 'TiSA is an attempt to institutionalise financialised imperatives globally'. Public services would be vulnerable to commercial involvement and accelerated risk of privatisation.

Public banks have clear associations with ideas of the social and solidarity economy (SSE), which has support from the UN, through its transformative potential towards providing decent work, environmental protection, sustainability and endorsement of equality. In contrast to the self-regulating market economy, the 'SSE sector is associated with democracy, with the argument that people should be able to shape their lives rather than only have the opportunity to sell their labour power in return for low wages and poor working conditions' (North, 2018: 75). Advocates concur that while the SSE project is both loosely defined and largely aspirational, the broader topic of economic democracy in which it is located has received revived attention in recent years through failure of free-market policies to address the climate emergency, provide decent and secure work or to engender equality.

Economic democracy and employee-owned enterprises

Economic democracy and employee-owned enterprises are linked but conceptually different aspects of economic behaviour. Employee-owned (EO) businesses operate at the micro (firm) level for most economies, and with the number and influence of employee-owned enterprises being limited, impact on major environmental concerns will always also be limited. Economic democracy concerns the operation of the economy more broadly. At this macro level, globally dominant free-market policies, whether fundamental or partially regulated, linked to the twin covert impacts of globalisation and financial control ensure that economic democracy as a practical exercise will remain a phantom rather than reality at a time when social and political democracy are themselves under threat from the same forces.

In considering EO businesses, first we need to recognise that these are different animals to the many companies that offer limited proportions of equity and no significant control to their employees. This is something that a recent endorsement of EO prepared on behalf of the UK government failed to distinguish. The 2012 *Nuttall Review of Employee Ownership* defined EO as a 'significant and meaningful stake in a business for all its employees' (2012: 20), with a 'significant and meaningful' ownership proportion identified at levels of 25 percent or even lower. To complicate matters further, the Review points out that: 'In practice, employee ownership takes many forms ... the literature reflects this diversity – each study uses its own definition of "employee ownership"' (*ibid.*: 23). This apparent diversity is confirmed by other writers. Gunderson (2019: 39), for example, contends that 'worker decision-making power' can extend from being notified about decisions to majority representation on major decision-making bodies.

A principal ambition of the Nuttall Review was to focus on the contribution of EO (however defined) on business performance and the contributions of employee-owners to it. There is an assumption of undisturbed organisational hierarchy under EO in the review with little or nothing to say about redistributed control over strategic decisions or links to the wider community. The orientation is strongly unitarist. The report reflects the perspective, promoted by governments, of voluntary adoption by enterprises to EO, supported by inducements, such as tax relief, with the prime aim to promote employee productivity and enterprise performance. At the same time, there are arguments from more radical sources that in common with cooperatives, fully employee-owned and directly controlled enterprises could be more in tune with environmental and community protection as well as providing employment security and meaning to employee-owners (Erdal, 2011). One problem of course is that employee-owned companies neither reflect nor represent the functioning of the economy. In these circumstances, fully employee-owned companies, whatever their positive implications for employee sovereignty, are (a) always likely to be in a minority in a market economy and (b) subject to the same financial constraints facing conventional enterprises though these constraints in terms of access to external capital investment and its terms for provision may be tighter. For this reason, many employee-owned companies are located in service and professional sectors rather than manufacturing (Stern, 2020: 11). Numbers of fully employee-owned and controlled enterprises will be modest, potentially confined to distinctive areas of the economy and despite the best efforts of participants, often at risk of degeneration, takeover or demise, especially under conditions of extreme competition when employees, harbouring fears of losing both jobs and equity, (re)turn their fully owned companies to the limited protection provided by conventionally organised capitalist enterprises (Spear, 1999).

With regard to economic democracy (ED), there are also conceptual and practical problems in that it has no agreed definition. As the author of a recent article on the topic acknowledges: 'there is no systematic analysis of the environmental impacts of economic democracy due to a lack of an actually existing democratic economy to analyse' (Gunderson, 2019: 39). There are substantial reasons for this situation. We saw earlier that the many experiments in nationalisation through public ownership failed to confer democratic control by workers or their representatives, even though renationalisation of key public services retains considerable popular support. We must also acknowledge the massive shifts noted by Richard Hyman, which have taken place in the 'cancer stage of capitalism' during which globalisation has loosened national sovereignty over economic policy-making while 'financial capitalism is one of the principal grave-diggers of social democracy' (2015: 17). The adverse effects on workers have been well-documented along with failure or refusal by companies or governments to confront the gravity of the environmental crisis. Nevertheless, models of ED have been authoritatively presented as system alternatives to free-market economies, with the central premise being that workers, either individually or through elected representatives, control the directions of the economy and its constituent parts. We saw in earlier chapters that more ambitious collective ownership schemes such as the Swedish Meidner Plan to democratise the economy through challenging capitalist ownership and control have been successfully neutered by the same powers that fail to give support to cooperative endeavour. Cumbers (2019; 2020) argues that the present free-market system has failed to enfranchise working people in terms of employment, community and environment and proposes instead an economic democracy founded on three 'interlocking pillars' comprising individual economic rights, democratic collective ownership of business, and public participation in economic decisions.

Taking his arguments further, Cumbers asserts that everyone has the right to be a fully informed and active economic citizen in a democratic society, for which people need adequate resources. A potential partial means of achieving this is through provision of a universal basic income (see Chapter 12) along with other progressive policies aimed at providing working people with greater economic independence, such as living wage provision and a progressive tax system. Public banks, described above, can also have important informative and participative qualities. The second pillar offers diverse forms of collective ownership, recognising that earlier forms of public ownership such as nationalisation failed to offer democratic control, whether to workers or to the communities served by the undertakings. Remunicipalisation exercises noted earlier along with cooperative models examined above provide examples of how this might be undertaken.

The third policy pillar proposal widens citizen participation in macro-economic decisions. Cumbers (2020) cites a number of experiments of this kind. He reviews participatory budgeting, in which a proportion of the budget is allocated to citizens' groups who meet to decide investment priorities. This experiment was first conducted in one Brazilian city, then spread to other cities in Brazil and North America. In Brazil, impacts have been generally positive despite the small proportion of budget allocated to the citizen groups. A second proposal has been to make macro-economic policy-making more open and accountable to citizens, possibly through establishing councils of citizens to deliberate and comment upon constitutional and economic matters and reforms. Recently, the parliament representing the small German-speaking region of Belgium has handed some powers to a citizens' assembly, chosen by lot, in a permanent arrangement to involve citizens in policy and decision-making. German-speaking citizens are empowered to put issues on the agenda, discuss potential solutions and, crucially, be able to monitor recommendations as they pass through parliamentary and governmental levels (Van Reybrouck, 2019). Probably the most ambitious citizens' assembly is the

Convention citoyenne pour le Climat convened in 2019 by President Macron, in which 150 members of the public have been selected by lot to assess the question of 'how to reduce greenhouse gas emissions in France by at least 40% by 2030, in the spirit of justice'. The assembly is representative of the country in members, chosen in a stratified sample according to age, gender, qualifications, location and occupation. To date, the assembly has met on four occasions. Measures proposed by the convention are to be enacted through a national referendum, parliamentary vote or adopted directly as executive orders (Involve, 2020).

Some countries operate systems of direct democracy exercised through referenda. Switzerland is probably the best-known example, though in recent years, referenda have been held in Catalonia, in an unauthorised plebiscite in pursuit of independence from Spain (2017) and the UK over Scottish independence (2014) and whether to remain in the EU (2016). Neither of the latter was successful but demonstrated the problem of distilling complex issues into a simple binary yes/no choice. The Swiss system is more firmly embedded within the constitution and referenda can be held at communal, cantonal and federal levels. Citizens can initiate a referendum to challenge legislation at any of these levels. At federal level, there are three types of referendum in Switzerland: mandatory, which must take place if Parliament intends to amend the constitution; optional, which is triggered if 50,000 valid signatures are gathered and operates if a parliamentary decision is to be contested; and popular initiatives to change the constitution that require 100,000 signatures. Of course, topics for referenda need not be progressive and progressive topics can be defeated, as was the case in 2016 when a popular initiative to introduce a universal basic income was easily defeated and in 2019, when a proposal that at least ten per cent of newly built homes in Switzerland should belong to non-profit developers such as housing cooperatives also failed to gain majority support.

Conclusions

For more than 100 years academic and political debates have continued about potential forms of economic or industrial democracy to challenge the subordinate role of labour in the market economy, though practical progress has been slow, patchy and reformist rather than radical: and any advances are perpetually under threat from reversals, helped by the impact of economic cycle fluctuations, government policy and more recently, globalisation. As numerous commentators have pointed out, capital has always possessed the dominant ideology and ultimately the power to defend and advance its interests, even if it occasionally has to offer political or economic concessions at times of resurgent labour movements and support by political sympathisers.

In recent years, authorities like Cumbers (2020), Hyman (2015) and Marois (2017a; 2018) have conducted thorough and critical re-examinations of the potential for models of economic democracy to challenge profit-driven modes of production at a time of crisis for the free-market economy. Whilst these authors do not focus exclusively on organisational strategies toward safeguarding the environment, there is clearly potential for shared or exclusive worker control to free organisations to pursue social policies as well as provide secure and dignified work lacking in the 'precarity' economies prominent in most countries. Hence Hyman recognises a central role for reformed and open trade unions in shifting toward a more democratised workplace. He argues that unions must first be all-embracing models of internal democracy in order to engage fully with membership and to ensure that they are attractive to the widest potential membership. He also recognises that unions must act in alliance with other national and international progressive movements, of particular import with regard to combatting the climate emergency. Cumbers (2020) argues for an economic democracy established along three pillars of legislated individual economic rights, diverse forms of collective ownership, articulated as Marois indicates (2017a; 2018) in cooperative, fully employee-owned and remunicipalised forms, and creating 'deliberative and knowledgeable publics', in which trade unions would figure strongly. It cannot be assumed, however, that employee or popular control will automatically lead to environmental enhancement or pursuit of socially desirable objectives in sectors such as arms production: preserving jobs will remain a priority for unions and their members. To effect more radical change will require the committed financial and strategic support of government. Critically, genuinely cooperative financial institutions and public banks can and do endorse environmentally sound practice through their investments, but to be effective must be supported through state policies that stop subsidies and dissuade investments in fossil fuels whilst promoting investment in clean production and technology (Oberholzer, 2019). The state's legislative and policy role in fostering organisational change toward emissions control is vital and failure of most countries to meet Paris 2015 GHG targets imperils the global environment. There is no longer any justification, if there ever was, for the state to maintain its regulation-free stance. To do so invites further climatic disaster.

Movement is underway, with international bodies like the UN and its ILO labour standards agency recognising the failures of profit-seeking economies to combat global harm, environmental degradation or to ensure decent and secure employment. The imperative now is for these and other authoritative institutions to insist on radical change in corporate governance. There is no longer any doubt that a state of emergency exists, which requires that governments stop subsidising fossil fuels; that finance institutions withdraw timeously from investing in fossil fuels; that the public interest be recognised in the democratic oversight of industry; that different models of workplace citizenship be explored through worker directors, trade union participation in line with national traditions and development of cooperatives; that liberalising union membership and recognition be treated as established governmental policy; that company boards expanded with worker director members be required to adopt ecological agendas that must meet with internationally-agreed standards. The positive impact of the interventionist state has been vividly demonstrated during the coronavirus epidemic emergency and by those European governments embracing corporatist and coordinated economic policies in combatting GHG emissions. As Greta Thunberg has starkly warned, the planet is on fire, and governments have just one final opportunity to reject the failed neo-liberal model of self-governing markets and do what governments are expected to do – to govern on behalf of their citizens.

Bibliography

Ackers, P. (2010), 'An industrial relations perspective on employee participation', in A. Wilkinson, P. Gollan, M. Marchington and D. Lewin (eds), *The Oxford Handbook of Participation in Organizations*, Oxford, Oxford University Press; 52–75.

Altuna-Gabilondo, L. (2013), 'Solidarity at work: the case of Mondragon', United Nations Research Institute for Social Development, www.unrisd.org/thinkpiece-altuna.

Bamburg, J. (2017), 'Mondragon through a critical lens', *Fifty by fifty*, https://medium.com/fifty-by-fifty/mondragon-through-a-critical-lens-b29de8c6049.

Barandiaran, X. and Lezaun, J. (2017), 'The Mondragón experience' in J. Michie, J. Blasi and C. Borzaga (eds), *Oxford Handbook of Mutual, Cooperative and Co-Owned Business*, Oxford, Oxford University Press; 279–294.

Basterretxea, I., Heras-Saizarbitoria, I., Lertxundi, A. (2019), 'Can employee ownership and human resource management policies clash in worker cooperatives? Lessons from a defunct cooperative', *Human Resource Management*, 58; 585–601. doi:10.1002/hrm.21957.

Blumberg, P. (1968), *Industrial Democracy: The Sociology of Participation*, London, Constable.

Burke, B. (2010), 'Cooperatives for "fair globalization"? Indigenous people, cooperatives and corporate social responsibility in the Brazilian Amazon', *Latin American Perspectives*, 37 (6); 30–52.

Buxton, J. (2019) 'Continuity and change in Venezuela's Bolivarian Revolution', *Third World Quarterly*, doi:10.1080/01436597.2019.1653179, 27 August.

Cheney, G., Cruz, I., Peredo, A. and Nazareno, E. (2014), 'Worker cooperatives as an organizational alternative: challenges, achievements and promise in business governance and ownership', *Organization*, 21 (5); 591–603.

Cooperatives Europe (2015), *Building People-Centred Enterprises in Latin America and the Caribbean*, Brussels.

Co-Operative Way Report (2019), *Report*, Manchester, Co-Operative Group, April.

Cumbers, A. (2019), 'Economic democracy: why handing power back to the people will fix our broken system', *The Conversation*, https://theconversation.com/economic-democracy-why-handing-power-back-to-the-people-will-fix-our-broken-system-126122.

Cumbers, A. (2020), *The Case for Economic Democracy*, Cambridge, Polity.

Dale, A., Duguid, F. *et al.* (2013), 'Cooperatives and sustainability: an investigation into the relationship', *International Cooperatives Association*, www.crcresearch.org/sites/default/files/u641/131124_ica_sustainability_scan_final_01.pdf.

Davidson, C. (2012), 'The Mondragon cooperatives and 21st century socialism: a review of five books with radical critiques and new ideas', *Perspectives on Global Development and Technology*, January; 229–243.

Defourny, J. and Develtere, P. (1999), 'The social economy: the worldwide making of a third sector', in J. Defourny, P. Develtere and B. Fonteneau (eds), *L'Économie Social au Nord et au Sud*, Brussels, De Boeck.

Dow, G. (2003), *Governing the Firm*, Cambridge, Cambridge University Press.

Erdal, D. (2011), *Beyond the Corporation: Humanity Working*, London, Bodley Head.

Errasti A. (2015), 'Mondragon's Chinese subsidiaries: Coopitalist multinationals in practice', *Economic and Industrial Democracy*, 36 (3); 479–499.

Errasti, A., Bretos, I. and Nunez, A. (2017), 'The viability of cooperatives: the fall of the Mondragon Cooperative Fagor', *Review of Radical Political Economics*, 49 (2); 181–197, https://doi.org/10.1177/0486613416666533.

Fossil Fuel Finance Report (2020), 'Banking on climate change', www.ran.org/wp-con tent/uploads/2020/03/Banking_on_Climate_Change__2020_vF.pdf.

Gasper, P. (2014), 'Are workers' cooperatives the alternative to capitalism?' *International Socialist Review*, 93, https://isreview.org/issue/93/are-workers-cooperatives-alterna tive-capitalism.

Gil, B. (2013), 'Moving towards eco-unionism: reflecting the Spanish experience', in N. Räthzel and D. Uzzell (eds), *Trade Unions in the Green Economy*, London, Routledge; 64–77.

Gosling, P. (2018), *The Fall of the Ethical Bank*, Manchester, Coop Press.

Gosling, P. (2019), 'The rise and fall of the "ethical bank"', *Ecologist*, https://theecologist. org/2019/jan/15/rise-and-fall-ethical-bank.

Gunderson, R. (2019), 'Work time reduction and economic democracy as climate change mitigation strategies: or why the climate needs a renewed labor movement', *Journal of Environmental Studies and Sciences*, 9 (1); 35–44.

Hallerberg, M. and Markgraf, J. (2018), 'The corporate governance of public banks before and after the global financial crisis', *Global Policy* 9 (1); 43–53.

Harnecker, C. (2009a), 'Workplace democracy and social consciousness: a study of Venezuelan cooperatives', *Science & Society*, 73 (3); 309–339.

Harnecker, C. (2009b), 'Main challenges for cooperatives in Venezuela', *Critical Sociology*, 35 (6); 8941–8962.

Hodgson, C. (2020), 'Climate activists accuse UK lender of financing fossil fuel', *Financial Times*, 17 March, www.ft.com/content/6bd38b42-652e-11ea-b3f3-fe4680ea68b5.

Hyman, R. (2015), 'The very idea of democracy at work', *Transfer: European Review of Labour and Research*, 22 (1); 11–24.

ILO (2011), 'Promoting co-operatives: a guide to ILO recommendation 193', Geneva, ILO.

ILO (2019), 'Transforming our world: a cooperative 2030: Cooperative contributions to SDG 13', Geneva, ILO.

ILO/ICA (2014), 'Cooperatives and the sustainable development goals', Geneva, ILO.

Involve (2020), 'Convention citoyenne pour le climat: what can we learn from the French citizens' assembly on climate change?', www.involve.org.uk/resources/blog/opinion/ convention-citoyenne-pour-le-climat-what-can-we-learn-french-citizens, January.

Kasmir, S. (1996), *The Myth of Mondragon: Cooperatives, Politics and Working-Class Life in a Basque Town*, New York, SUNY Press.

Kasmir, S. (2016), 'The Mondragon cooperatives and global capitalism: a critical analysis', *New Labor Forum* 25 (1); 52–59.

Larrabure, M., Vieto, M. and Schugurensky, D. (2011), 'The new cooperativism in Latin America: worker-recuperated enterprises and socialist production units', *Studies in the Education of Adults*, 43 (2); 181–196.

Logue, J. and Yates, J. (2001), *The Real World of Employee Ownership*, Ithaca, Cornell University Press.

Logue, J. and Yates, J. (2005), 'Productivity in cooperatives and worker-owned enterprises: ownership and participation make a difference!', Geneva, ILO.

Marois, T. (2017a), 'How public banks can help finance a green and just energy transformation', Transnational Institute, Amsterdam, November.

Marois, T. (2017b), 'TiSA and the threat to public banks', Transnational Institute,
Amsterdam, April.

Marois, T. (2018), 'Towards a green public bank in the public interest', United Nations
Research Institute for Social Development, Working Paper 2018–2013, Geneva.

Marois, T. (2019), 'Public banks and a just and green transition', *Canadian Dimension*,
https://canadiandimension.com/articles/view/public-banks-and-a-just-and-green-transi
tion, July.

McMahon, C. (2019), 'The political economy of worker cooperative development: Mei-
theal and sustainability', Unpublished doctoral dissertation, National University of
Ireland.

Mondragon (2018), 'Annual report', www.mondragon-corporation.com/wp-content/
themes/mondragon/docs/eng/annual-report-2018.pdf.

North, P. (2018), 'Transitioning towards low carbon solidarity economies?', in P. North
and M. Scott Cato, *Towards Just and Sustainable Economies*, Bristol, Polity Press;
73–95.

North, P. and Scott Cato, M. (2018), 'Introduction: new economies North and South –
sharing the transition to a just and sustainable future', in P. North and M. Scott Cato,
Towards Just and Sustainable Economies, Bristol, Polity Press; 1–12.

Nuttall, G. (2012), 'Sharing success: the Nutall review of employee ownership', The
Department for Business, Innovation and Skills, London.

Oberholzer, B. (2019), 'Can sustainable finance really help solve the climate crisis?',
Social Europe, www.socialeurope.eu/can-sustainable-finance-really-help-solve-the-
climate-crisis, December.

Olsen, E. (2014), 'The relative survival of worker cooperatives and barriers to their
creation' in T. Kato (ed), *Advances in the Economic Analysis of Participatory &
Labour-Managed Firms*, 14; 83–107.

Oseen, C. (2016), '"It's not only what we say but what we do": pay inequalities and
gendered workplace democracy in Argentinian worker cooperatives', *Economic &
Industrial Democracy*, 37(2); 219–244.

Palmer, T. (2019), 'State of the sector: US worker cooperatives in 2017', *Journal of Participation and Employee Ownership*, 2 (3); 190–201.

Pateman, C. (1970), *Participation and Democratic Theory*, Cambridge, Cambridge University Press.

Pencavel, J. (2002), *Worker Participation: Lessons from the Worker Co-ops of the Pacific
Northwest*, New York, Russell Sage Foundation.

Pendleton, A. (2005), 'Employee share ownership, employment relationships and corporate governance', in B. Harley, J. Hyman and P. Thompson (eds), *Participation and
Democracy at Work*, Basingstoke, Palgrave Macmillan; 75–93.

Pérotin, V. (2012), 'The performance of worker cooperatives', in P. Battilani and H.
Schröter (eds), *The Cooperative Business Movement 1950 to the Present*, Cambridge,
Cambridge University Press; 195–221.

Russell, R., Hanneman, R. and Getz, S. (2011), 'The transformation of the kibbutzim',
Israel Studies, 16 (2); 109–126.

Singer, P. and Primavera, H. (2018), 'Solidarity economy policy dialogue in Latin
America: transferring Argentine experience of social currency to Brazil', in P. North
and M. Scott Cato, *Towards Just and Sustainable Economies*, Bristol, Policy Press;
195–212.

Spear, R. (1999), 'The rise and fall of employee-owned UK bus companies', *Economic and Industrial Democracy*, 20 (2); 253–268.

Stern, S. (2020), 'Redesigning the CEO: how employee ownership changes the art of leadership', *Ownership at Work*, March.

UN (2017), *Report of the Secretary-General*, 'Cooperatives in social development', General Assembly, A/72/159.

UN (2019), *Financing for Sustainable Development: Report of the Inter-agency Task Force in Financing for Development*, New York, United Nations.

Utting, P. (2015), 'Introduction: the challenge of scaling up social and solidarity economy', in P. Utting (ed), *Social and Solidarity Economics: Beyond the Fringe*, London, Zed Books; 1–37.

Van Reybrouck, D. (2019), 'Belgium's democratic experiment', *Politico*, 25 April, www.politico.eu/article/belgium-democratic-experiment-citizens-assembly

Vásquez-Léon, M. (2010), 'Walking the tightrope', *Latin American Perspectives*, 37 (6); 3–11.

Weinberger, E. (2020), 'California breathes new life into public banking movement', *Bloomberg Law*, https://news.bloomberglaw.com/banking-law/california-breathes-new-life-into-public-banking-movement.

Wolff, R. (2012), 'Yes, there is an alternative to capitalism – Mondragon shows the way', *The Guardian*, 24 June, https://www.theguardian.com/commentisfree/2012/jun/24/alternative-capitalism-mondragon.

12 Towards sustainable work

New production paradigms

As we examined in Chapter 4, the way we currently produce goods and services, and the neo-liberal economic ideology that sets the conditions under which such production takes place, is fundamentally environmentally unsustainable. Most of our raw materials are finite, the planet's capacity for absorbing our waste products is close to saturation point, and we are already experiencing the climatic consequences of two centuries of industrialisation powered by burning locked up carbon. A recent report by the Institute for Public Policy Research warns that the window of opportunity to avoid catastrophic outcomes in societies around the world is rapidly closing and that such outcomes include economic instability, large-scale involuntary migration, conflict, famine and the potential collapse of social and economic systems (Laybourn-Langton et al., 2019).

Following from this, it is becoming impossible to continue to take economic growth as our index of progress and prosperity and, for the planet to continue to be habitable, a good deal of current economic activity will have to be curtailed and replaced by more environmentally positive alternatives: this will entail significant structural shifts in the economy as we move away from unsustainable economic activity, which is reliant on fossil fuel for either its energy or its raw materials.

While there will be an inevitable decline in resource-extractive industries, traditional sectors of the economy such as manufacturing, construction, food and agriculture, retail and transport will still be required. However, it is increasingly argued that the way they are organised will have to change, with a shift from what has been termed the current and conventional 'linear' model of production, best summed up as 'take-make-use-lose': extract minerals, biomass and fossil fuels, make them into products, consume them, then throw them away (Raworth, 2017; 221). The proposed alternative is the 'circular economy', which would run on renewable energy, eliminate waste and toxic materials by design and see waste as either biological waste that can be returned to nature, or technical waste that can be recycled or re-formed; the defining characteristic of circular production gives a high priority to durability, repairability and recyclability. Several companies are currently claiming to have adopted 'circular manufacturing' through the recovery of components from their products for either refurbishment and re-sale or for re-manufacturing. Welcome as this is, the problem, as Raworth points out, is that each initiative is top-down and internal to each corporation with companies seeking control over their own used products and thus circularity, even in the same industry, is fragmented into disconnected parts (*ibid.*: 228); this is some way from being the basis for developing an ecologically sustainable industrial economy.

In view of this, it is argued that to achieve anything approaching sustainability there will have to be a substantial sectoral restructuring of contemporary economies, away from resource-heavy and high waste producing industries and activities. As such structural shifts in the past have seldom resulted in the smooth transition of employment from declining to new economic sectors, such trends could suggest an increase in structural unemployment. Conversely it is argued that, if we take the case of energy production, renewable energy tends to be more labour intensive than traditional modes of energy production: globally more than ten million people now work in renewables. In the UK, 126,000 people were employed in renewable heat, power and transport in 2015/2016, an increase of 2.5 percent over 2014/2015 (Green New Deal Group, 2018: 8). Similarly, the GND proposals for retrofitting housing or expanding public transport networks offer the advantages of being labour intensive, hard to automate and impossible to relocate abroad; in addition, such restructuring has to take place where these services are supplied, meaning that job creation is localised.

Most of the proposals for a sustainable green economy have limited themselves to this sort of sectoral level analysis. In this book we want to try to operate at the level below this – that of the jobs people would do. We need to ask, what would *jobs* in this sustainable space look like and how would they be experienced?

Not just green jobs but good jobs

We have examined the multiple components of contemporary work degradation, including in-work poverty, job insecurity and precariousness, physical and mental ill-health at work, enhanced surveillance and managerial control, bullying work cultures and the threat of automation and AI. As we indicated, the continued existence of such features in the labour markets of the major industrial nations is *economically* unsustainable: quite apart from growing societal inequality (Dorling, 2019: 82; Brown, 2017), low productivity (Taylor, 2017) and the contradictions of the downward implications for a government's tax income and the upward pressure on its benefit system, the global mobility of capital and its search for ever cheaper labour has meant that, for many sectors of the economy, the emiseration of employment will continue in what is quite literally a transnational 'race to the bottom'. If a transition to a more sustainable, less growth-dominated, economy is to be matched by sustainable work that offers respect and responsibility to all workers, alternative models are needed (Piketty, 2020: 967; Wilkinson and Pickett, 2010: 226; Raworth, 2017).

The mood for change is certainly evident: a YouGov survey of over 2,000 adults conducted in May 2020 found that more than 60 percent think that the UK should prioritise improved social and environmental outcomes over GDP growth once the pandemic is over, which links to calls for quarterly economic growth statistics to be abandoned and replaced by a system of well-being indicators (including health, education and carbon emissions), which should instead become stated government targets (Positive Money, 2020).

Similarly, there is no shortage of awareness that something is fundamentally wrong with the labour market and its ability to deliver quality jobs. The Royal Society of Arts, for example, launched a Future Work Centre, which has produced a *Four Futures of Work* report with scenarios based on projected variations in variables such as robotics and AI, the gig economy and surveillance (Dellot *et al.*, 2019). While the authors have clearly focused on several major current and emergent trends in labour market practice and include other significant uncertainties such as the state of the global economy, the future of worker voice and the state of immigration, nevertheless there is a whiff of technological determinism about such forecasts. Of more concern to us is the relegation of the consequences of climate crisis to a passing reference to 'non-tech drivers' in their list of such other 'uncertainties': this failure to address the climate crisis means that we have no way of assessing whether jobs in any of their scenarios are sustainable.

The other missing element is any suggestion of worker agency: scenarios such as these present what is being done *to* workers and there is no suggestion that employees might have responses of their own. In contrast, JT involves 'a process of change ... to a more sustainable society, based on a low carbon economy [which requires] ... proactive policy-making and implementation' (Snell and Fairbrother, 2013: 147) and fundamentally seeks to extend worker, community and union roles in the change process. Genuine 'good work' requires that the interests of employees, communities and the environment be recognised and acted upon. Unlike contemporary human resource models, which mask employer and managerial control objectives through imposing hyped but potentially exploitative employee empowering and engagement exercises, genuine collective participative and cooperative working will need to be part of a radically revised economic programme, especially in those countries where market domination has been strongest and social protection weakest.

What is good work?

Before we can try to identify the characteristics of sustainable jobs, we need to understand the nature of work itself. Sociologists of work such as Sheila Allen have long pointed out that it is about more than just putting bread on the table as work has always carried with it a legacy of moral and political values and ideas about social worth (Allen, 1997: 57). From the very start of the industrial period early philosophers and later social scientists have debated the 'meaning of work', these debates revolving around two polar positions. The first is that work will always be experienced as unpleasant drudgery with little intrinsic meaning for the worker and performed solely for monetary reward, while the opposite view holds that, while work has often been so experienced, it does not *have* to be like this and that work can and does contain many non-monetary rewards. In reality, there is a nugget of truth in both the critical and the optimistic standpoints. We know that, in addition to its primary function as the source of economic livelihood, there *can be* a range of non-monetary factors that are valued by employees – security, a recognition of worth and source of personal identity (Du Gay, 1996), pride in work and a degree of control and individual or collective agency (Braverman, 1974), the presence of an occupational community (Brown *et al.*, 1972), teamwork and membership of social groups (Hodson, 2001).

At various times in the last century management initiatives claimed to be able to remediate the fact that job design under capitalism has all too often resulted in meaningless tasks: stripping away all non-monetary rewards from work, reducing its meaning down to the basic 'cash nexus', has contributed to a lack of employee commitment to the organisation and in the past to increased levels of industrial conflict. The most recent of these initiatives was the human resource management offensive from the 1980s, which aimed at replacing employees' collective identity with individualised commitment to the organisation though a menu of high commitment strategies. Instead of confrontation there was to be involvement, instead of collective bargaining there was to be a personal individualised sense of belonging to the organisation, instead of management diktat there was to be 'empowerment'. For the subsequent decades, such catechisms of HRM as 'people are our greatest asset' were dutifully recited as an essential component of the new management orthodoxy.

It is now clear that 30 years of such HRM discourse has resulted in little visible improvement in the quality of most employees' working experience (Taylor, 2002) and the reason is not hard to find. To link the use of human resources with the strategic objectives of the company does not necessarily result in better or more satisfying work. For example, if the goal is to compete on the basis of cost-minimisation, the organisation may simply wish to have the cheapest and most disposable resources they can get – and not necessarily in the home country. Our analysis in Chapter 4 indicated that for LMEs with shareholder value as their main short-term priority, this has all too often been the case, with the result that in many organisations HRM has become the most disliked (and now often outsourced) management function. This should warn us that, as the basic characteristics of employment are as yet unchanged, we should not assume that a future 'green economy/society' will automatically generate non-alienating and self-fulfilling jobs.

Academic disciplines that focus on employment tend to define and emphasise 'good' work differently: economists focus on paid work and favour attributing goodness to attached or extrinsic rewards in terms of remuneration and conditions of employment in return for productive contribution. At its neo-liberal extreme, an 'economic human' is assumed to be 'self-reliant, self-interested, singly focused on economic gain, the organic seat of "human capital" and ultra-rational in its conduct' and motivated by 'egocentric competitiveness' (Fleming, 2017: 98). Occupational psychologists and health specialists tend to emphasise the positive influence of social relationships at work on mental and physical well-being (and hence on performance) and the negative impacts of insecurity and absence of personal control. Sociologists focus on variations in occupational status and worker autonomy and the alienating power and control dynamics of economic and managerial systems that tend to deny workers these attributes (Anderson, 2017; Hodson, 2001; Bolton, 2007; Pettinger, 2019; Frayne, 2015).

Tempered by workplace reality, observations from workers themselves suggest modest expectations and ambitions: to be offered security, a reasonable wage, to be treated with respect and without discrimination; and to be offered a measure of responsibility (Bloodworth, 2018; Ehrenreich, 2002). Surveys often indicate that for workers, irrespective of status, occupational 'meaningfulness' is a prized characteristic (Dromey, 2014; Alfes *et al.*, 2010), and indeed occupies 'a fundamental human need' (Bailey and Madden, 2017: 3). However, defining meaningfulness can be problematic (*ibid.*, 2017; Frayne, 2015; Baldry *et al.*, 2007). For some occupations it is associated with personal identity and a sense of contributing positive social utility to community and to society alongside the freedom to exercise independent decision-making. Typically, these comprise professional or quasi-professional occupations populated by doctors, teachers, social workers and academics who traditionally enjoy autonomy over their work, based on an assumption of authority derived from their professional training and expertise and independent disciplinary oversight by their professional institutions. Also, because entrance to professions is secured through these same institutions, numbers introduced into the occupation and numbers actively practicing can be restricted, safeguarding both rewards and status. Though not enjoying the same professional status, occupations such as nursing and nursery teachers also derive meaningfulness not primarily from financial reward but rather from group solidarity (Bolton and Laaser, 2020) and the satisfaction of contributing to civic education and individual welfare (Cohen *et al.*, 2019).

An alternative construction for meaningfulness is that, rather than presenting meaning *in* work, the remunerative rewards acquired *through* work provide a way to offer meaning to private and family life by giving opportunities to acquire products and services that enable a comfortable and coveted lifestyle or class association. Work, however degraded, becomes the instrument for satisfying other, external needs. And, of course, many people in order to live do undertake routine work with little intrinsic satisfaction.

These different interpretations also raise a further question: meaningful to whom? For classical economists and their acolytes, rewards associated with work derive from its supposed market value in a competitive economy. In turn, people performing the lowest-paying jobs are assumed to have the lowest 'human capital' and market value, and hence are easily disposable, live precariously and enjoy little social prestige. But the experience of the Covid-19 pandemic has shown that low ascribed market value need not necessarily equate to low public value. Whilst many highly paid employees and managers with undecipherable job titles but questionable social value have kept a low profile during the Covid-19 crisis, the truly essential, and potentially hazardous, work has been performed by previously unnoticed delivery drivers, refuse collectors, postal workers, rail and bus personnel and supermarket check-out operators, and possibly above all, by nursing and care home staff, many recruited from among the BAME community.

Tasks, jobs, work and employment

This is where we need to broaden our understanding of work in the sense of any task performance that creates value. In the above discussion of 'work' we are actually talking about *employment* – the physical and mental labour that is bought and sold in the labour market. While conventional economics quantifies the total reward going to labour as the figure for the national income, this ignores all those other forms of work that are not marketised, that is, not performed for a monetary reward but that contribute to the overall welfare of communities and society and, arguably, without which a society could not function; these include domestic labour, voluntary and community work, and unpaid caring (such as within a family). It can be argued that these essential activities are actually central to the core economy but because they are unpaid they are currently given low status and exploited; their value is not included in the official count of the national product as the market has not allocated a price for them. Perhaps the most significant of these is, and always has been, domestic labour. When the tasks undertaken in domestic work are undertaken by employees in the labour market we give them labels such as child-care and early years education, social care, cleaning, cooking and food preparation, laundry, garment repair. When done in the home, as part of the domestic rather than public economy, they are given no monetary value. To return to Marçal's example, Adam Smith could not have produced *Wealth of Nations* without his mother's unpaid domestic labour (Marçal, 2015).

There are clearly specific satisfactions to be obtained from the performance of tasks, as evidenced by the fact that we may choose to spend our leisure (that is, non-employed) time in maintaining cars, building bookshelves, cooking or dressmaking, all of which in a marketised context can provide intense dissatisfactions to their performers. This indicates that there is nothing special about the tasks undertaken in non-marketised work, as the same tasks (cooking, child-care, cleaning, laundry work and so on) can be performed as wage or salaried employment. The UK Office of National Statistics in fact uses this criterion to value unpaid work; within what it terms the 'general production boundary' unpaid activity is deemed as productive if it could be contracted out to a market provider and its recent calculations valued unpaid work at £1tn for the year 2014 (ONS, 2017); this compares to gross domestic product (GDP), or the money economy, which was £1.8tn in current prices in the same year. Within overall figures for unpaid work, the work of an estimated 6.8 million unpaid carers in the UK economy was valued at £140bn in 2015, almost as much as the NHS budget (*ibid.*, 2017; FitzRoy and Jin, 2018), while the value of formal volunteering was put at £23bn (ONS, 2017). It is not surprising that some economists have referred to this unremunerated economy as the 'core economy'.

Work in this sense is therefore in itself value neutral – it depends on the social context in which the activity takes place as to whether it is regarded as good or bad, acceptable or unacceptable. The fact that we can find the same tasks undertaken in both the market and non-market sectors suggests that this division between 'employment' and non-market 'work' is a social rather than an economic construct and it is of course heavily gendered. Despite decades of change in social attitudes and legal acceptance of gender equality, it is still the case that the burden of domestic work and unpaid caring falls on women (who account for 69 percent of the value of unpaid child-care and 59 percent of unpaid adult care).

The major difference is that employment is a relational concept – it reminds us that labour power is employed in the specific context of a social and economic relationship between its buyers and sellers. The two distinctive features of this relationship are its monetary basis – workers *sell* their labour power in the labour market to earn the means of subsistence – and the fact that it is an authority relationship – the employer, having purchased the labour power of the worker, is then deemed to have the right to control and organise how it is used – in other words, the right to manage. Inherent in both of these aspects is a latent (and often manifest) conflict of interest: the terms of the employment exchange may be disputed by either party over, for example, what is felt to be an appropriate level of reward, over what is felt to be an appropriate level of effort, over the hours that are worked and the conditions under which work is performed. In resolving such conflicts, the employer of course, through having the power to permanently terminate the relationship, has by far the upper hand compared to the individual worker; only where employees can act collectively can this power discrepancy be challenged.

On a more abstract level, to be employed is a powerful strand in the dominant societal value-systems of most advanced economies. Subsequent interpretations of the Protestant Work Ethic, first analysed by Max Weber in the nineteenth century, imbued being in employed work with a moral value that has made 'getting a job' something that everyone not only has to do to earn a living but *should* do as a good citizen or as religious duty (Bendix, 1963; Anthony, 1977). This ideology of work means that, in addition to the need for subsistence, most people would prefer to be in work (employed) than unemployed; as only marketised work is given economic value, those unable to gain employment are likely to feel value-less.

Towards sustainability

We have argued in this book that the twin contemporary concerns of the degradation of work and the degradation of the planet are in fact intimately connected and intertwined, such that any positive resolution towards sustainability will have to meet both environmental and economic criteria to be within the ring of Raworth's 'doughnut' model (Raworth, 2017). The concept of a JT from fossil fuel-dependent and resource-heavy economies to productive activity that is both ecologically and economically sustainable implies that the transition should not result in the persistence of inequality and injustice. Thus, any transition to sustainability would have to overcome both the economic and social pressures to seek employment when the opportunities for doing so may be declining, and the historical undervaluing of work in the unpaid economy. This posits an additional challenge to any JT policies: in addition to 'greening' current jobs and economic sectors, it may be that we have fundamentally to reconsider how we have traditionally viewed work and non-work to include a recognition of the contribution to the overall welfare of a society made by *all* forms of labour. Also, to investigate sustainable alternatives to the current degraded state of so many jobs, we will have to look at the satisfactions we can get from both employment and un-waged work.

Tim Jackson argues that productive activities in a sustainable economy should be characterised by three operational principles: a positive contribution to well-being, the provision of decent livelihoods and low material and energy throughput (Jackson, 2011: 196). Based on the above discussion, we can expand on these to speculate on what the ideal features would look like for jobs that are both socially and environmentally sustainable.

For work to be *socially* sustainable it should not lead to exhaustion of an individual's physical energy, mental energy (burnout) or sense of self-worth (alienation) and the work should also contribute to the wider life of the community in which it is situated. Ideally, then:

a   Work should be rewarded equitably: it should provide monetary compensation for effort sufficient to offer a standard of living to sustain the worker and family.

b   Workers should be treated with respect and decency.

c   A job should involve a degree of control over work-related decisions, both over the performance of the task or tasks and wider decision-making. The product should be something that the worker can identify with as his/her own product.

d   Work should be meaningful: the nature of the work should be such that the worker wants to continue doing it and the nature of the organisation should want the worker to continue to do it via a reciprocated commitment.

e   There should be built-in opportunities for sociability – the non-monetary rewards from working with others.

f   The production of a good or service should contribute to the general welfare of the society.

Keeping the employment/work distinction in mind, we can see significant differences in those of the above factors that characterise employed and non-employed work respectively. Employed work is of course characterised by monetary reward (a) and the history of employment relations is the history of the degree to which this has or has not been perceived as equitable by employers and employees. Employment was often traditionally characterised by (c) and (e), although it is noticeable from our earlier examples of degraded work that the design of contemporary jobs specifically *and deliberately* removes just these non-monetary rewards from work. Following the contribution of Braverman (1974), much employment research has focused on (c) and the observed trends towards deliberately eliminating employee skill and control over the job. There has been a similar trend in minimising the sociability of working with others (e) often the redeeming feature of otherwise monotonous or repetitive work. These trends have been taken to extremes by the use of digital platforms for the allocation of tasks to workers: as we saw in Chapter 9, Amazon's Mechanical Turk and similar designs create a workforce of atomised individuals who not only have minimal contact with each other but have little idea of the end product that their input is contributing to.

In contrast, non-market work such as that in the voluntary sector, while not enjoying the monetary reward of (a) is often characterised by (b) (c) (d) (e) and (f). There are of course non-market activities not associated with these factors such as those that we undertake that could be (or once were) done by paid labour such as self-service garage forecourts and self-service supermarkets (Standing, 2017).

For work to be *environmentally* sustainable it should ideally:

a   Not involve natural resources (including power sources) that cannot be replaced; all resource use should have the aim of recycling or re-use.
b   Not involve the polluting of air, water or land, whether through the production process or through creation of unusable waste products.
c   Work should not result in products, services or waste that inflict irreparable damage on either people (tobacco, armaments, particulates) or the environment (plastics, heavy metals, carbon).

This is more challenging as the whole way in which the production of goods and services has been organised up to now has contributed to negating all of the above. To meet these criteria would involve changing the parameters for socio-economic success from measuring productivity in terms of labour productivity but rather to look at resource productivity (Raworth, 2017: 267); as Daly and Cobb (1994: 296) remind us, the means of producing more with fewer people (the conventional definition of productivity) has historically involved the substitution of fossil fuel energy for human labour. This cautions us that many aspects of 'good work' are not necessarily sustainable work: a clear example would be the degree of work-centred group cohesiveness and integration with community that mid-twentieth sociologists found in coal mining communities (Dennis *et al.*, 1956). Yet, such work was always physically dangerous and, when the coal seams became uneconomical to work and the burning of coal itself, the original fossil fuel, had become almost a pariah activity, those communities were broken and abandoned.

If we broaden our evaluation of the contribution that work makes to the welfare of society to include both employed and non-employed work this also suggests we will have to examine how we reward such work and how work is organised, in terms of the patterns of working time and the most effective organisational forms that will meet the above goals. One proposed answer, currently receiving a good deal of attention, is the suggestion to pay everyone in society a Universal Basic Income or UBI. Further, according to supporters, by helping to break the link between work and consumption, UBI could encourage us to work less and consume less, with environmentally beneficial consequences for economic activity and growth (Maslin and Lewis 2019).

Universal Basic Income

According to the Basic Income Earth Network (BIEN, 2020), UBI is defined as 'a periodic cash payment unconditionally delivered to all on an individual basis, without means-test or work requirement'. It is thus paid at regular intervals (such as weekly or monthly) not as a one-off grant, and paid in money rather than in kind or vouchers dedicated to a specific use (such as food vouchers). It is paid to everyone usually resident in a given community or country, and on an individual basis rather than to households (unlike current welfare payments) (Mason, 2016). These characteristics present a number of both practical and philosophical questions (De Wispelaere and Stirton, 2004) but in essence UBI is seen as a modest amount of money whose purpose is to provide basic economic security rather than total security or affluence. It would be a predictable regular amount and be non-withdrawable.

This is not a new idea and has its roots in historical campaigns for social justice. By the twentieth century variations of the idea of a basic income for all had been espoused among others by religious groups, by philosopher Bertrand Russell, political economist GDH Cole (who may have coined the phrase 'basic income'), and economists such as James Meade, James Tobin and J K Galbraith who saw it as less likely to create welfare dependency than means-tested systems (for a full history of the concept of UBI see BIEN, 2020). Writing very much in the social justice tradition, the most radical justification for UBI is given by the economist Guy Standing, one of the founders of BIEN. Standing argues that the most equitable way of seeing UBI, and one that overcomes many of the objections, is to see it as a social dividend from the collective wealth of society that has been built up by preceding generations. Viewed in this way, basic income would be a *right* of citizenship that could not be taken away; it is not 'welfare' by another name – it is income (Standing, 2017). This idea of basic income as a right of citizenship is put forward by several of the campaign groups for UBI, such as the Citizen's Income Trust in the UK and BIEN, who argue that every member of society has a right to basic subsistence and to live in a way that does not subject them to social exclusion (Hirsch, 2015).

Unusually, proposals for basic income have been supported by those on the right and the left, albeit for different reasons. Free-market advocates, from

Milton Friedman onwards, have seen UBI as a way of scrapping all existing welfare programmes and replacing them with one single payment, cutting all other forms of state intervention and essentially leaving the provision of existing social goods (education, health) to the market. In contrast, for those on the left it is seen as a way of mitigating poverty and, because of the universal and unconditional basis to UBI, getting rid of the old distinction between the 'deserving' and 'undeserving' poor. It is recognised that UBI would not eradicate poverty on its own and does not have to replace existing anti-poverty programmes, which would continue as a top-up to basic income (De Wispelaere and Stirton, 2004).

Growing support for UBI is driven by recognition that, as we have seen, current economic and social policies are resulting in gross inequalities and a growing precariat, which increasingly fragmented welfare systems are structurally unsuited to support. Whereas in the late twentieth-century economy, characterised by long-term secure employment, the risks of ill-health, accidents or unemployment could be calculated and built into systems of social insurance, this is increasingly no longer possible in economies characterised by temporary, part-time and fluctuating employment and in which the post-war consensus on the distribution of national income no longer holds, with a declining share going to labour. Similarly, minimum wage legislation, often introduced by governments following the decline of collective bargaining and trade union activity, is not suited to a flexible labour market. Paradoxically, a minimum wage floor expressed in a rate per hour works best in an economy with full-time regular employment; it does not offer much support to temporary or agency labour or where workers may be juggling several jobs, making it almost impossible to calculate an hourly rate. Neither does it cover non-employees such as those designated as 'self-employed' and 'independent contractors'.

Looking to the future, Painter and Theung (2015), point out that the ageing demographic profile of the developed economies means that the 'caring economy' is bound to expand: as we have noted above much of this work is undertaken either by the low-paid agency sector or as part of the unpaid work economy. Here a basic income would both give recognition to valuable but unpaid work and also mitigate the dilemma faced by those having to choose between employment and caring for a vulnerable relative.

There have of course been several major objections, both economic and 'moral', to any proposals for introducing UBI. While most objections have focused on the cost and impracticability of the policy, there are more deep-rooted ideological objections. The whole idea of UBI runs counter to one of the basic philosophical assumptions built in to the liberal free-market model: that individuals basically do not want to work and will only seek it if there is a 'price' or financial reward for their labour (or if there is a threat to cut whatever welfare benefits they may currently be receiving). Therefore, it is argued, to give people 'something for nothing' would disincentivise working and jeopardise labour supply. Despite the fact that, as we have seen, the high level of *unpaid* labour in the economy immediately refutes this assertion, it is undoubtedly true that, because of the dominance of this ideology, any UBI proposal, particularly if financed through taxation, would have to overcome deeply entrenched populist objections based on the 'free riders' argument of 'my money going to those who don't contribute anything themselves'. (In contrast, public responses during the Covid pandemic have demonstrated that self-organised mutual aid, drawing voluntary and supportive solidarity across communities, exists and can flourish.)

This connects to what one might term 'popular moral' objections: the view that people only have a right to be supported if they are unable to support themselves and that a basic income for all sends the wrong message that society owes you a living whether or not you make an effort. This perspective has been developed by authors on the libertarian right such as Charles Murray (2008) to claim that tax revenue money given to all would only be spent on morally 'bad' activities such as drinking and gambling. This is not borne out by the experience so far of basic income pilots and similar schemes in India and Africa where it was found that the schemes could actually lead to a reduction in the consumption of 'therapy bads' like drugs and alcohol and that the money was more likely to be spent on personal goods such as food for children, healthcare and schooling. The paternalist view, that poor people in receipt of 'public money' should only spend on bare essentials, ignores the fact that taxpayers have never had the right to determine how their money is spent; for example, the current authors may object to the continued financing of nuclear weapons but have to respect decisions made by the elected legislature. Standing argues that the question of whether UBI should be introduced cannot be reduced to empirical tests of consequent behaviours; if it is seen as a *right* then 'asking if it "works" makes no sense, any more than the abolition of slavery' (Standing, 2017: 275).

UBI proponents have raised several more concrete counter-arguments to these objections. Firstly, the free rider argument is seldom raised against the many other forms of income that are not related to productive activity: inherited wealth, tax breaks and subsidies, and 'rentier' income derived solely from possession of assets. Secondly, the effects of UBI on the supply of labour are more complex than the crude 'paying people to be lazy' argument. On the one hand labour supply could be increased as UBI would reduce the 'poverty trap' whereby someone moving from benefits to a low paying job can both lose income and have to cover the costs of transport and child-care: the combination of the cut in means-tested benefits like universal credit and new liabilities for income tax, means that a substantial part of any additional income from a new job at present goes back to the state. However, although the supply of labour at the low end of the labour market might be increased, so might the bargaining position of workers, as a basic income would enable more people to refuse disliked or unpleasant jobs, or demand more pay for doing them. This in turn would refute the objection to UBI often made by trade unions, that it would enable employers to reduce wages (which is exactly what working tax credits can encourage).

The 'my taxes' argument by opponents of UBI is usually based on the assumption of 'strict revenue neutrality', that is that UBI would be funded solely by adjusting personal income tax rates and allowances, plus savings on current welfare spending. In arguments over the affordability of, and various mechanisms for funding UBI there are complicated calculations on both sides (for detailed examples see Standing, 2017; Painter and Theung, 2015; Hirsch, 2015; Torry, 2015; OECD, 2017) but supporters point out that alternative financing could come from reductions in such non-welfare spending as subsidies and regressive tax breaks or from new sources of finance such as a sovereign wealth fund, a carbon tax or financial transaction tax, plus the elimination of tax avoidance by the super-rich. The whole debate over basic income clearly raises questions over what exactly are society's fiscal, social and moral priorities, and how can we determine these.

It is the suggestion of alternative financing that has boosted environmentalist support for UBI as it can be seen as a way of encouraging carbon reductions via various sorts of carbon tax. The US Citizens' Climate Lobby calculate that a $15 per ton carbon tax applied to domestic fossil fuel production and imports could raise $117bn a year, which could finance a dividend of $323 per person with highly progressive distributional effects as nearly 90 percent of households living below the Federal Poverty Level would benefit (Ummel, 2016). Although not strictly UBI as it would be taxable, the benefits could, in addition to an individual's basic income, include changing consumption patterns towards more environmentally friendly goods and services and away from resource and energy-intensive activities, whose prices would rise in any case due to the carbon tax.

Lastly, to return to the desirability of re-evaluating how we assess all work, the 'free riders' model illustrates the problems caused by the conventional treatment of paid labour as the only measurable means of adding value to the economy. For example, those who currently reduce their paid hours in order to care for children or vulnerable family members are judged to be doing less 'work'. If we widen the analysis to include all forms of work, marketised and non-marketised, we get some different conclusions. A basic income system would encourage a shift from valuing only paid 'labour' to valuing *all* 'work' – child-care, care for the elderly and vulnerable, voluntary work, community work and personal development. Ironically these arguments were given increased momentum by the Covid-19 pandemic, during which most European governments were forced rapidly to introduce job retention schemes (such as the UK 'furlough' scheme under which the government paid 80 percent of employees' wage costs) in order to prevent an immediate spike in unemployment; in the UK this amounted to around £10bn per month. At the same time social and community networks were kept going by a huge increase in unpaid voluntary work.

There have by now been several pilots and trials of UBI-type schemes in both developed economies (Canada, Alaska, Finland) and in economies in the global South (Madhya Pradesh in India, Namibia, Kenya). Although Standing (2017) points out that few of these have met all the criteria for full UBI, in being neither universal nor paid to individuals nor unconditional, a number of positive consequences have been noted. In the Manitoba trial, in the 1970s, where a guaranteed income was given to those out of work, there was a recorded reduction in hospitalisation, especially for admissions related to mental health and to accidents and injuries (*ibid.*: 257). Trials in Madhya Pradesh State in India from 2011 to 2013 resulted in improved nutrition, and increased school attendance and performance; as the basic income grants enabled small-scale investments in such things as seeds, sewing machines, and repairs to equipment, there was a reduction in debt and, because payment was to individuals rather than households, women gained economically and in terms of self-esteem (*ibid.*: 235).

Significant attention within Europe was given to Finland's two-year experiment, starting January 2017, in giving 2,000 unemployed citizens their unemployment benefit on an unconditional basis. Rather than being means-tested, they received €560 per month irrespective of whether they were living with another wage earner, or whether they got or were actively looking for a job. Unfortunately, the experiment was threatened after the first year by a new right-wing government who introduced harsher penalties for benefit recipients who failed to find a set number of hours of work or pursue suitable training. However, while days in employment increased in the second year in both the pilot group and a control group, it was the pilot group who worked on average six more days a year, despite that fact that they were less affected by tighter conditions. As van Parijs (2020) has pointed out, this cannot be taken as a test of the effects of a basic income as it is not known what effect it would have on those currently in work and who might be enabled to cut their hours on receiving basic income in addition to their wage or salary. Other results from the Finnish experiment are interesting – recent immigrants to Finland in receipt of the benefit worked more days than either the rest of the pilot or the control group, and basic income recipients as a whole reported more positive feelings about health and stress.

The USA has shown an example of a partial basic income scheme in operation in the Alaska Permanent Fund Dividend. This is funded from oil and gas returns and pays out to every Alaskan citizen, a variable amount, averaging around $1,100 each year. This deviates in several key respects from the ideal UBI model described above: it is a variable and unpredictable amount and gives only a fraction of an adequate basic income, despite the huge oil wealth of a sparsely populated state. It can, however, be seen as a sort of social dividend for future Alaskan generations rather than providing basic income security. Only Finland has introduced a nationwide pilot scheme in 2017, and Canada, Scotland, Spain and the Netherlands have tried or contemplated more localised trials.

Whatever the structural mechanics of such a scheme, we can see from the above discussion that the introduction of a basic income would require significant changes in the way society views work, employment, income and the role of the state, in particular the acceptance of the idea that everybody should receive a basic level of support with no work-based conditions (Hirsch, 2015). Politicians are unlikely initially to see such value-shifts as being acceptable to voters, a view supported by evidence of social attitudes; the 2011 British Social Attitudes Survey demonstrated a lessening of support over the past two decades for those on benefits and, for example, a belief by over half of those surveyed, that unemployment benefit was too high and 'discouraged work'. There had also been a decline in the numbers believing that government should redistribute income in favour of 'the less well-off' (Clery, 2012). Painter and Theung (2015) argue that the public's dislike of anything that suggests 'something for nothing' should be recognised and some form of reciprocity built into any future UBI scheme. For example, they suggest that to be in receipt of the basic income, every 18–25-year-old would be expected to sign a public 'contribution contract' with their local community, which will commit the recipient to contribute to the extent they are able through earning, learning, caring or setting up a business. To avoid basic income being seen as simply another impersonal welfare institution, this would tie its receipt, particularly in early adulthood, to a socially embedded contribution.

Reduction in working time

In addition to income, the other component of employment that has significant implications both for the quality of working life and for the environment is the amount of time we spend at work (and the amount of time we do not). This has been one of the key areas of contention and dispute between employer and employee since the start of the factory system and is one of the major components of the 'wage/effort bargain' between the two. The number of hours worked by men, women and children went unregulated in most industrial societies for decades and early attempts from the 1830s onwards by the growing trade union movement to limit or reduce working hours were met with stiff and outright opposition by employers and politicians alike. A consistent feature of the history of working time is the opposition by British governments to the setting of any limits that might apply across the labour market. The first reform proposed by the new International Labour Organization after the First World War was the Washington Forty-Eight Hours Convention but this was not ratified by the British government; even when a Labour government attempted to ratify it in the early 1930s it was prevented by fierce opposition in Parliament. By 1939 the government still refused to ratify the 1919 Forty-Eight Hours Convention and the ILO's Forty Hours Convention of 1936–1937.

Because of such institutional opposition, reductions in working time have invariably come about through trade union campaigns and collective action during periods of union strength. Notably, in the decades after both World Wars the unions were able to exploit their wartime growth to secure large reductions in hours for industrial workers (Skidelsky, 2019). In the UK of the 1950s and 1960s, one of the features of the 'post-war consensus' was the trade-off in which, in return for higher productivity, the unions were able to negotiate both shorter working hours and growth in real wages. However, as the steam began to run out of the post-war boom, further reductions in working time became harder to come by. The major attempt to limit working hours in the late twentieth century came from outside the UK. The EU Working Time

Directive establishing a maximum 48-hour week was implemented in 1998 but was resisted by the British government (following their deep dislike of all aspects of the EU Social Dimension). Even when adopted by a Labour government in 2003 it had two major weaknesses: employers could 'ask' workers to opt out of coverage and it exempted those, such as managerial and professional grades, who could claim a degree of autonomy and control over their hours.

The Working Time Directive seems to have had minimal impact on limiting the long hours worked by British employees (by 2004 nearly one-fifth of British workers had signed the opt-out), or the fact that in many cases the extra hours consist of unpaid overtime. By the first decade of the new century 36 percent of men and 32 percent of women were working more hours than they were contracted for and by 2019 the TUC calculated that 5.1 million employees provided 39 million hours of unpaid overtime each week, working out at a weekly average of 7.6 hours per person. (Bunting, 2004; TUC, 2020). Paradoxically, the problem is worse at the upper ends of the labour market where long hours for managerial and supervisory groups are driven by an organisational culture that stresses 'going the extra mile' and constantly evaluates commitment to the organisation by the amount of time spent there.

The most comprehensive review of current patterns of working time in the UK is provided by Skidelsky (2019). Using ONS and Eurostat data, his report found that British full-time employees (74 percent of the workforce) still work longer weekly hours than full-time employees in all other EU countries except Greece and Austria; the EU average being 41.2 while the UK's is 42.5: 17.1 percent of all employees and nearly 26 percent of the self-employed were working over 45 hours.

A combination of socio-economic circumstances since the turn of the century has resurrected arguments and campaigns for working time reduction (WTR), although not all of them from the same perspective. British working hours have remained comparatively long because the conditions that enabled post-war hours reductions have not applied for several decades. Union strength, particularly in the private sector, has been weakened by successive tranches of restrictive legislation, and without 'wages push' it has been easier for employers to go for the short-term option of relying on cheap and 'flexible' labour rather than invest in productive capital, with the result that productivity, even before the 2008 recession, had been virtually flatlining (Skidelsky, 2019). The ending of progressive reductions in working time is not solely a British phenomenon but since the 1980s has been shared with most other OECD countries (Devetter and Rousseau, 2011).

The reorganisation of work around information and communications technology has changed many of the certainties and assumptions about work and time and has also altered patterns of consumption and demand for products and services. Consumers are now advised by organisations that they can make retail purchases, organise their finances, take out insurance or book a holiday at virtually any time of the day or night, resulting in extended working hours and fluctuating shift patterns, often set by peaks and flows in demand for the service, for the employees at the receiving end of these requests (Baldry *et al.*, 2007). In addition, as we saw when looking at the relationship between technology and patterns of work, mobile devices and the internet have resulted in many employees being 'on call' or reachable at any time and, in the case of zero-hours working, often being unable to predict the amount of work-time required of them. These trends have raised significant question marks over what has become known as work-life balance (Hochschild, 1997; Baldry *et al.*, 2007; Hyman *et al.*, 2003) and the blurring or dissolving of the demarcation between what we have conventionally regarded as two spheres of social activity.

The contemporary UK working culture that encourages long hours and often results in employee exhaustion has resurrected historical arguments for reducing the hours of work to improve employee well-being. The 2017 Skills and Employment Survey found employees having to work harder than they have for the past 25 years and reported that 55 percent of women and 47 percent of men said they 'always' or 'often' went home from work exhausted (quoted in Skidelsky, 2019). The main health consequences of long hours working have been shown to include increased stress, risks of impaired mental health, an increase in cardiovascular disease, risk of miscarriage, headaches, musculoskeletal disorders and insomnia (Devetter and Rousseau, 2011)

The promise of increased productivity through the diffusion of IT and the consequent threat of technological unemployment has given further traction to the traditional trade union demands to share out any reductions in the amount of employed work. The threat of a technologically induced reduction in the demand for labour prompted the French government in the 1990s to undertake one of the few state-backed experiments in cutting working time by introducing a 35-hour week in the private sector, with the twin goals of reducing unemployment and improving employee well-being. Although these goals were weakened by employers and subsequent legislation (Skidelsky *op. cit.*: 32–32) there was initially a ten percent reduction in hours worked by full-time workers in the first eight years of the legislation. What is interesting is that employers reported no drop in overall productivity, partly confirming the claim by unions that workers are actually more productive when they are working shorter hours. Those workers who reported an improvement in the quality of working life tended to be the more skilled as they had more choice about when to reduce their hours, such as taking a longer weekend. Lower-skilled workers tended not to have this degree of choice and also reported an increase in work intensification (undoubtedly another factor in maintaining levels of productivity).

While much of the past focus on the effect of WTR has been employee health and welfare and the sharing of employment opportunities, the climate emergency has prompted claims that WTR can contribute to a more sustainable economy. One calculation (in King and van den Bergh, 2017) is that a 20 percent reduction in working time could result in 16 percent reduction in energy use. The environmental case for WTR starts from the opposite perspective from previous approaches; rather than see time reductions as a result of increased productivity (and therefore economic growth), the environmental argument is that, to sustain the climate at acceptable levels, economic activity and growth will very likely have to be reduced, thereby reducing the demand for both raw materials and labour and a reduction in waste emissions.

Environmental models of WTR usually assume that the increased non-work or 'leisure' time will be seen as increased time for creative pursuits, volunteering, or for increased social interaction such as caring or child-care, whereas in reality the way we spend our leisure time at present is predominantly characterised by consumption, so that a counter-argument might assert that increased leisure could actually lead to increased consumer demand and increased carbon emissions. But French economists Devetter and Rousseau (2011) have argued that consumerism and its associated values are themselves partly a product of long working hours. Using French household expenditure figures, they show that where households as a whole work long hours this not only increases consumption but actually changes the type of consumption towards those areas that have been found to have a greater environmental impact, in particular housing and household equipment, clothing, furniture, expenses associated with food (mainly meals eaten out), and behaviour relating to transport and leisure. The reasons seem to be complex: long hours working (even where not a voluntary choice) represents a trade-off between income and free time. There is a suggestion that where there are no longer enough time opportunities to engage in self-expressive social activities; one way of establishing or maintaining social identity is through what Veblen termed 'conspicuous consumption' in 'a race for status', resulting in over-consumption and its associated negative environmental impacts. In support of this, studies on the French 35-hour-week law showed that its main impact on everyday life seemed to be its clear effect on non-commercially related activities (time with children, rest, sport, etc.). Devetter and Rousseau concluded: 'In terms of consumption, these results indicate that the main effects of working time reduction are much more qualitative (consuming differently) than quantitative (consuming less)'.

The limited empirical data so far suggest that WTR does have the potential to reduce emissions (Pullinger, 2014). A review of data from OECD countries between 1970 and 2007 found that countries with lower average hours of work have lower resource usage and lower carbon footprints (Skidelsky, 2019) and work by Rosnick (2013) and others has found that, while the relationship between working hours and carbon emissions is complex and not clearly understood, shorter work hours are associated with lower GHG emissions and therefore less global climate change, mainly through lowering consumption.

Although WTR is usually seen in terms of the length of the working day, other means of achieving reductions include a shorter working week or increased holidays. King and van den Bergh (2017) looked at five different ways of reducing working time and evaluated their chances of reducing carbon emissions, basing their calculations on a putative 20 percent reduction in UK hours (which, as this would bring UK hours down to the level of Germany or Netherlands, they point out is hardly radical). The usual work pattern in OECD countries of a five-day working week with a two-day weekend could be shortened for most sectors by adding Friday to the weekend, although sectors such as hospitality, entertainment, retail and sport might wish to take a different day. An alternative method of achieving a four-day work week could be a free Wednesday. Keeping a five-day working week, a 20 percent reduction in hours could also be achieved by reducing the current average length of 7.5 hours to a six-hour day. Holiday entitlement in the UK is currently around 25 days paid holiday plus eight public holidays; a 20 percent reduction in working hours would increase holidays to around 70 calendar days. Lastly the authors looked at what they call workforce minimisation: using hot-desking to stay open for five days but every employee only working for four, resulting in ability to reduce office space and save on overheads.

As an addition to this observation, we may add that the experience of the changes to work patterns created by the Covid-19 pandemic has demonstrated that possible energy savings from dispersed or home working are complex and difficult to calculate. Spending time at home increases individuals' energy needs when compared to workers sharing an office together and using the same lighting and heating systems, while on the other hand there are significant savings in commuting and travel-based energy consumption.

While King and van den Bergh calculated that the options of a three-day weekend, a free Wednesday and more efficient use of office space could lead to greater emissions reductions than more holidays and shorter working days, they point out that it is extremely unlikely that the same sorts of reduction could be applied across the whole labour market. Public services such as health and education are likely to find such reductions difficult, and industries such as chemicals and some manufacturing, many of them highly energy-intensive and based around continuous production and a heavy reliance on shift-work, are unlikely to reduce operational hours. They also note that in the UK, due to high proportions of part-time workers, reducing hours of full-time employees by 20 percent would not reduce overall hours worked by this amount.

Rosnick (2013) has calculated that a combination of shorter workweeks and additional vacations to reduce average annual hours in the US by just 0.5 percent per year over the rest of the century would eliminate about one-quarter to one-half of the global warming that is not already locked in. However, in a more pessimistic prediction, Frey (2019), using OECD data on carbon productivity per sector, national predictivity in terms of GDP per hour worked and a calculation of the per capita carbon budget needed to keep warming below 2°C, has calculated that the level of working hours that would be necessary to be considered sustainable would have to be significantly below current levels. The results are quite startling: for the OECD on average a sustainable working week would be something like six hours, compared to the current average 40-hour week, while for the UK a sustainable working week would be around seven to eight hours, compared to the current average of around 43.

It should be noted from the above discussion that, unlike previous working time campaigns, which were based on maintaining income levels, calculations about the environmental benefits of WTR assume that reduced energy use will primarily occur through reduced incomes and reduced consumption. Rosnick (2013) notes that this is likely to be much more difficult in an economy where inequality is high and/or growing (as is the case in the USA and UK); without a more socially equitable spread of the benefits, the majority of workers would have to take an absolute reduction in their living standards in order to work less.

Conclusions

The above discussion suggests that societies, in order to meet the JT twin goals of avoiding a climate catastrophe and improving the welfare of all their citizens, will need to 'change the social logic' (Jackson, 2011). With productive activity oriented towards a more 'circular' economic system, societies should take as a measure of socio-economic progress the contribution to societal and environmental welfare of all forms of work, both marketised and unpaid. Instead of trying to squeeze either UBI or WTR into a 'business as usual' model, they should be examined together as a major ingredient in a concerted alternative to the current broken economic system (Cojocaru, 2019). It also seems likely that meeting the need to provide work that is both environmentally sustainable and economically and psychologically fulfilling will require new forms of organisation that are characterised by worker voice and collective participation.

Bibliography

Alfes, K., Truss, C., Soane, E. *et al.* (2010), *Creating an Engaged Workforce*, London, CIPD.

Allen, S. (1997), 'What is work for? The right to work and the right to be idle', in R. Brown (ed.) *The Changing Shape of Work*, Basingstoke, Macmillan; 54–68.

Anderson, E. (2017), *Private Government*, Princeton, NJ, Princeton University Press.

Anthony, P. (1977), *The Ideology of Work*, London, Tavistock.

Baldry, C., Bain P., Taylor P. *et al.* (2007), *The Meaning of Work in the New Economy*, Basingstoke, Palgrave/Macmillan.

Bailey, C. and Madden, A. (2017), 'Time reclaimed: temporality and the experience of meaningful work', *Work, Employment & Society*, 31 (1); 3–18.

Bendix, R. (1963), *Work and Authority in Industry*, New York, Harper and Row.

BIEN (2020), 'About Basic Income', Basic Income Earth Network, https://basicincome. org/basic-income.

Bloodworth, J. (2018), *Hired: Six Months Undercover in Low-Wage Britain*, London, Atlantic Books.

Bolton, S. (ed.), (2007), *Dignity in and at Work*, Oxford, Butterworth-Heinemann.

Bolton, S. and Laaser, K. (2020), 'The moral economy of solidarity: a longitudinal study of special needs teachers', *Work, Employment & Society*, 34 (1); 55–72.

Braverman, H. (1974), *Labor and Monopoly Capital: The Degradation of Work in the Twentieth Century*, New York, Monthly Review Press.

Brown, R., Brannen, P., Cousins, J. and Samphier, M. (1972), 'The contours of solidarity: social stratification and industrial relations in shipbuilding', *British Journal of Industrial Relations*, 10 (1); 12–41.

Brown, R. (2017), *The Inequality Crisis*, Bristol, Polity Press.

Bunting, M. (2004), *Willing Slaves: How the Overwork Culture is Ruling our Lives*, London, HarperCollins.

Clery, E. (2012), 'Welfare: Are tough times affecting attitudes to welfare?', *British Social Attitudes*, 29, London, NatCen Social Research.

Cohen, L., Duberley, J. and Smith, P. (2019), 'Losing the faith: public sector work and the erosion of career calling', *Work, Employment & Society*, 33 (2); 326–335.

Cojocaru, A. (2019), *Basic Income and Working Time Reduction: what is their environmental impact?*, London, Royal Society of Arts. www.thersa.org/fellowship/fellow ship-news/fellowship-news/new-basic-income-report–basic-income-and-working-time-reduction.

Daly, H. and Cobb, J. (1994), *For the Common Good*, Boston, Mass., Beacon Press.

Dellot, B., Mason, R. and Wallace-Stephens, F. (2019), *The Four Futures of Work: Coping With Uncertainty in an Age of Radical Technologies*, London, Royal Society of Arts.

Dennis, N., Henriques, F. and Slaughter, C. (1956), *Coal is Our Life*, London, Tavistock/SSP.

Devetter, F-X. and Rousseau, S. (2011), 'Working hours and sustainable development', *Review of Social Economy*, 69 (3); 333–355.

De Wispelaere, J. and Stirton, L. (2004), 'The many faces of Universal Basic Income', *The Political Quarterly*, 75 (3); 266–274.

Dorling, D. (2019), *Inequality and the 1%*, 3rd edn, London, Verso.

Dromey, J. (2014), 'MacLeod and Clarke's concept of employee engagement: an analysis based on the Workplace Employment Relations Study', London, ACAS.

Du Gay, P. (1996), *Consumption and Identity at Work*, London, Sage.

Ehrenreich, B. (2002), *Nickel and Dimed: Undercover in Low-Wage USA*, London, Granta.

Fleming, P. (2017), *The Death of Homo Economicus*, London, Pluto Press.

FitzRoy, F. and Jin, J. (2018), 'Basic income and a public job offer: complementary policies to reduce poverty and unemployment', *Journal of Poverty and Social Justice* 26 (2); 191–206.

Frayne, D. (2015), *The Refusal of Work*, London, Zed Books.

Frey, P. (2019), *The Ecological Limits of Work: on Carbon Emissions, Carbon Budgets and Working Time*, Crookham, Hampshire, Autonomy Research.

Green New Deal Group (2018), 'Jobs in every constituency', www.greennewdealgroup.org/wp-content/uploads/2018/09/GNDJobsReport9-18.pdf.

Hirsch, D. (2015), 'Could a "citizen's income" work?' Joseph Rowntree Foundation, March 2015. www.jrf.org.uk/report/could-citizens-income-work.

Hochschild, A. (1997), *The Time Bind*, New York, Henry Holt.

Hodson, R. (2001), *Dignity at Work*, Cambridge, Cambridge University Press.

Hyman, J., Baldry, C., Scholarios, D. and Bunzel, D. (2003), 'Work-life imbalance in call centres and software development', *British Journal of Industrial Relations*, 41 (2); 215–239.

Jackson, T. (2011), *Prosperity without Growth: Economics for a Finite Planet*, London, Earthscan.

King, L. and van den Bergh, C. (2017), 'Worktime reduction as a solution to climate change: Five scenarios compared for the UK', *Ecological Economics*, 132; 124–134.

Laybourn-Langton, L., Rankin, L. and Baxter, D. (2019), *This is a Crisis: Facing up to the Age of Environmental Breakdown*, Initial Report, London, Institute for Public Policy Research.

Marçal, K. (2015), *Who Cooked Adam Smith's Dinner?*London, Portobello Books.

Maslin, M. and Lewis, S. (2019), 'We've declared a climate emergency – here's what UBI could do to help the planet', *The Conversation*, 17 May.

Mason, P. (2016), *Postcapitalism: A Guide to our Future*, London, Penguin/Random House.

Murray, C. (2008), *Guaranteed Income as a Replacement for the Welfare State*, Oxford, The Foundation for Law, Justice and Society. www.fljs.org/sites/www.fljs.org/files/publications/Murray.pdf.

OECD (2017), 'Basic Income as a policy option – can it add up?' Policy Brief on the future of work, May, www.oecd.org/employment/emp/Basic-Income-Policy-Option-2017.pdf.

ONS (2017), 'Changes in the value and division of unpaid volunteering in the UK: 2000 to 2015', www.ons.gov.uk/economy/nationalaccounts/satelliteaccounts/articles/changesinthevalueanddivisionofunpaidcareworkintheuk/2015.

Painter, A. and Theung, C. (2015), *Creative Citizen, Creative State: The Principled and Pragmatic Case for a Universal Basic Income*. London, Royal Society of Arts, www.thersa.org/discover/publications-and-articles/reports/basic-income.

Pettinger, L. (2019), *What's Wrong with Work?*, Bristol, Polity Press.

Piketty, T. (2020), *Capital and Ideology*, Cambridge, Mass., Harvard University Press.

Positive Money (2020), 'The tragedy of growth', http://positivemoney.org/wpcontent/uploads/2020/05/Positive-Money-The-Tragedy-of-Growth.pdf.

Pullinger, M. (2014), 'Working time reduction policy in a sustainable economy: criteria and options for its design', *Ecological Economics*, 103; 11–19.

Raworth, K. (2017), *Doughnut Economics: Seven Ways to Think Like a 21st Century Economist*, London, Penguin/Random House.

Rosnick, D. (2013), *Reduced Work Hours as a Means of Slowing Climate Change*, Washington D. C., Center for Economic and Policy Research, https://cepr.net/documents/publications/climate-change-workshare-2013-02.pdf.

Skidelsky, R. (2019), *How to Achieve Shorter Working Hours*, London, Progressive Economy Forum.

Snell, D. and Fairbrother, P. (2013), 'Just transition and labour environmentalism in Australia', in N. Räthzel and D. Uzzell (eds), *Trade Unions in the Green Economy*, London, Routledge; 146–161.

Standing, G. (2017), *Basic Income: and How We Can Make It Happen*, London, Pelican/Penguin Books.

Taylor, M. (chair) (2017) Good Work: The Taylor Review of Modern Working Practices https://assets.publishing.service.gov.uk/government/uploads/system/uploads/attachment_data/file/627671/good-work-taylor-review-modern-working-practices-rg.pdf.

Taylor, R. (2002), 'Britain's world of work – myths and realities', *ESRC Future of Work Commentary Series*, Swindon, Economic and Research Council.

Torry, M. (2015), 'Research note: A feasible way to implement a citizen's income', *Citizen's Income Newsletter*, 2015 (1), London, Citizen's Income Trust, https://citizensincome.org/wp-content/uploads/2016/02/CIT_Newsletter_2015_Issue_1.pdf.

TUC (2020), www.tuc.org.uk/blogs/work-your-proper-hours-day-lets-stop-working-free.

Ummel, K. (2016), 'Impact of CCL's proposed carbon fee and dividend policy: A high-resolution analysis of the financial effect on U.S. households; Citizens' Climate Lobby Working Paper v1.4', https://citizensclimatelobby.org/wp-content/uploads/2016/05/Ummel-Impact-of-CCL-CFD-Policy-v1_4.pdf.

van Parijs, P. (2020), 'Basic income: Finland's final verdict', *Social Europe*, 7 May, www.socialeurope.eu/basic-income-positive-results-from-finland.

Wilkinson, R. and Pickett, K. (2010), *The Spirit Level*, London, Penguin.

13 A sustainable world?

Introduction

This book was motivated by recognition that for many years a global existential environmental crisis was mounting, but that much had to be done by corporations, governments and the international community to address the oncoming disaster. Across the world, economic growth has been the policy priority of choice, accompanied by insatiable consumerism and unsustainable production. At the same time, enhanced global competition has resulted, particularly in liberal-market economies, in strategies of short-term cost reduction, aimed primarily at maintaining and enhancing shareholder value. The experiential result of this in the lives of literally millions of workers world-wide has been steady deterioration in both job quality and the ability of employment to sustain non-work life. As we have seen, low-paid, temporary, precarious, uncertain, and often unhealthy, work frequently characterises both ends of global value chains and all stages in between.

When we set out on this book, the world was facing one global environmental emergency. We hardly expected it to be written in the depths of another, more immediate, global crisis presented by the coronavirus. Yet, as environmentalists and scientists have pointed out, the two crises are closely related in causation and, some would say, in impact. Whilst evidence for interconnected causation between environmental despoliation and pandemic is strong (FAIRR, 2020), it has become clear that the pandemic has affected different ethnic groups, communities and countries very differently: poorer countries and poorer communities within prosperous countries have suffered most and longest (Platt and Warwick, 2020; Public Health England, 2020; Hall and Taylor, 2020). Workers who are mobile between countries have been essential to maintain food supply chains and to serve public health during the pandemic, but receive little protection from their host countries, even within the EU (Rasnača, 2020). Equally, with the environment: famines, flood risk, enforced migration, health risk and job degradation and loss are to be found among and between the poorest societies.

Nevertheless, extreme crisis provides opportunities for reflection and potentially the pre-conditions for major economic and social change (Scheidel, 2017). Though the Covid-19 pandemic has devastated communities across the world, the rapid decline in atmospheric pollution through lower carbon and NO_2

emissions has alerted us to the potential health and psychological benefits of cleaner air, what some commentators are calling the 'silver lining' of the viral crisis (Teale, 2020). A British Lung Foundation survey indicates that nearly two million Britons with lung conditions reported health improvement with the decline in air pollution (Lovett, 2020). Similar reductions in air-borne pollutants have been reported across Europe (CAMS, 2020). But experience from China demonstrates how quickly this improved situation can reverse. In March 2020, air pollution in China plunged by about a third compared with the previous year, but levels rapidly rebounded to pre-crisis levels, largely driven by industrial emissions, and especially in coal-burning regions of the country, where commercial activities swiftly resumed (CREA, 2020; Hepburn *et al.*, 2020). Not surprisingly, CREA's senior analyst made the plea that 'it is essential for policymakers to prioritise clean energy'. More broadly, these crises confirm that long-assumed human dominance over nature is a chimera: rather than control nature, we must learn to live by its rules. Progress toward sustainability can no longer be considered a desirable option: it is imperative.

This chapter examines potential routes that a sustainable economy and society may take, and the many challenges faced in implementation. Calls are being made by statutory advisory bodies (CCC, 2020), trade unions (TUC, 2020) and researchers for new – and urgent – directions in economic, transport and employment policies that are consistent with an environmentally sustainable future. In particular, the two crises have stimulated assessment of the impact of potential Covid-19 recovery packages on sustainability and climate change. Recent research notes common features between the two: market failures, externalities, breakdown in international cooperation, complex scientific questions, political leadership, system resilience and reliance on public support (Hepburn *et al.*, 2020). The severity of the climate crisis is clear: GHG emissions are expected to fall by an unprecedented eight percent in 2020 as a consequence of reduced global polluting activities (Le Quéré *et al.*, 2020), but to keep temperature increase below 1.5°C, emissions must decrease by 7.6 percent *annually* for the next ten years, reductions far higher than those achieved by any country following the Paris Agreement. Hepburn *et al.* argue that urgent post-viral *rescue* programmes compatible with longer-term *recovery* programmes are needed to address the twin crises. Rescue packages are needed to effect rapid support to beleaguered economies while recovery packages aimed at decoupling GHG emissions from economic activity demand deeper consideration and are likely to differ between higher-income and lower- and middle-income countries. Hepburn and colleagues argue that two factors affect the outcomes of fiscal recovery packages, the speed of impact and the longer-term multiplier effect, where, for example, benefits of public investments in clean energy infrastructure or domestic insulation retrofits induce further investment or filter through to other aspects of economic and social life, such as lower fuel bills (Allan *et al.*, 2020: 7). On the basis of their findings from a global survey of 231 key economic and finance actors covering over 50 countries, Hepburn *et al.* contend that these programmes are primarily the responsibility of individual

governments but should be overlaid with international coordination and cooperation as necessary. Potential national recovery programmes emerging from this
and other research are considered below.

We stress that while national governments are critical in setting the terms for
sustainability, they must also collaborate effectively with social partners,
including employers, local authorities and unions. Though the record of governments subsequent to the Paris accord has been patchy and generally inadequate, there was expressed intent by the signatory countries to take action to
keep global temperature rise below 2°C. Similarly, the 17 goals contained in the
UN's 2015 'action plan for people, planet and prosperity', signed by 193 Heads
of State, are accompanied by a range of specified far-reaching sustainability
targets. Securing sustainability through meeting these agreed targets provides
ample justification for intervening in and regulating economic and corporate
behaviour and also for meeting a comprehensive and effective international
response to the emergency.

However, there are negative warning signs based on earlier experiences. In
her book, *The Shock Doctrine*, Naomi Klein (2007: 405–413) demonstrated how
free-market zealots, politicians and economists alike, in spurious pursuit of
'freedom', have exploited crisis situations to apply their fundamentalist market-
informed 'shock treatment' to civic calamities when citizens are at their most
vulnerable, as happened in the aftermath of the 2005 New Orleans floods and
following the 2007 financial crash. A second, and from a business perspective,
very appealing option is for governments to promote policies to restore economic activity, growth and inevitably, emissions, lost during the pandemic.
Alternatively, to realistically confront the ecological crisis, there are growing
calls for transformational shifts in political direction towards social and economic democracy, genuine 'system changes', proposed by Thomas Piketty, Paul
Adler and many of the authorities cited in this volume. Within the same ideological framework lies the 'doughnut' economics advocated by Kate Raworth,
in which economic activity should be directed to benefit all in society whilst
operating within boundaries that do not imperil the earth's resources.

Based on the evidence accumulated in this book, our stance is clear. Neither
fundamentalist Friedmanite nor business-as-usual approaches will serve the
cause of planetary salvation or the interests of the public who work for and buy
from enterprises eschewing green development. We place ourselves firmly
within the social-democracy camp but recognise the many constraints that need
to be confronted and overcome at company, sector, national and international
levels for sustainable transformation to be achievable. As we argue in Chapter
4, any reassessment of national priorities should take into consideration that
growth-based GDP, founded on ever-increasing consumption, is an inadequate
and potentially damaging measure of the economic health of a country or the
well-being of its citizens. Life satisfaction measures are less growth dependent
(and can be positively associated with environmentally sustainable degrowth)
and provide the substance of the UN's sustainable development goals and to
GNDs.

Governments and sustainability

Countries have different systems of government and economic trajectories, but all face ecological and sustainability challenges. For this reason, we contend that national governments must take appropriate steps and take them rapidly to avert a disaster that threatens to be more profound even than the Covid pandemic. We therefore support fiscal measures that subordinate economic growth to sustainable growth; advocate policies that offer accountability and wider control over corporate interests; and argue that principles and practices of corporate governance through economic democracy should be developed and observed. Progress of these projects in meeting environmental and social objectives should be formally monitored at national level and supported through international cooperation.

Companion papers by Allan *et al.* (2020) and Hepburn *et al.* (2020) focus on governmental fiscal recovery programmes aimed at delivering both economic and climate goals. Historically low interest rates provide an impetus for Gudgin *et al.* (2020) to advance similar public investment-driven recovery proposals, recommending in addition, that smaller projects up to the value of £500 million, be devolved to control at metropolitan and local authority levels. Though the three studies cover similar territory and are directed purposefully at policy makers, Allan *et al.* and Gudgin *et al.* are more UK-specific while Hepburn *et al.* adopt perspectives that may vary according to national context. The latter propose a set of actions offering both strong multiplier and positive climate impact. Five policy items are identified, comprising: investment in clean physical infrastructure; building efficiency retrofits; education and training to offset pandemic-linked unemployment and decarbonisation-related structural unemployment; natural capital regenerative investment; and clean research and development investment. They also add that for lower- and middle-income countries, rural support expenditure could have high priority. The authors are clear that affordability of these initiatives would vary but in a period of low interest rates and potentially high multiplier returns, they are confident that the interventions should be financially sustainable.

The starting point for Allan *et al.*'s interventions is the UK's binding commitment to achieve net zero carbon emissions by 2050, though some commentators question whether the UK government is taking sufficient strategic action to make climate action a priority (Sasse *et al.*, 2020) or devoting sufficient funds to meet the emergency (Beament and Montague, 2020). To achieve this and based on examination of the government response to the 2008 financial crisis, Allan *et al.* draw up a recovery package of ten potential fiscal interventions, which 'collectively ... will strongly influence whether the targets agreed at COP21 in Paris are met' (2020: 3), whilst also helping to ensure post-Covid economic recovery. The projects largely overlap with those of Hepburn *et al.*, with the addition of investment in broadband connectivity, incentives for electric vehicles manufacture and use and improved cycling and pedestrian infrastructure. Company bailouts should be made conditional on meeting measurable climate-positive criteria (see below) and social outcomes. The authors insist that 'investments must collectively target a net-zero emissions future' (2020: 7) for which effective monitoring must be in place. They suggest that national oversight on progress could be built through establishing a Climate Change Emergency Committee alongside a Net Zero Delivery Plan, which would detail 'the UK's pathway to net zero by 2050' with broad representation from government, the Committee on Climate Change and other bodies, though apparently with no space specifically for the CBI or TUC (2020: 13). Finally, the global nature of the sustainability crisis is recognised, and an international Sustainable Recovery Alliance is suggested for establishment at the postponed COP26, now scheduled to meet in Glasgow in 2021. A major report produced by the highly respected International Energy Agency (2020) in association with the IMF, recommends investment of around $1tn annually through joint public and private expenditure over the next three years to oversee a global sustainable recovery plan that would secure economic growth, preserve nine million jobs a year and significantly reduce energy-related GHG emissions, along with a range of other positive health and well-being outcomes. With analysis based on factors including individual country sustainability circumstances, overall emphasis would be on clean energy expansion and energy-saving projects similar to those identified above. The Report also commends much closer international collaboration in pursuit of the UN's SDGs and Paris targets.

Though these proposals, stimulated by the Covid pandemic, enjoy obvious merit, the UK manufacturing base has been progressively hollowed out over the past 40 years, production has increasingly been outsourced and offshored, and considerable rebuilding of the country's industrial, technological and skills capacity is required in order to contemplate implementation (TUC, 2020). Moreover, investment in fossil fuels has continued unabated. The UK's largest pension fund, the Universities Superannuation Scheme (USS), continues to heavily fund the fossil fuel sector, while its single largest holding of £500 million is invested in oil company Shell (Spence, 2020), despite plummeting demand for oil-based products and failure by oil companies to invest seriously in renewable energy. BP, for example, with an annual investment budget of $12bn, invests just $500 million in green energy. Nevertheless, while oil production continues to offer profitable returns, the sector will continue to attract funding. It is therefore a crucial role for governments to use taxes, subsidy withdrawal and other disincentives to ensure that products and processes that destroy people's health and welfare are not profitable whilst ensuring that sustainable assets are (Oberholzer, 2019).

These observations lead to two further issues. First, the fiscal plans concerned principally with public revenues and expenditures do little to address the power and influence of the largely unregulated global finance sector in shaping national restructuring policies (Pettifor, 2019; 2020). And, second, while few would argue against progressive, albeit top-down, fiscal programmes, these also need to be monitored and supervised from below by the people directly most affected, workers and local communities prominently, for whom shared control over the financial and commercial organisations whose activities affect them is imperative for sustainability and equality to progress (Piketty, 2020: 972). We will recall from earlier chapters that few corporations have taken strategic moves to reduce their emissions or to raise levels of sustainable employment. GNDs, trade unions, environmental and elected local body participation are needed to ensure that national and corporate policies are in harmony with stated environmental objectives.

During the pandemic it became clear that many of those whose work had been previously dismissed as peripheral and treated accordingly, whether through no pay or little recognition, as with domestic and voluntary labour, or treated at the margins of the labour force through zero-hours, fake self-employment and agency contracts in fact made significant contributions to maintaining the economy, and to protecting citizens' health and welfare, often at considerable risk to their own. As we indicated in Chapter 12, a UBI can offer recognition to those whose work is socially and domestically valuable but not financially compensated and can also reduce pressures on carers squeezed between the twin demands of employment and care. UBI can also support transitions between occupations, providing opportunities for reskilling for work in climate-friendly or new technology sectors and UBI funded through carbon taxes could be directly associated with decarbonification. Nevertheless, the road to UBI would not be straightforward: there is no single formula or mechanism for delivery and there are political and ideological objections associated with perceptions of freeloading. However, the pandemic has revived support for UBI (Nettle *et al.*, 2020) and we believe that it is incumbent on governments to consider its merits in terms of reducing inequality, helping to lubricate a just employment transition for workers from unsustainable industries and through its potential to support the environment.

The control of corporations

Hence, the challenges to be faced should not be underestimated. Economic activity is at the core of any sustainable strategy and corporations, private and public, are at the heart of economic activity. Corporations exist to satisfy footloose shareholders and, to retain their support, executives can go to extreme lengths to defend profit-making capabilities and cost minimising activities. Many public sector organisations have also been channelled into dysfunctional market-based patterns of corporate behaviour where they are expected to meet imposed targets through adopting private-sector managerial practices. There is ample evidence that under competitive market conditions, regulations providing employee and civic protections are treated by enterprises as cost-raising obstacles to be avoided or even resisted. In consequence, products and processes known to cause environmental and personal damage are backed by aggressive campaigns, often supported by dubious think tanks sponsoring equally dubious scientific justifications while launching corrosive personal attacks against opponents. Since the 1960s, independent investigators have forensically deconstructed the claims and activities of international chemical, pharmaceutical, tobacco industries and major atmospheric polluters and all have been faced with orchestrated attempts to undermine their authority. Nevertheless, many enterprises continue to oppose or ignore regulation over their activities or dispute the need for environmental controls. Recourse even to law-breaking has been vividly revealed in David Michaels' contemporary accounts of corporate misbehaviour by nominally reputable international enterprises, many of which proudly flaunt their well-publicised CSR and green credentials. Diesel cars designed to cheat emissions controls, alcohol and soft drink producers disputing negative health impacts and fossil fuel companies denying scientific climate evidence are all too prominent, yet many nonetheless continue to be successful in their commercial activities (Michaels, 2020). Clearly, not all enterprises act, in Bakan's words, as psychopathic creations, though by single-minded pursuit of profit-seeking interests and of externalising costs, their antisocial tendencies all contribute to 'the routine and regular harms caused to ... workers, consumers, communities, the environment' (2004: 60). Corporations as commercial entities not only wield considerable financial power but as Michaels points out, corporate wealth is deployed, often covertly through powerful lobbyists (Geoghegan, 2020), to strengthen their grip on media and to sway political influence. Big corporate donors, often with fossil fuel interests, anonymously fund climate-change denying Republican Party campaigns in America and hold sway over government policy through their vast financial contributions. Their 'successes' include generous government subsidies for oil exploration and extraction, disabling USEPA and in persuading Trump to withdraw from the Paris Agreement (Michaels, 2020: 193–198).

Evidence suggests then that big corporations in particular, pressured by financial interests, are reluctant to alter their behaviour voluntarily, yet it is equally clear that to meet environmental and JT challenges, corporate power and influence must be restrained and made answerable to a wider constituency than shareholders and senior executives alone. The role of national governments in controlling corporate behaviour is therefore crucial and to do this governments must be detached from undemocratic corporate influence (Adler, 2019: 33). Depending upon domestic legislation, conventions and traditions, specific steps to be taken may vary, but core issues such as requiring companies to pay appropriate taxes for environmental costs would be expected to be applied universally to achieve the needed impact and to prevent social dumping or to remove threats to national governments by international investors to shift capital from high to low tax regimes in a process termed 'abusive transfer pricing' (Pettifor, 2019: 80). Costs deriving from commercial activity have frequently been borne not by the companies responsible but by public authorities charged with personal and community welfare. One tragic, but not isolated, example is that of the DuPont corporation, which knowingly adopted the most rational (i.e. organisationally cost-effective) means for disposal of toxic waste-products by dumping them into the local West Virginia environment to the health and welfare detriment of local inhabitants, farm animals and wildlife (Shapira and Zingales, 2017). Minimising private costs continues to be a prime driving force in corporate behaviour, and under the 'polluter pays' principle, these costs must be representative of the full damage attributable to a company's or sector's commercial activities. Such a policy would have multiple benefits. For the rational employer, an imposed tax fully representative of environmental costs would be expected to act as a signal to companies and their investors to engage in sustainable activities not attracting a tax levy (especially if these activities were to attract a subsidy: the carrot to accompany the stick). Second, local and wider environments would benefit from reduced pollution levels and improved worker and societal health. Third, there would be reduced consumption of higher priced antisocial products, as has been experienced following tax increases on sugary beverages in a number of countries (Roache and Gostin, 2017; Sorensen *et al.*, 2017) and price increases on tobacco (Michaels, 2020: 213).

Implementing economic democracy

As we noted in earlier chapters, economic democracy can be established through different routes. The crucial factor is to ensure that the control and accountability of corporations extend beyond those of equity-holders and senior executives alone or that essential services be returned to public ownership and jurisdiction. In Piketty's words, a 'just society' provides absolute equal access to fundamental goods, in 'such a way as to allow its least advantaged members to enjoy the highest possible life conditions' (2020: 968). One significant step that governments should take in this direction is to reverse privatisation of core services such as water provision, utilities and transport systems. As Chapter 6 has demonstrated, privatisation, even in terms of profitability, has rarely been the unqualified success claimed by its free-market ideologues and has often been reliant on continuing generous government subsidies. At the same time, privatisation has provided least benefit to the most disadvantaged in society. However, it is acknowledged that top-down nationalisation has not proved to be an exemplar whether in terms of economic dynamism or worker democracy. Conversely, remunicipalisation projects have grown in recent years in essential services such as water and energy provision. A recent survey by the Transnational Institute revealed more than 1,400 remunicipalisation projects in 2,400 cities covering 58 countries with services extending into an expanding range of projects, including waste management and telecommunications. The study confirmed that there had been a positive ecological transition in at least 119 of these initiatives (TNI, 2020). As well as wider consideration to environmental issues, the remunicipalisation process allows for greater local public and worker input into decision-making with potential for both greater efficiency, improved working conditions, and relief from fuel poverty. Moreover, it is a policy that is readily applicable to economies in different stages of development, as we see in its growth in countries as diverse as Germany and Canada and those of South America (*ibid.*, 2020).

Further, in Chapter 11, we demonstrated that worker cooperatives in both developed and emergent economies often focus on sustainable agriculture and green products whilst making positive contributions to the environmental health of workers and communities. Formally, cooperatives offer equal opportunities for all members to contribute to and benefit from decisions. As we have seen, there are high numbers of agricultural and rural cooperatives in Latin American and Caribbean countries, often anchored in wider social objectives expressed through the social and solidarity economy movement. In principle, cooperatives aim for member inclusiveness and equality and, in recent years, growing attention to environmental protection. As long as equality and environmental safeguards are upheld, there is no reason why worker cooperatives should not contribute to the just society noted earlier and therefore should be supported by governments, both financially and through support services.

The critical role of trade unions

We consider that trade unions are vital participants to meet the challenge of sustainable commercial activity, sustainable employment and reducing inequality in accordance with the Paris Agreement, in ensuring just transitions and in meeting UN 2030 sustainability goals. There are convincing justifications for a central union role. First, we have indicated in earlier chapters that trade unions have historically campaigned for safe and healthy working conditions and in recent years for action against polluting activities by corporations that compromise the welfare of their communities. This protective role has again been demonstrated during the pandemic (TUC, 2020) with, for example, teaching unions offering guidance to protect children and teachers. Second, we have seen that unions have been more active than employers in developing viable plans for green production and for more extensive GND and equality reforms. It will be recalled that the original New Deal owed much of its success to the strengthened role offered to trade unions. Also, union members, and potential members, are most vulnerable to job loss with associated deterioration in mental and physical health brought about through changes in industrial activity, such as enforced cessation of fossil fuel mining (McIvor, 2017), and there is a convincing case that unions be involved in plans to switch to alternative sources of energy, founded on provision of secure employment and necessary skills training for displaced positions. Moreover, affected workers are likely to live in communities most touched by atmospheric and other forms of pollution and unions alongside community groups should be offered full disclosure on hazards and the authority to intervene in production processes to ensure a clean local environment. A recent study of nearly three million births in California found that proximity to active rural oil and gas production facilities was associated with adverse birth outcomes and specifically low birth weight and with babies small for their gestational age (Tran *et al.*, 2020).

Though unions are vital actors in support of transformational environmental change, there are problems that must be acknowledged and, where appropriate, confronted. Unions can be faced with conflicting priorities. As Gingrich (2013) noted, members expect them to act in defence of their jobs and when they do attempt to obstruct environmental progress, it is invariably through rational fear of job and industry loss, as in Germany's Lusatia region, where 100,000 mining jobs have dwindled to 6,000 over the past 30 years. Remaining miners are struggling to maintain their employment, being offered few alternative opportunities. Their persisting influence can be seen in that, while many European countries including Sweden, Belgium and Austria are now coal free and others expect to end coal dependence before 2025, Germany, which aims to start retreating from nuclear power by the end of 2022, still has a minority working population but high dependency on extreme-polluting lignite for powering energy. The country only expects to exit from coalmining by 2038, delays largely attributable to local union fears for alternative employment. We can see a similar picture in aviation and vehicle manufacture, where thousands of jobs have been threatened following the pandemic, leading to union calls for bailouts, some of which, as we show below have been successful. Where equivalent employment alternatives are offered and negotiated, union cooperation rather than obstruction is more evident (Hess, 2012; TUC, 2019).

As Chapter 8 demonstrated, workers and collective labour institutions also face profound global challenges. Over the past 30 years, union membership and influence across the world have steadily declined and in recent years, unions have faced mounting legal and physical restraints on their activities (ITUC, 2020). At the same time, and linked to unions' enforced decline, problems facing workers in terms of security, pay, equality, participation and meaningfulness have worsened. Even before the pandemic, an atmosphere of crisis for labour has long permeated political and economic systems (Pettifor, 2019: 8–10). Further, though unions are formally democratic institutions, answerable to their members, Richard Hyman (2016) reminds us that formal claims to practising democracy are insufficient: it needs to be exercised and refreshed through engagement with the widest possible membership constituency on a regular and informed basis. Democratisation of the workplace requires that workers themselves organise in ways that represent the interests of all.

Trade unions are also ideologically internationalist in orientation, though globalisation has served to severely undermine international worker solidarity as companies seek out the cheapest and most flexible labour options. Nevertheless, foundations for international labour cooperation, essential to confront an environmental crisis exacerbated through globalisation, are present. Confederations such as the ETUC and ETUI aim to coordinate and serve unions across Europe. More broadly, the International Trade Union Confederation represents over 200 million workers in over 160 countries. There are also a number of specific sectoral unions, including IndustriALL, a global confederation representing 50 million workers in mining, engineering and manufacturing in over 140 countries and the Industrial Transport Workers Federation of 700 unions representing workers in some 150 countries. Further, despite shortcomings noted in Chapter 6, the International Labour Organization, part of the UN system, aims to work with governments and employers alongside worker organisations to secure good working conditions. Also, we have seen that unions often work effectively in cooperation with sympathetic governments and supportive community groups.

How should unions be involved?

From a longer-term strategic perspective, unions need to be involved as equal partners alongside national and local government, employer organisations and community groups, so that ways to meet objectives can be agreed and progress monitored, with prime emphasis on meeting Paris and sustainable development objectives. While the horrors of the pandemic have demonstrated the significance of global cooperation and the resources needed to combat transnational crises, the virus has exposed the need for more immediate domestic initiatives to support an environmentally sensitive and equality orientated economic recovery. Those in precarious occupations have borne a large burden during the pandemic as many of these workers face exposure to the virus. There is also the reality that vulnerable workers are vulnerable because they have few legal protections or collective representation. We argue that there is no room for zero-hour contracts in a modern economy and that worker interests would be well served by making union representation easier to achieve through permissive legislation. Transition to green energy and sustainable production is essential but reports of poor working conditions in some new and green sectors indicate that transition to these sectors must be associated with legislative safeguards for workers and requirements for union intervention. Vulnerable workers have benefitted when unions are able to act on their behalf (TUC, 2017) and so have employers. Therefore, working time reductions, which can help the environment as well as productivity and worker quality of life, should be undertaken through joint employer-trade union agreement. Another aspect exposed by the response to the virus and that can have positive implications for both environment and worker well-being is the increase in homeworking for some occupations, which can relieve pressure on transport systems and help ensure an effective balance between work and home life (Pratt, 2020). Again, like all proposed changes to employment, intentions to introduce elements of homeworking should not be imposed unilaterally by management but discussed, agreed and reviewed between employer and employee representatives.

Using taxpayers' contributions, governments are being encouraged to support key and strategic sectors where local economies and large numbers of employees have been disadvantaged by the pandemic. Governments could take an equity stake in companies, as Germany intends to do with Lufthansa subject to confirmation from Brussels, despite resistance from environmental groups opposed to unconditional support for the heavily polluting aviation sector. It is also proposed as a viable policy option in France, with €7bn initially allocated to Air France (Noisette, 2020) as well as contingent financial support for the domestic car industry. Though the mitigation requirements demanded for the Air France bailout are mild, a more stringent condition for government intervention should be that supported enterprises produce detailed and formal operational plans for engagement in product and employment sustainability. For example, for the aviation sector, this would require formal commitment to better designed aircraft and much-reduced carbon-emitting fuel systems. At present most of the agreed aviation bailouts have no binding climate conditions (Transport Environment, 2020). Financial support should also be conditional on establishing enhanced employee participation in enterprise decision-making to ensure that agreed sustainability objectives are being met. This could be achieved by introducing worker directors in those industries with a governmental stake and also insisting on full recognition and strong bargaining rights for trade unions, along with policies prohibiting the use of precarious work and zero-hours contracts.

More generally, worker directors for companies employing 500 or more employees should be mandatory unless national legislation already provides for representation in smaller enterprises and worker directors, elected from the unionised workforce should occupy half the seats on boards of directors. For international companies, parity board representation should be on the main board of directors as well as local boards. UK employers have traditionally opposed worker directors on grounds that unlike European countries with dual-level supervisory and executive boards (with only minority worker representation on the executive board), the UK has adopted a single unitary board system. Nevertheless, a number of European countries with unitary boards successfully operate with worker directors (Conchon, 2015) and there is no reason why countries like the UK should not do the same. There would be concerted resistance from employer interests, resenting interference in their affairs, but without intervention and monitoring, there are few signs that employers would adopt and follow the commercial and employment sustainability strategies that the crisis demands.

For organisations with European connections, a case can be made for reinforcing and redirecting European Works Councils. As we saw in Chapter 8, these bodies were originally established 25 years ago to provide information and consultation arrangements over cross-national issues for employees working in transnational companies, but notwithstanding revisions to the Directive, EWCs still face obstacles that limit their representative impact. One problem is that only about half of eligible companies have established an EWC. A second is that representative influence has generally been low, compounded by cases of managerial lack of cooperation, for example, in providing adequate information or training for representatives. Representatives also complain that consultation often follows management decisions, rather than informs them. Companies with interests in accession states have rarely established EWCs, though these countries tend to be more dependent on fossil fuel operations than other European states. The original EU Directive and its 2009 revision were passed before wide acknowledgement of the global impact of the environmental crisis. The situation has now radically changed and we propose that all transnational companies with operations in Europe be formally required to establish an EWC; that a core remit of the EWC be to advise on and monitor operational and costing plans for transition to environmentally friendly conditions prepared by (joint) boards of directors; and that progress be evaluated at regular meetings of the EWC. Full and continuous training and appropriate facilities should be provided for participation by employee representatives, as should facilities for dissemination of progress to employees. Failure by companies to comply with decisions or to offer full representation should be reported to the EU who should have powers to force company compliance if breaches are confirmed. As we propose that union membership and recognition be actively encouraged, it would also be expected that the majority of employee representatives be drawn from recognised unions.

Therefore, for a green movement to gather momentum, we would advocate that trade union recognition and active collective participation should take place through co-determination, EWCs and enhanced collective bargaining at organisational levels. Legislation should be enacted to provide for union recognition and voice. Unions should also be fully engaged as partners where strategic decisions are made at sectoral levels, such as in France and Germany, and most importantly in tripartite economic and environmental management at governmental levels, as demonstrated in the corporatist systems found in Nordic countries. A role here would be especially important in establishing and monitoring government-backed GNDs and sustainability targets. Where unions and their members are at risk of abuse and discrimination, as occurs in many South American and Asian countries, legislated protection by state authorities is vital and if abused, countries should be subject to internationally designated sanctions.

Prospects for Green New Deals

GNDs represent ambitious and radical proposals for combatting the environmental and equality crises though some establishment-promoted GNDs are GND in name only, consisting of little but empty rhetoric. With their history and interests, it is perhaps not surprising that unions have been instrumental, along with supportive political parties and community partners, in promoting and drawing up GNDs and JTs involving social interventions to accommodate worker protections when shifting to sustainable production. There are concrete issues to be faced, though. It is one thing to draw up a GND, it is another to implement one. Because of their radical and transformative intent, political and commercial opposition has been vigorous, and as yet GNDs remain largely defined as social-democratic, mature economy, manifestos rather than practice. Though several GNDs have emerged with different national (USA) or international (EU) emphases, genuinely transformative programmes share common characteristics. First is that economic and social reform plans need to be subject to wide consultation. Second is to create high quality, sustainable and unionised jobs, based on decarbonisation targets set by the UN. Third, more ambitious programmes aim to confront poverty and inequality to secure economic and social change.

As might be expected, these objectives are considerably less attractive to conservative governments and their financial supporters. Despite stated commitments to meet UN and Paris targets, there has been little political willingness to engage with the more progressive or universal elements of GNDs, notwithstanding acknowledgement that decarbonisation is making inadequate progress, that inequality remains a stain on modern society, that governments must adopt more forceful sustainability policies and that they must cooperate with one another and with international institutions, in ways they signally – and fatally – have failed to do during the coronavirus pandemic. One objection (though one that neglects compensatory job creation and health benefits) is that the costs of implementing even restricted GNDs would be substantial, especially at a time when most economies are suffering from the impact of the pandemic. At the heart, as economist Ann Pettifor, one of the prime backers of GNDs in the UK, acknowledges, is the need for structural change to subordinate the 'financial sector to the interests of society and the planet' (2019: 8). How might this be done, especially on a global stage, and bearing in mind the massive disruptive power of global financial interests? She argues for a number of steps, commencing with increasing public understanding of the dependence of the finance sector on government spending of public money, though the lack of public protest following the banks' bailout following the recent financial crisis might sound a warning note. As indeed might Trump's election in 2016, the *gilets jaunes* protests in France over fuel tax and British enthusiasm for a Brexit that is most likely to cause catastrophic economic damage. Nevertheless, the success of a GND movement rests initially on demands that 'public financial assets be used for public, not private benefit, and be deployed in the service of humanity and the ecosystem' (*ibid.*:162). Though people-powered demands could provide the impetus for such changes, these would surely falter without the support of united political forces to convert demands into established policies, policies that could (indeed, *would*) be interpreted and portrayed by opponents and their media as damaging to public interest rather than promoting it. We fear that general public support could be a long time coming without significant political changes in direction.

Naomi Klein (2019) also pins her hopes for a GND in the USA on political change underpinned by a powerful social movement, arguing that politicians not being actively corrupt is insufficient – they need to be 'up for the political fight of the century' (2019: 261). Before the 2020 election the Democrats were active through the select committee of the House of Representatives Democrats in promoting sustainability, releasing a massively detailed and scientifically informed official report and recommendations for 'solving the climate crisis' based on a 'clean energy economy and a healthy and just America' (House Select Committee, 2020; Roberts, 2020). The plan closely follows the GND initiative presented by Congresswoman Alexandria Ocasio-Cortez and Senator Ed Markey, though the report does not offer job guarantees, universal healthcare or proposals for nationalisation.

Though less radical than the Ocasio-Cortez proposals, President Joe Biden enters the White House with a very different environmental agenda to that of his predecessor, who denied climate change, enthusiastically embraced fossil fuel development, challenged environmental science, significantly reduced the staffing and diminished the protective remit of the Environmental Protection Agency. Hostility to multilateral initiatives was reflected in withdrawal from the Paris accord, intention to pull out of the WHO, undermining the UN and willingness to blame other countries, notably China, for the economic and pandemic crises faced by the USA. Biden has signalled a more progressive US approach toward sustainability, pledging to re-join the Paris accord and aiming for carbon-free electricity production by 2035 and net zero-emissions by 2050, joining other major economies like the EU, China, UK and South Korea. He is also prepared to spend heavily on upgrading buildings, public transport and electric cars, though all these ambitions may be tempered by Senate and Supreme Court opposition. Nevertheless, under more enlightened and collaborative US leadership, the path may be open for essential closer international cooperation and action in reducing carbon emissions.

Possibly the best hope for regional progress is presented by the European GND (EGD), powered by European Commission President, Ursula von der Leyen, which has a number of identified mitigation goals and targets, applicable to all 27 EU member states, though objections from some coal-dependent accession states and hesitancy from others, may limit progress in a European union composed of diverse political colours. On the other hand, the EGD will benefit from considerable funding from the European Investment Bank (€1tn. of climate-friendly investments planned for 2021–2030), the European Fund for Strategic Investment (potentially €650bn) and estimates that one-quarter of all EU programme expenditure will be specified for climate change projects (Watt, 2020).

Clearly, the success of all GNDs will depend on the popular support that can be built up for them and the extent to which this support is developed by political parties across national states. The role of governments is therefore vital. Nevertheless, major polluters like India, Russia and conservative administrations in Anglo-Saxon economies have yet to demonstrate genuine interest in strongly interventionist GNDs and powerful neoliberal opponents have already mounted strong oppositional campaigns.

Multilateral cooperation and coordination

There is little doubt that the coronavirus pandemic severely exposed the frailties of international cooperation and as analysts have warned, it is essential that in dealing with the sustainability crisis the international community learn from the errors and missteps that have proved to be so costly in human and economic terms. Essentially, global problems demand global solutions but as Chapter 7 shows, the main deficiencies, demonstrated by national responses to the pandemic, can be summarised as: (i) lack of confidence by national governments in international support bodies, up to and including the UN, (ii) failures in domestic responses to the crisis attributed to alleged partisan behaviour by international bodies such as the WHO, rather than to failures in domestic policy and application, (iii) failure by countries to provide and share adequate, accurate and timeous information on progress of the outbreak, (iv) failure or refusal by countries to learn from the experience of others and (v) determination by governments to act independently of one another, irrespective of evidence available from other countries (Stockton *et al.*, 2020). Diminishing confidence in international agencies can also result in declining finances, which further undermine their potential for positive intervention (Maxmen, 2020).

A major fulcrum for this multilateral failure has been the fracturing of the world into power blocs dominated by China and the USA. International relations have been increasingly polarised by what an analyst at the Asia Research Institute describes as a Trump-inspired 'geopolitical contest against China'. And as China becomes more successful in increasing its intervention in global affairs, the greater the antagonism expressed by the USA and its populist allies while other Western countries, some with growing economic ties with China, try to maintain a semblance of neutrality. Nevertheless, with the support of many lower-income countries, Chinese international influence is on the ascendancy. The Chinese supported the campaign of Tedros Adhanon Ghebreyesus to head the WHO and Chinese nominees also lead four UN specialist agencies. Tension increased following Trump's attempts to divert public attention from the massive USA coronavirus infection rate and associated subsequent economic decline by blaming China for its approach to the crisis and its perceived influence over international agencies: American withdrawal from these multilateral bodies opened the door for further Chinese influence.

Our argument is clear. A coordinated multilateral response to the sustainability crisis is essential, and with the election of Joe Biden, hopes have been raised of a return to more stable and progressive international relations. Concern about access to power supplies under Trump led China to resume domestic fossil fuel production and carbon emissions have again risen – and China already accounts for nearly one-third of global CO_2 emissions. If multilateral relations do not resume on a cooperative and coordinated basis, it seems difficult to know how the UN's SDGs or Paris targets can be met, amidst mutual suspicion and economic aggression between major powers and their acolytes. A fundamental question is how to return to less abrasive international relations and restore trust between major powers, bearing in mind the rapidly closing time window for taking remedial action (International Energy Agency, 2020).

One potential opportunity is provided by Biden's election and another by the COP26 meeting rearranged for November 2021. Following expert counsel, previous environmental 'summits' helped to draw attention to the urgency of the situation to world leaders, culminating in the COP21 Paris meeting, which led to establishing the 2015 targets and commitment to them by individual countries. Since then, the UN's IPCC and other international bodies have endeavoured to maintain some sort of unity of direction among otherwise disputatious nations, though as we have seen, individual governmental actions have been too ill-defined, or restricted, to make the necessary environmental impact. During the virus pandemic, government officials throughout the world have claimed that their actions were led by the 'science': though faced with a new and unfamiliar threat, the status of the science itself was always being questioned and rightly so. There is no credible scientific dispute over the causes and effects of global heating and climate change, and there are no justifications for governments not to follow the explicit guidance that the international scientific community offers for countries to meet their agreed environmental and sustainability obligations. Further obfuscation and delay by national governments endanger planetary survival. The pandemic, terrible as it has been, does now allow for governments to follow economic and social routes that can not only serve to prevent further environmental catastrophes but also reduce inequality and enhance the livelihoods and welfare of millions of workers, their families and communities. We are confident that the strategies outlined in this book will help to ensure positive practical steps in these directions. But, and it is a massive *but*, there will be opposition, potentially covertly driven, from financial and commercial interests threatened by shifts to more democratic accountability both at the workplace and in society. It is initially essential, therefore, as Pettifor argues, for the global financial system 'to be returned to its role as the servant, not the master of the economy and ecosystem' (2019: 9). Writing before the pandemic, Pettifor pointed out that prospects of regaining public authority over national and international monetary systems might be utopian unless and 'until a global shock makes system change inevitable' (*ibid.*: 29). That global shock *has* occurred, interest rates are minimal, allowing governments to examine ways to refloat economies in sustainable directions, moves that would be welcomed by high proportions of the public. The programmes outlined above provide means whereby green investment, green new deals and appropriate fiscal instruments could set economies and societies along more equitable and sustainable routes. It is questionable whether any further opportunities will present themselves.

And finally ....

In many ways this has become a very personal book for us. We have spent much of our working lives teaching and researching the elements of good employment relations; many of our students have gone on to become managers, policy specialists and trade union representatives. With colleagues nationally and internationally we have worked with management and unions researching employment issues in the workplace such as occupational health and safety, technological change and opportunities for employee voice. It is therefore dispiriting, to put it mildly, to see the employment clock being viciously turned back to the nineteenth century with the re-emergence of sweatshops, low-paid and insecure work, increase in work-related ill-health and absence of worker representation or protection.

At the same time, as parents and grandparents, we have viewed with alarm the increasingly stark warnings about the vulnerability of the planet's ecosystem and despair at the failure of governments and organisations to come together to take appropriate action; we have a deep concern about the world we are leaving to our children.

We feel that the same misguided philosophy underlies both these trends: it is a mind-set that can view both people and environment as 'resources' to be used and then disposed of when not wanted. When value can no longer be extracted from employees' labour power, those workers will be laid off until further notice or will become unemployed. When anything of value has been extracted from the earth's raw materials, those elements with no market value – gaseous emissions, mining spoil, slag and general waste – will be dumped back into the environment. Indeed, it could be argued that the very word 'environment' sends out the wrong signals to anyone trying to understand our place in the world, in the sense that it suggests something 'out there', separate from us, the stage on which our actions take place but, at the same time, over which we have dominion. In reality, of course, we are just as much a part of the global ecosystem as the blue whale, the redwood and the bee, and the thing about an ecosystem is that it is a *system*, whose interconnected elements affect, support and threaten each other. However, what makes us, as *homo sapiens*, different, and as far as we know unique, is that unlike the whale, the redwood and the bee we have the cognitive ability to both perceive the consequences of our actions in advance and to choose to do something about them. But we do have to choose.

Bibliography

Adler, P. (2019), *The 99 Percent Economy*, Oxford, Oxford University Press.

Allan, J., Donovan, C., Ekins, P. *et al.* (2020), 'A net-zero emissions economic recovery from COVID-19', Oxford Smith School of Enterprise and the Environment, Working Paper No. 20-01.

Bakan, J. (2004), *The Corporation*, London, Constable.

Beament, E. and Montague, B. (2020), 'Boris green industrial revolution "a rehash"', *The Ecologist*, 18 November, https://theecologist.org/2020/nov/18/boris-green-industrial-revolution-rehash.

CAMS (2020), 'Providing air quality information in support of COVID-19 research', Copernicus Atmosphere Monitoring Service (CAMS), https://atmosphere.copernicus.eu/providing-air-quality-information-support-covid-19-research.

CCC (2020), 'Reducing UK emissions: Progress report to Parliament', Committee on Climate Change, 25 June, www.theccc.org.uk/wp-content/uploads/2020/06/Reducing-UK-emissions-Progress-Report-to-Parliament-Committee-on-Cli.._-002-1.pdf.

Conchon, A. (2015), 'Workers' voice in corporate governance; a European perspective', *Economic Report Series*, London, Trades Union Congress (TUC).

CREA (2020), *Centre for Research on Energy and Clean Air*, 'China's air pollution overshoots pre-crisis levels for the first time', https://energyandcleanair.org/china-air-pollution-rebound-briefing/.

FAIRR (2020), 'An industry infected: animal agriculture in a post-COVID world', London, Jeremy Coller Foundation.

Geoghegan, P. (2020), *Democracy for Sale: Dark Money and Dirty Politics*, London, Head of Zeus.

Gingrich, M. (2013), 'From blue to green: a comparative study of blue-collar unions' reactions to the climate change threat in the United States and Sweden', in N. Räthzel and D. Uzzell (eds.), *Trade Unions in the Green Economy*, London, Routledge; 214–226.

Gudgin, G., Lightfoot, W., Lyons, G. and Zeber, J. (2020), 'Why the Government should spend more on capital', London, Policy Exchange.

Hall, P. and Taylor, R. (2020), 'Pandemic deepens social and political cleavages', *Social Europe*, 22 June.

Hepburn, C., O'Callaghan, B., Stern, N. *et al.* (2020), 'Will COVID-19 fiscal recovery packages accelerate or retard progress on climate change?', Oxford Smith School of Enterprise and the Environment, Working Paper, No. 20-02.

Hess, D. (2012), *Good Green Jobs in a Global Economy: Making and Keeping New Industries in the United States*, Cambridge, MA, MIT Press.

House Select Committee on the Climate Crisis (2020), *Solving the Climate Crisis*, https://climatecrisis.house.gov/sites/climatecrisis.house.gov/files/SCCC%20summary.pdf.

Hyman, R. (2016), 'The very idea of democracy at work', *Transfer*, 22 (1); 11–24.

International Energy Agency (2020), 'Sustainable Recovery: World Energy Outlook Special Report', June, www.iea.org/reports/sustainable-recovery.

ITUC (2020), *Global Rights Index 2020*, Brussels, International Trade Union Confederation.

Klein, N. (2007), *The Shock Doctrine*, London, Penguin.

Klein, N. (2019), *On Fire: The Burning Case for a Green New Deal*, London, Allen Lane.

Le Quéré, C., Jackson, R., Jones, M. *et al.* (2020), 'Temporary reduction in daily global CO2 emissions during the COVID-19 forced confinement', *Nature Climate Change*. https://doi.org/10.1038/s41558-020-0797-x.

Lovett, S. (2020), 'Lockdown: nearly two million Britons with lung conditions report improvement as air pollution drops', *The Independent*, 4 June, www.independent.co.uk/independentpremium/uk-news/lockdown-air-pollution-drop-health-lung-coronavirus-a9546996.html#gsc.tab=0.

Maxmen, A. (2020), 'What a US exit from the WHO means for COVID-19 and global health', *Nature*, 27 May, www.nature.com/articles/d41586-020-01586-0.

McIvor, A. (2017), 'Deindustrialization embodied: work, health, and disability in the United Kingdom since the mid-twentieth century', in S. High, L. MacKinnon and A. Perchard, *The Deindustrialized World*, University of British Columbia, UCB Press; 3–22.

Michaels, D. (2020), *The Triumph of Doubt*, Oxford, Oxford University Press.

Nettle, D., Johnson, E., Johnson, M. and Saxe, R. (2020), 'Why has the pandemic increased support for the Universal Basic Income?', www.danielnettle.org.uk/download/188-UBI-pandemic-preprint-v1.pdf.

Noisette, T. (2020), 'Faut-il des contreparties écologiques aux aides publiques?' No. 2899, *L'Obs*, 28 May, https://www.nouvelobs.com/debat/20200601.OBS29573/faut-il-des-contreparties-ecologiques-aux-aides-publiques.html.

Oberholzer, B. (2019), 'Can sustainable finance really help solve the climate crisis?', *Social Europe*, 10 December, https://socialeurope.eu/can-sustainable-finance-really-help-solve-the-climate-crisis.

Pettifor, A. (2019), *The Case for the Green New Deal*, London, Verso.

Pettifor, A. (2020), 'Rebuild the ramshackle global financial system', *Nature*, 17 June, www.nature.com/articles/d41586-020-01507-1.

Piketty, T. (2020), *Capital and Ideology*, Cambridge, Mass., Harvard University Press.

Platt, L. and Warwick, R. (2020), 'Are some ethnic groups more vulnerable to COVID-19 than others?', London, Institute for Fiscal Studies.

Pratt, L. (2020), 'Two-thirds of UK staff feel they can work effectively from home', *Employee Benefits*, https://employeebenefits.co.uk/67-uk-work-effectively-home.

Public Health England (2020), 'Disparities in the risks and outcomes of COVID-19', London, June.

Rasnača, Z. (2020), 'Essential but unprotected: highly mobile workers in the EU during the Covid-19 pandemic', *ETUI Policy Brief*, No. 9, Brussels.

Roache, S. and Gostin, L. (2017), 'The untapped power of soda taxes: incentivizing consumers, generating revenue, and altering corporate behavior', *International Journal of Health Policy and Management*, 6 (9); 489–493.

Roberts, D. (2020), 'House Democrats just put out the most detailed climate plan in US political history', *Vox*, www.vox.com/energy-and-environment/2020/6/30/21305891/aoc-climate-change-house-democrats-select-committee-report.

Sasse, T., Rutter, J., Norris, E. and Shepheard, M. (2020), 'Net Zero: how government can meet its climate change target', London, Institute for Government.

Scheidel, W. (2017), *The Great Leveler: Violence and the History of Inequality from the Stone Age to the Twenty-First Century*, Princeton, NJ, Princeton University Press.

Shapira, R. and Zingales, L. (2017), 'Is pollution value-maximizing? The DuPont case', *Working Paper 23866*, Cambridge, MA, National Bureau of Economic Research.

Sorensen, C., Mullee, A. and Duncan, H. (2017), 'Soda-taxes: old and new', *The Tax Adviser*, www.thetaxadviser.com/issues/2017/jun/soda-taxes.html.

Spence, B. (2020), 'Why is the UK's largest pension scheme still investing in fossil fuels?', *Times Higher Education*, 11 June, 28.

Stockton, B., Schoen, C. and Margottini, L. (2020), 'Crisis at the Commission: inside Europe's response to the coronavirus outbreak', *The Bureau of Investigative Journalism*, www.thebureauinvestigates.com/stories/2020-07-15/crisis-at-the-commission-inside-europes-response-to-the-coronavirus-outbreak?utm_source=Nature+Briefing&utm_campaign=27ccd738e2-briefing-dy-20200716&utm_medium=email&utm_term=0_c9dfd39373-27ccd738e2-44223933.

Teale, C. (2020), 'COVID-19 may sport the thinnest silver lining: a cleaner climate', *Smart Cities Dive*, March 19, www.smartcitiesdive.com/news/coronavirus-impact-cities-climate-change-efforts/574450.

TNI (2020), 'The future is public', Amsterdam, Transnational Institute, www.tni.org/files/publication-downloads/futureispublic_online_def_15_june.pdf.

Tran, K., Casey, J., Cushing, L. and Morello-Frosch, R. (2020), 'Residential proximity to oil and gas development and birth outcomes in California: a retrospective cohort study of 2006–2015 births', *Environmental Health Perspectives*, 128 (6); 1–13.

Transport Environment (2020), 'Bailout tracker', https://www.transportenvironment.org/what-we-do/flying-and-climate-change/bailout-tracker.

TUC (2017), 'What have trade unions ever done for us? Ask 3000 workers at Sports Direct', London, Worksmart.

TUC (2019), 'How industrial change can be managed to deliver better jobs', London, Trades Union Congress.

TUC (2020), 'A better recovery: Learning the lessons of the corona crisis to create a stronger, fairer economy', London, Trades Union Congress.

Watt, A. (2020), 'The European Green Deal: will the ends, will the means?', *Social Europe*, 14 January, www.socialeurope.eu/the-european-green-deal-will-the-ends-will-the-means.

Wolff, R. (2012), 'Yes, there is an alternative to capitalism – Mondragon shows the way', *The Guardian*, 24 June, https://www.theguardian.com/commentisfree/2012/jun/24/alternative-capitalism-mondragon.

Index